IN VITRO–IN VIVO CORRELATIONS

ADVANCES IN EXPERIMENTAL MEDICINE AND BIOLOGY

IN VITRO–IN VIVO CORRELATIONS

Edited by

David Young
University of Maryland at Baltimore
Baltimore, Maryland

John G. Devane
Elan Corporation, plc
Athlone, Ireland

and

Jackie Butler
Elan Corporation, plc
Athlone, Ireland

SPRINGER SCIENCE+BUSINESS MEDIA, LLC

Library of Congress Cataloging in Publication Data

In vitro-in vivo correlations / edited by David Young, John G. Devane,
  and Jackie Butler.
       p.   cm. -- (Advances in experimental medicine and biology ; v.
  423)
    Includes bibliographical references and index.
    ISBN 978-1-4684-6038-4   ISBN 978-1-4684-6036-0 (eBook)
    DOI 10.1007/978-1-4684-6036-0
    1. Drugs--Solubility--Testing. 2. Drugs--Bioavailability-
  -Testing.   I. Young, David, 1952-   . II. Devane, John G.
  III. Butler, Jackie. IV. Series.
  RS189.I438 1997
  615'.19--dc21   97-17156
                                                              CIP

Proceedings of a workshop on In Vivo–In Vitro Relationships,
held September 4 – 6, 1996, in Baltimore, Maryland

© 1997 Springer Science+Business Media New York
Originally published by Plenum Press, New York in 1997
Softcover reprint of the hardcover 1st edition 1997

http://www.plenum.com

10 9 8 7 6 5 4 3 2 1

PREFACE

This book represents the invited presentations and some of the posters presented at the conference entitled "*In Vitro-In Vivo* Relationship (IVIVR) Workshop" held in September, 1996. The workshop was organized by the IVIVR Cooperative Working Group which has drawn together scientists from a number of organizations and institutions, both academic and industrial. In addition to Elan Corporation, which is a drug delivery company specializing in the development of ER (Extended Release) dosage forms, the IVIVR Cooperative Working Group consists of collaborators from the University of Maryland at Baltimore, University College Dublin, Trinity College Dublin, and the University of Nottingham in the UK. The principal collaborators are:

Dr. Jackie Butler, Elan Corporation
Prof. Owen Corrigan, Trinity College Dublin
Dr. Iain Cumming, Elan Corporation
Dr. John Devane, Elan Corporation
Dr. Adrian Dunne, University College Dublin
Dr. Stuart Madden, Elan Corporation
Dr. Colin Melia, University of Nottingham
Mr. Tom O'Hara, Elan Corporation
Dr. Deborah Piscitelli, University of Maryland at Baltimore
Dr. Araz Raoof, Elan Corporation
Mr. Paul Stark, Elan Corporation
Dr. David Young, University of Maryland at Baltimore

The purpose of the workshop was to discuss new concepts and methods in the development of *in vitro-in vivo* relationships for ER products. The original idea went back approximately 15 months prior to the workshop itself. For some time, the principal collaborators had been working together on various aspects of dosage form development. It was obvious that more and more of our time was spent discussing issues and aspects of IVIVR development since it plays such a key role in dosage form development, particularly for ER products. We felt it was important to provide a forum where scientists from industry, academia, and regulatory authorities could come together and discuss how best to develop methods in this complex area.

It is also important to emphasize and appreciate the background and context of the workshop. One can readily identify major milestones and initiatives over the last 10 years that have contributed significantly, either directly or indirectly, to the development of IVIVR methodology. In the last 5 years, the pace of these initiatives has accelerated, very much driven by the FDA. In particular, one can identify the SUPAC IR and ER Workshops of the early 90s, resulting in the 1995 issuance of the SUPAC IR guidance, with additional draft guidances prepared during 1996. In the context of IVIVR this has culminated in the issuance of the FDA draft guidance on IVIVC, which was completed just prior to the workshop.

At the point where we set a date for this workshop, we did not anticipate being so timely in relation to the issuance of the draft guidance. However, we believe that the discussions and ideas generated during the workshop provided a better understanding of the guidances and gave valuable feedback to the agency in what was the first public forum since the issuance of the guidance.

As one point of clarification, the Cooperative Working Group, who organized this meeting, talk about IVIVR. Many of the speakers and, indeed, the FDA draft guidance, referred to IVIVC. For the purposes of this workshop, these terms were viewed as interchangeable.

CONTENTS

EXAMPLES OF *IN VITRO–IN VIVO* RELATIONSHIPS WITH A DIVERSE RANGE OF QUALITY

Russell J. Rackley

Biopharmaceutics
Purepac Pharmaceutical Co., a subsidiary of Faulding, Inc.
200 Elmora Avenue
Elizabeth, New Jersey 07207

The purpose of this chapter is to present an introduction of *in vivo - in vitro* relationships (IVIVRs) and illustrate a number of examples which IVIVRs may be used to evaluate oral, modified-release formulations. Examples given are meant to portray applications to extended-release formulations, which may be defined as a dosage form allowing for a reduction in dosing frequency as compared to that drug presented as a conventional dosage form (1). "Controlled-release" throughout this chapter refers to an extended-release dosage form. In this chapter, in accordance with the meeting, IVIVR is used interchangeably with IVIVC (*in vitro - in vivo* correlation).

BACKGROUND

In vitro-in vivo correlation (IVIVC) has been defined by the United States Pharmacopeia (USP) Subcommittee on Biopharmaceutics as: "the establishment of a relationship between a biological property produced by a dosage form, and a physicochemical characteristic of the same dosage form" (2). A Food and Drug Administration (FDA) interpretation of *in vitro-in vivo* correlation has been cited (3) as: "To show a relationship between two parameters. Typically a relationship is sought between *in vitro* dissolution rate and "*in vivo*" input rate. This initial relationship may be expanded to critical formulation parameters and "*in vivo*" input rate." As suggested by the either of the cited definitions of IVIVC, physicochemical properties of a dosage form other than dissolution should not be overlooked as an *in vitro* measurement. However, with respect to quality control testing, more weight tends to be placed on the cumulative dissolution of a dosage form over time as an *in vitro* indicator of *in vivo* performance. *In vivo* performance may typically be assessed in man by rate and extent of absorption of an oral dosage form. For controlled-release dosage forms, it is especially desirable to determine the cumulative absorption-time profile. Although, other endpoints of *in vivo* performance of a dosage form might be investigated, such as measurement of drug efficacy. The ultimate goal of an IVIVC should be to establish a meaningful relationship between *in vivo* behavior of a dosage form and *in vitro* performance of the same dosage form, which would allow *in vitro* data to be used as a surrogate for *in vivo* behavior.

Skelly and Shiu (4) have inferred that dissolution testing evolved as a tool for biopharmaceutical investigation of *in vitro-in vivo* correlation throughout the 1950's and 1960's. The United States Pharmacopeial Convention has greatly influenced the standardization and general acceptance of dissolution since this time (5). As a result of the introduction of different generic forms of digoxin around the early 1970's, the use of dissolution testing began to be widely recognized throughout the industry as a quality control test (4). This was in fact due to the *in vitro-in vivo* correlations that were demonstrated between different digoxin products (6–8). Although the evolution of *in vitro-in vivo* correlation may be rooted in conventional immediate-release dosage forms, the concepts are applicable toward the development and support of controlled-release dosage forms.

RATIONALE OF IVIVC

For controlled-release products, an *in vitro-in vivo* correlation should be based on an intra-product comparison. That is, only variations of the drug-release rate for a specific formulation controlled-release mechanism may be considered in the correlation of a given product. Because different controlled-release products generally employ different controlled-release mechanisms, it may not be possible to make a comparison between different controlled-release products. It must be emphasized that *in vitro-in vivo* correlations for controlled-release formulations should be considered to be product-specific.

In vitro-in vivo correlation for controlled-release dosage forms would be of benefit if utilized in one or more of the following ways:

- surrogate to bioequivalency studies which might typically be required with scale-up or minor post-approval changes (SUPAC), where minor post-approval changes may include site of manufacture, formulation, or strength;
- support and/or validate the use of dissolution testing and specifications as a quality control tool for process control, dissolution specifications in quality control ranges may be shown to be relevant to *in vivo* data;
- predict *in vivo* performance of a formulation based on *in vitro* dissolution data, termed "biorelevant dissolution" (9), which may be used in the justification of dissolution specifications and may aid in the design of formulation release-time profiles resulting in optimal plasma concentration-time profiles;
- identify appropriate dissolution conditions for a formulation which result in data relevant to *in vivo* performance.

CONSIDERATIONS

A number of rational considerations should be taken into account before attempting an *in vitro-in vivo* correlation for solid oral dosage forms (3,10–12). Factors affecting the success of an *in vitro-in vivo* correlation may include one or more of the following:

- if a drug has a fairly narrow therapeutic window, an *in vitro-in vivo* correlation may still not be acceptable as a surrogate for bioequivalency testing (13)
- the pharmacodynamic properties (therapeutic or adverse) of the drug have been evaluated and there is minimal lag (or hysteresis) for effect vs. plasma concentration (12)
- degree of linearity in pharmacokinetics of the drug may limit degree of correlation, and variation in the rate and extent of absorption or disposition may limit correlation
- the physicochemical nature of the drug is not foreseen as a limiting factor potentially leading to absorption variability *in vivo*
- the physical absorption of drug should not be the rate-limiting step in the absorption process (i.e. is not permeation rate-limited)
- release from the dosage form is the rate limiting step in the overall absorption process (i.e. is dissolution rate-limited)
- ideally, the formulation is relatively insensitive to the range of variation expected within the *in vivo* environment (effect of pH, surfactants, agitation, etc.)
- dose (mg) and solubility (mg/mL) will determine the volume of dissolution media necessary to evaluate the formulation *in vitro*, sink conditions may be difficult to maintain for large volumes and may indicate problems *in vitro* and *in vivo*; however, in these cases the USP Apparatus III and USP Apparatus IV may be investigated (12).

Logically, a higher degree of correlation may be expected with controlled-release formulations, since release from the dosage form tends to become the rate-limiting step in absorption, overcoming permeation rate limitations. Also, the time-frame allowed to characterize the profile of dissolution or absorption is much longer than with immediate-release dosage forms, which permits a greater degree of accuracy and precision in characterizing the dissolution and absorption profile.

It has been recommended that validation of a Level A correlation for a controlled-release formulation would be accomplished by preparing one or more batches of drug formulation which release at different rate(s) by varying critical formulation components that are likely to vary during normal manufacturing (1). A small pharmacokinetic study (N=6) may then be conducted to determine whether the Level A correlation is still supported by the new batches.

It should be realized that success with one drug formulation may not be extrapolated to all drugs or necessarily to different formulations of the same drug. All formulation development efforts for new drugs should be treated on a case-by-case basis.

DECONVOLUTION AND CONVOLUTION

Scheme I summarizes the relationships of deconvolution or convolution to *in vitro-in vivo* correlation or biorelevant dissolution. Based on a knowledge of the pharmacokinetic system for a drug, the plasma concentration-time profile resulting from administration of an oral dosage form may be taken apart, or deconvoluted, to give an absorption-time profile for the oral dosage form *in vivo*. Also, based on the assumption that release of drug from the controlled-release formulation is the rate-limiting factor in the absorption process, then the absorption-time profile resulting from deconvolution may be considered to be indicative of *in vivo* dissolution. It is necessary to determine the absorp-

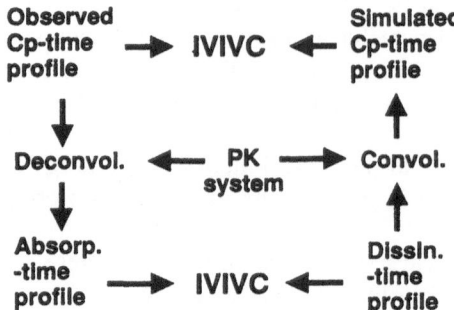

Scheme 1. Relative approaches to IVIVC may be based on deconvolution or convolution methodologies.

tion-time profile in the assessment of Level A Correlations. The more common examples of deconvolution are the Wagner-Nelson equation (14) for one-compartment model drugs and the Loo-Riegelman equation (15) for two-compartment model drugs. Numerical deconvolution may have an advantage in that *in vivo* dissolution may be determined, using a program such as PCDCON (16). In general, the use of deconvolution methods (14–17) requires a fairly good understanding of a drug's pharmacokinetics and pharmacokinetic principles. These methods theoretically imply investigation of the drug administered intravenously; but, in reality the use of oral solution or an immediate-release form of the drug might be an acceptable alternative.

Pharmacokinetic modeling for a drug may be necessary to use some deconvolution techniques correctly. Plasma concentration-time data may be analyzed with standard programs such as NONLIN (18), SAS (19), or NONMEM (20) in the determination of an appropriate model and parameter estimates for the model. Here it is important to keep in mind the drug-related considerations listed previously for potential evaluation of an *in vitro-in vivo* correlation.

A knowledge of the pharmacokinetic system of a drug may be combined with, or convoluted with, the dissolution-time profile of an oral dosage form to simulate a plasma concentration-time profile for administration of the controlled-release dosage form (Scheme I). Leeson et al. (21) used this process as a method to evaluate prototype controlled-release formulations *in vitro* without performing more costly and time-consuming *in vivo* testing, and the concept has been termed "biorelevant dissolution." It is essential that sensitivity of the oral controlled-release dosage form be investigated by testing over a diverse range of *in vitro* testing conditions in order to simulate the *in vivo* environment (see Dissolution Specifications). As with deconvolution, convolution may only be used when a valid pharmacokinetic system for the drug has been established. Simulations resulting from the combination of dissolution data and the pharmacokinetic system are only relative to the *in vivo* behavior of the reference form of the drug on which the pharmacokinetic system is based (21).

A relatively simple method of convolution may be based on the numerical integration of a dissolution profile, transformed to rate, and input into the absorption compartment of an appropriately defined pharmacokinetic model. The simulation software known as STELLA II (22), provides a quick and efficient means of convoluting data, making model modifications, or testing different dosing regimens; however, meaningful parameter estimates must be provided to the model and both drug- and formulation- related consid-

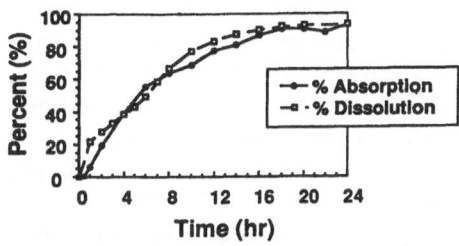

Figure 1. Qualitative, visualization of in vitro - in vivo relationship.

erations previously mentioned must be taken into account in order to generate meaningful simulations of *in vivo* performance.

LEVEL A *IN VITRO-IN VIVO* CORRELATION METHODS

• Qualitative:

The first test of a Level A correlation is to visualize the dissolution and absorption profiles together to assess the degree of superimposition. If the profiles do not appear to superimpose each other with respect to rate and extent of drug release, then pursuit of a more quantitative correlation may not be warranted. If, however, the profiles to appear to superimpose each other in this manner, possibly with a lag-time for absorption, then a more quantitative correlation may be explored.

This formulation represents a developmental formulation of highly soluble drug salt in an oral osmotic controlled-release system with an outer immediate-release component. Visualization of Figure 1 demonstrates a qualitative *in vitro-in vivo* correlation in which it may be concluded that there is a fairly good Level A correlation.

• Quantitative:

Linear Relationship of Absorption as a Function of Dissolution. Using x-y data pairs representing estimates of dissolution and absorption at common time points, linear regression is performed. By far, this has been one of the more popular methods used initially to investigate a quantitative relationship between dissolution and absorption data. Superimposable data should have a one-to-one relation; therefore, linear regression of absorption

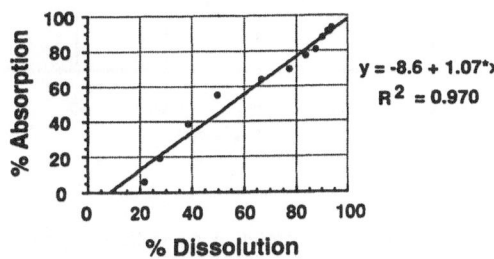

Figure 2. Quantitative, linear regression of in vitro - in vivo relationship.

versus dissolution at common time points should demonstrate a linear relationship with a slope of one, an intercept of zero, and a coefficient of determination (r-squared) of one. It has been suggested that a y-intercept less than zero might be explained by a lag-time in absorption, whereas a positive y-intercept may require additional evaluation (3).

Quantitatively, the slope of this apparent relationship is close to unity; the intercept is slightly negative, indicating an overall lag of absorption relative to dissolution; and a fairly high coefficient of determination (r-squared) is observed (Figure 2). This further supports the notion of a Level A correlation for the data presented in Figure 1, although there is systematic variation about the regression.

Hwang et al. (23) have mathematically demonstrated that the lag of absorption-time data in relation to dissolution-time data would be eliminated if *in vivo* dissolution-time data were estimated using numerical deconvolution methods. In cases where drug-release is constant (zero-order), then a simple time-shift in the absorption-time profile, equivalent to the reciprocal of the first-order absorption rate constant, may be used to superimpose dissolution-time data and establish a Level A correlation (23). When drug release is a slow first-order process, the same approximation appears to apply. In order to accommodate the

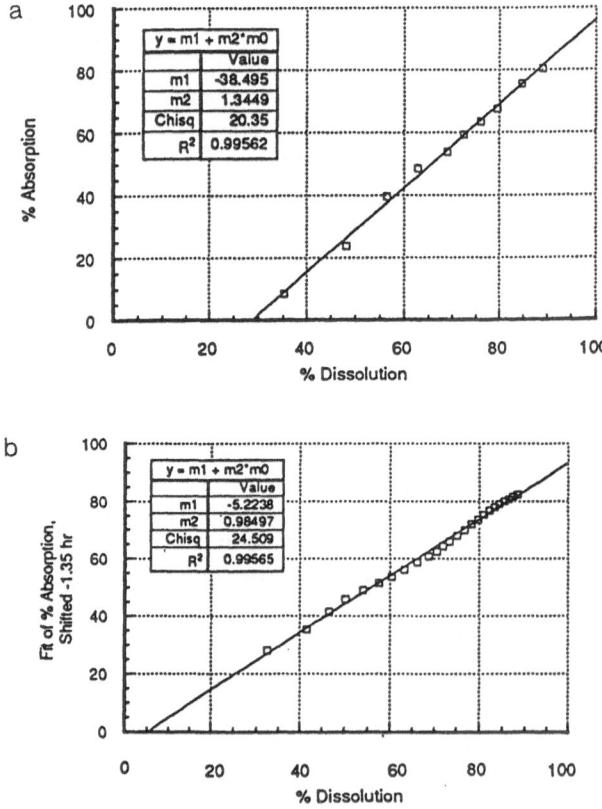

Figure 3. a - Linear regression of absorption vs. dissolution, with a large negative y-intercept. This type of regression may suggest a lag in estimated absorption, relative to dissolution. b - Linear regression for cubic spline fits of absorption (-1.35 hr) vs. dissolution. Shifting the absorption data by 1/ka and correcting for lag in estimated absorption demonstrates an improved linear relationship with dissolution data.

time shift, interpolation of data may be used if there is a high degree of confidence in the profile or process (i.e. with dissolution in a controlled system).

For this example, a different set of data initially indicated that the dissolution and absorption profiles were quite similar, with the exception that the absorption appeared to lag behind the dissolution profile approximately 1 to 2 hr. Linear regression of absorption versus dissolution at common time points indicates a fairly large negative y-intercept and a slope greater than one, although there appears to be a strong linear relationship (Figure 3a). Based on the concepts presented by Hwang et al. (23), an investigation of a time shift was conducted in an effort to improve the correlation. The first-order absorption rate constant for the drug in this formulation was previously estimated as 0.741 hr^{-1}, based on administration of a rapidly releasing oral dosage formulation. Therefore, an appropriate time shift of absorption data should be 1/0.714 or 1.35 hr. The absorption time profile was then shifted by a time interval or -1.35 hr to account for lag of absorption estimates to *in vivo* dissolution. Linear regression was again performed using absorption data shifted by -1.35 hr and the slope was nearly one with a y-intercept close to zero, and a strong linear relationship remained (Figure 3b). The linear regression was weighted to $1/y^2$ to account for the higher distribution of data near 100% and m0 is a variable representing the abscissa (x-axis).

NON-LINEAR RELATIONSHIP OF ABSORPTION AS A FUNCTION OF DISSOLUTION

There are a variety of possibilities in which non-linear functions may be investigated and the utility of this approach appears to be somewhat empirical. When linear regression does not indicate a good correlation, a non-linear function may indicate an excellent relationship. Higher order polynomial equations may be used in many cases to describe data with curvature. Unlike a linear relationship for superimposable data, the parameter estimates for higher order polynomial equations are more difficult to interpret. Nevertheless, this approach might be considered as a valid method of expressing a relationship between dissolution and absorption data.

When a third order polynomial is explored for the data in Figure 1, a very good relationship may be defined between the absorption and dissolution data (Figure 4); however, the coefficients should be reproducible for a Level A correlation.

Other methods suggested for investigation of nonlinear relationships include the use of Emax equation and its variations (24), the Weibull equation (25), and exponential or Gompertz equations (13).

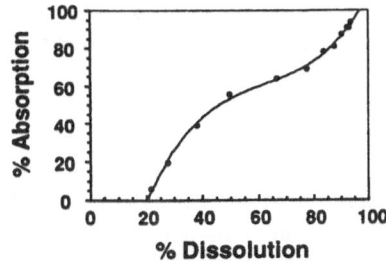

Figure 4. Quantitative, non-linear in vitro - in vivo relationship.

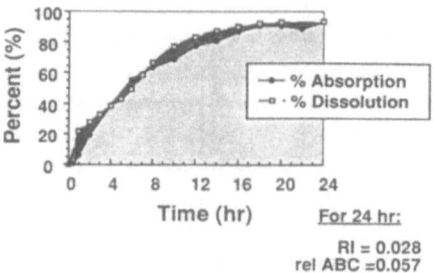

Figure 5. Quantitative, area between the curves; Resigno Index (RI) was 0.028, based on unweighted data up to 24 hr. Relative area between the curves (rel ABC) was 0.057, which means the area between the dissolution and absorption curves represents a deviation of 6% relative to the area under the dissolution curve up to 24 hr, which serves as the reference.

RESCIGNO INDEX

It has been previously proposed by Dr. J. Powers (13) that an index reported by Rescigno (26) be considered for comparing the similarity of dissolution- and absorption-time curves. The method was originally proposed for the assessment of concentration-time profiles for bioequivalency. However, cumulative absorption and dissolution tend to plateau at some asymptote; therefore, the time up to which absorption and dissolution data are collected should be specified. For the example given in Figure 1, index of Rescigno (26) was calculated for data up to 24 hr with an exponent of one without weighting and the resulting index was 0.028.

RELATIVE AREA BETWEEN THE CURVES

A variation of the Rescigno Index is to relate the difference of area between the cumulative dissolution and absorption curves to the area under the cumulative dissolution curve, as dissolution is assumed to be the independent variable. When the index is calculated in this way, it directly gives the fractional difference of absorption to dissolution. Again, since cumulative data are analyzed, the final time point used in the evaluation should be specified. The relative area between the curves for the data in Figure 1 was calculated as 0.057, indicating that the area between the curves up to 24 hr was less than 6% of the area under the dissolution-time curve up to 24 hr (see Figure 5).

If these methods do not appear to demonstrate a relationship of absorption to dissolution, then an alternate method of dissolution should be investigated, which generates a dissolution-time profile resulting in a closer match of the absorption-time profile. In any case the goal of an *in vitro* test should be to generate results that are indicative of *in vivo* performance.

F2 CALCULATION

With the possible exception of the F2 calculation (27), quantitative evaluation of dissolution data, as indicator of bioequivalence, has rarely been used. However, just in the same manner that the Rescigno Index was applied to dissolution and absorption data, it is conceivable that the F2 calculation be applied in a similar manner. For comparisons of absorption and dissolution data, it may be appropriate to correct for lag in the absorption profile.

Except for the concept of "mapping", the issue of variability in dissolution data and its relation to absorption variability has not been taken into account in the discussion of these methods. Thus, there appears to be a need for more statistically based guidelines in the establishment of equivalency of dissolution and absorption profiles. Future developments of *in vitro-in vivo* correlation and biorelevant dissolution will probably include the use of nonlinear mixed effect modeling in the evaluation of dissolution and absorption data, for solid oral formulations.

IVIVR FOR FOOD EFFECTS

There has been a growing interest in the ability to demonstrate *in vitro - in vivo* correlations for extended-release formulations, which may be subject to food effects. A somewhat simplistic approach has been to examine the absorption time profiles for fasting and fed conditions, relative to the dissolution profile. As seen in Figure 6, this approach may only demonstrate a correlation for one condition or the other. Figure 6a represents relatively good linear correlation for absorption vs. dissolution for a fasting study of a modified-release formulation. However, for the fed study phase it is apparent that a linear correlation would not be appropriate (Figure 6b).

A modern approach is exemplified by that of McCall et al. (28), where these authors attempted to optimize an *in vitro* test which mimicked post-prandial *in vivo* performance (Figure 7a-7d). This type of approach may be useful for screening and evaluation of extended-release formulations that may be sensitive to food effects.

DISSOLUTION SPECIFICATIONS

With controlled-release formulations, there is an inherent need to "profile" the release over time. Dissolution specifications define the acceptable range of dissolution-time data and should be representative of the profile and variability associated with a controlled-release dosage form. The USP (29) offers a guide suggesting the time over which the dissolution profile be related to the labeled dosing interval. And, there should be a minimum of three dissolution time points for dissolution testing of controlled-release dosage forms (12, 13): the first time point should assess dose dumping, the second or more time points should "profile" the dissolution-time curve, and the last time point should provide information as to the recovery of drug in the dosage form.

An investigation of the dependence of the formulation on pH and surfactants is recommended in media of various compositions (12). Also, a dependence on dissolution equipment, and range of equipment settings, should also be considered in the investigation. However, for controlled-release formulations that are sensitive to changes in the dissolution environment, dissolution should be examined to determine the *in vitro* conditions which achieve an optimal IVIVC (1). Again, one or more batches of drug formulation which release at different rate(s) could be examined in an *in vivo* study to determine whether the correlation is supported. Once a formulation has been finalized for clinical use, dissolution data should be collected and tracked. The dissolution data should include batch-to-batch variation. Dissolution specifications have implications related to critical manufacturing parameters. Batches manufactured near the limit of a critical manufactur-

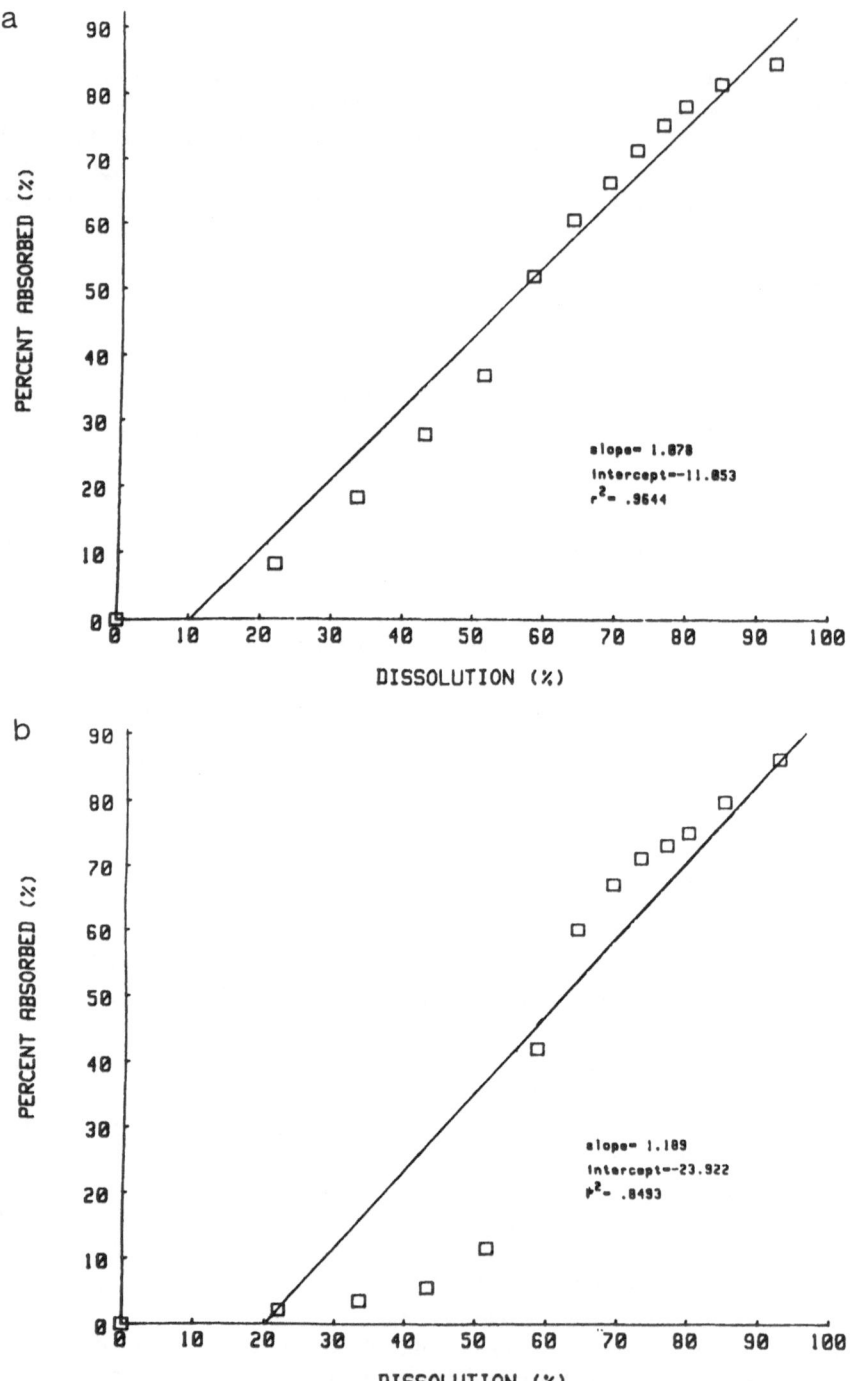

Figure 6. a - Approximated linear relation of a modified release formulation administered under fasting conditions. b - Approximated linear relation of the same modified release formulation administered under fed conditions. Nonlinear nature relationship becomes more apparent under fed conditions.

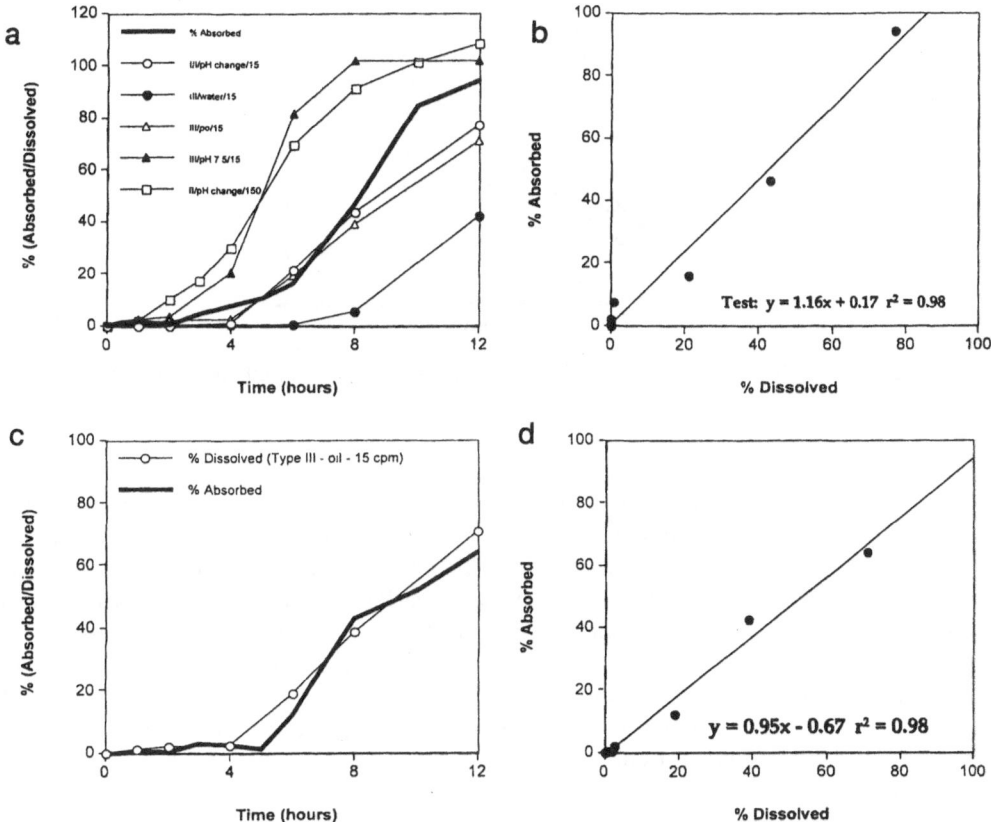

Figure 7. a - Qualitative assessment of in vitro - in vivo relationship for TIMERx formulation administered under fasting conditions, for absorption compared to dissolution in different apparatus and media. b - Quantitative linear relationship of in vitro - in vivo relationship for TIMERx formulation administered under fasting conditions, for absorption vs. dissolution optimized for fasted conditions. c - Qualitative assessment of in vitro - in vivo relationship for TIMERx formulation administered under fed conditions, for absorption compared to dissolution optimized for fed conditions. d - Quantitative linear relationship of in vitro - in vivo relationship for TIMERx formulation administered under fed conditions, for absorption vs. dissolution optimized for fed conditions.

ing parameter must pass dissolution specifications. Extension of the range of a critical manufacturing parameter may be justified if the batch will pass dissolution specifications. The USP acceptance tables (29) appear to be based on an empirically developed "decision tree." These USP tables are used as a guide for setting specifications and possibly have roots in the evaluation of digoxin, mentioned previously.

Typically, the historically-based average plus or minus some measure of variation has been used to set dissolution specifications (Figure 8). Experience has demonstrated that the average ± 3.0 SD's may be used as a guide to setting dissolution specifications for some modified release formulations. A more meaningful method of developing dissolution specifications would be through the use of *in vitro-in vivo* correlation or biorelevant dissolution based on a Level A correlation. Two ways have been suggested to go about this (1,12):

Deconvolution. Limits of variation observed in the actual plasma concentrations observed in clinical studies are assessed to determine absorption-time profiles (Scheme I, *in*

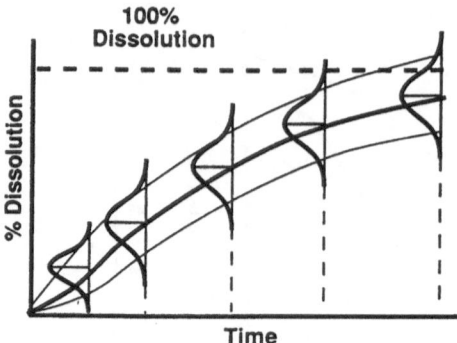

Figure 8. Example of dissolution-time data for a modified-release formulation with typical variation in dissolution at set time points.

vitro-in vivo Correlation). For example, the 95% confidence intervals of observed plasma concentrations may be deconvoluted to provide absorption-time profiles that may be used as a guide to set dissolution specifications. This approach, however, has potential problems associated with it because not all sources of variation in plasma concentration-time profiles are taken into account, and this approach tends to over-estimate the degree of variation that should be allowed in the dissolution specifications (Figure 9).

Rather, it would be more appropriate to deconvolute the variation associated with pharmacokinetics, which generally constitute a large portion of the overall variation observed in estimation of absorption.

Sources of Variation

$$
\begin{aligned}
\text{Var(abs)} &= \text{F[Var(diss), Var(pk), Var(error)]} \\
\text{Var(diss)} &= \text{F[Var(intra-batch), Var(inter-batch), Var(assay), Var(error)]} \\
\text{Var(pk)} &= \text{F[Var(intra-subj), Var(inter- subj), Var(assay), Var(error)]}
\end{aligned}
$$

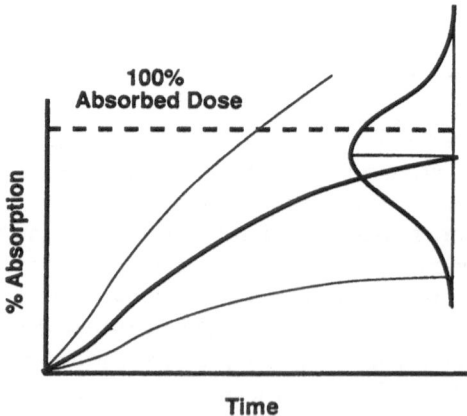

Figure 9. Example of absorption-time data with 95% confidence intervals as deconvoluted from mean plasma concentration-time data and respective 95% confidence intervals.

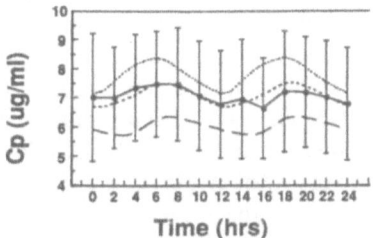

Figure 10. Average, steady-state plasma concentrations (±SD) observed for patients administered a control-release formulation compared to simulations based on convolution of low, mid, and high dissolution specifications. This convolution approach justifies proposed dissolution specifications.

Here Var() represents a some function of variation for either absorption (abs), dissolution (diss), or pharmacokinetics (pk). For each of these there may be further sources of variation, which are defined in the function of F[].

Convolution. Limits of dissolution are used to simulate drug input to a pharmacokinetic model associated with the active ingredient of the dosage form (Scheme I, Biorelevant Dissolution). This is essentially the "biorelevant dissolution" approach, coined by Dr. L. J. Leeson (9). The resulting simulations of plasma concentrations are compared to variation observed in the clinical studies.

Wagner-Nelson method (14) of deconvolution was used to estimate the absorption-time profile, which could be compared to cumulative dissolution-time data. Excellent linear correlations of absorption versus dissolution at common time points were observed. The Rescigno index (calculated with an exponent of one and no weighting) was found to be 0.043 up to 24 hr, while the relative area between the curves was 0.072 up to 24 hr. Dissolution data from 5 clinical batches were pooled (individual tablets, n=60), and specifications were established based on the average ± 2.5 SD at 3, 6, 12, and 24 hr. The upper and lower limits of the dissolution specifications were then convoluted, based on an appropriate pharmacokinetic system for actual patients, and compared to variation observed in clinical, steady-state studies (Figure 10).

Although variation in resulting simulations are within the variation observed for the actual plasma concentration-time profiles, this approach still fails to take into account all pertinent sources of variation. That is, variation associated with the pharmacokinetic system have not been convoluted in with the simulations, giving plasma concentrations which are dependent on absorption.

A PROPOSED APPROACH TO VALIDATION OF DISSOLUTION SPECIFICATIONS

Newer methods for establishing dissolution specifications for controlled-release formulations have now been proposed in the recent guidance for development, evaluation and application of *in vitro - in vivo* correlations for extended release solid oral dosage forms (30). Certainly, any two lots of a product on the market should bioequivalent. However, batches manufactured with averages centered at the limits of dissolution specifications should not necessarily be required to be bioequivalent. A biostudy to investigate IVIVR over this *in vitro* range would provide some insight as to whether a correlation was valid over this range. Traditionally, dissolution specifications have been set to encompass the limits of variation associated with lot-to-lot manufacture. And, the specifications are set with sensitivity to detect production of units outside the normal lot-to-lot variation. However, a lot with average dissolution centered on the lower limit (or upper limit) of dissolution specifications may very well not pass the dissolution specifications themselves and would theoretically not be on the market. For appropriate drug formulation candidates, side-batch formulations representing the limits of bioequivalence should be determined. These formulation side-batches would then be used to determine the appropriate limits of dissolution. Comparison of side-batches representing dissolution specifications to a manufacturing target reference, or mid-point batch, may be a more reasonable approach, but it too may suffer from some of the same limitations just mentioned.

CONCLUSION

Applicability of *in vitro-in vivo* correlation has currently evolved to include demonstration of a relationship of dissolution release specifications, based on variation in dissolution, to variation in plasma concentrations resulting from administration of a controlled-release formulation. Ideally, an *in vitro-in vivo* correlation should demonstrate a relationship between critical formulation parameters and dosage form performance, *in vitro* and *in vivo*. The prospect of using an *in vitro-in vivo* correlation as a surrogate for bioequivalency studies appears to have come of age, especially for situations involving scale-up or post-approval changes of a controlled-release dosage form. Inference of optimal biological performance should come from a Level A correlation. The most practical use of IVIVCs in support of dissolution specifications would be to demonstrate a relationship holds true over the limits of bioequivalency. However, actual specifications should be expected to be extend above and below the specific range validated so that the dissolution specifications will be useful as a tool for monitoring process control.

REFERENCES

1. AAPS/USP/FDA Workshop on Scale-Up of Extended-Release Dosage Forms; Crystal Gateway Mariott Hotel, Arlington, VA; Sept. 8–10, 1992; and Skelly JP, Van Buskirk GA, Arbit HM, et al.; Pharm Res 10(12):1800–1805, 1993.
2. Pharmacopeial Forum; July-August, 1988; pg. 4160.
3. Cardot JM and Beyssac E, Eur J Drug Metab Pharmacokin, 18(1):113–120, 1993.
4. Skelly JP and Shiu GF; Eur J Drug Metab Pharmacokin, 18(1):121–129, 1993.
5. Cohen JL, Hubert BB, Leeson LJ, et al.; Pharm Res, 7:983–987, 1990.
6. Lindenbaum J, Butler VP, Murphy JE et al.; Lancet, June 2(1):1215–1217, 1973.
7. Johnson BF, McCrerie J, Greer H, et al.; Lancet, June 30(1):1473–1475, 1973.
8. Shaw TRD, Raymond K, Howard MR, et al.; Br Med J, 4:763–766, 1973.
9. Leeson LJ; LJL Associates, Inc., personal communication.
10. Siewert M; Eur J Drug Metab Pharmacokin, 18(1):7–18, 1993.
11. Oosterhuis B and Jonkman JHG; Eur J Drug Metab Pharmacokin, 18(1):19–30, 1993.
12. Pharmacopeial Forum, 19(3):5366–5379; May-June, 1993.
13. Generic Drugs Advisory Committee, Open Session; Rockville, MD; Jan. 11–12, 1994.

14. Wagner JG and Nelson E; J Pharm Sci, 52(6):610–611, 1963.
15. Loo JCK and Riegelman S; J Pharm Sci, 57:918–928, 1968.
16. Gillespie WR; PCDCON: Deconvolution for Pharmacokinetic Applications; The University of Texas at Austin, Austin, TX.
17. Chan K, Langenbucher F, and Gibaldi M; J Pharm Sci, 76(6):446–450, 1987.
18. PCNONLIN; Statistical Consultants, Inc.; Lexington, KY.
19. SAS; SAS Institute, Inc.; Cary, NC.
20. NONMEM; NONMEM Project Group, University of California; San Francisco, CA.
21. Leeson LJ, Adair D, Clevenger J, et al.; J Pharmacokin Biopharm, 13(5):493–514, 1985.
22. STELLA II; High Performance Systems, Inc.; Hanover, NH.
23. Hwang SS, Bayne W, Theeuwes F; J Pharm Sci, 82(11):1145–1150, 1993.
24. Holford NHG and Sheiner LB; Clin Pharmacokin, 6:429–453, 1981.
25. Langenbucher F; Pharm Ind, 38(5):472–477, 1976.
26. Rescigno A; Pharm Res, 9(7):925–928, 1992.
27. Moore JW and Flanner HH, Pham Tech, June 1996, pg 64–74.
28. McCall TW. Diehl D, Baumgarner C, et al.; 11th Annual AAPS Meeting, October, 1996, Poster PDD 7279 [TIMERx is a trademark of TIMERx Technologies].
29. USP XXII; Chapter <711>, pg 1578–1579; and Chapter <724>, pg 1580–1581.
30. Malinowski H, Marroum P, Uppoor VR, et al.; Guidance for Industry; Extended Release Solid Oral Dosage Forms; Development, Evaluation, and Application of *in vitro - in vivo* Correlations; Center for Drug Evaluation and Research (CDER), July 1, 1996.

2

DISSOLUTION ASSAY DEVELOPMENT FOR *IN VITRO–IN VIVO* CORRELATIONS

Theory and Case Studies

Brian R. Rohrs, John W. Skoug, and Gordon W. Halstead

Pharmaceutical Development
Pharmacia and Upjohn
Kalamazoo, Michigan 49001

1. THEORY

1.1. Introduction

To obtain an *in vitro-in vivo* relationship (IVIVR), two sets of data are needed. The first set is the *in vivo* data, usually a pharmacokinetic metric derived from a plasma concentration profile (e.g., AUC, Cmax, % Absorbed, etc.). The second data set is the *in vitro* data. This is usually drug release data from a dissolution test and most often takes the form of percent dissolved as a function of time. A mathematical model describing the relationship between these data sets is then developed. While fairly obvious, it should be pointed out that when trying to develop an IVIVR, the *in vivo* data is fixed. Once this data is generated, it establishes the relevant performance of a particular dosage form or series of formulations. The *in vitro* drug release profile, on the other hand, may be modified through changes in the dissolution test conditions. It is the goal of the pharmaceutical scientist to vary *in vitro* conditions such that a dosage form or formulation series behaves in a manner similar to that found *in vivo*.

In the ideal world, *in vivo* data would be available prior to developing the dissolution assay. The *in vivo* drug release performance of a formulation or series of formulations would then be established, and the dissolution test would have a defined optimization endpoint. In the real world however, the drug development timeline requires choosing initial dissolution test conditions prior to release of clinical formulation lots. *In vivo* data is gathered, and only then is the information available to assess whether the *in vitro* test conditions are appropriate. If the test does not provide the correct relative relationship, the *in vitro* test conditions should be modified.

This sequence of events puts a great deal of emphasis on the initial dissolution test. Although the test conditions may be altered during drug development, there is usually a significant amount of inertia associated with the initial dissolution test. A data base of formulation performance and stability has often been established. Changing the test involves re-validation of the assay and analyst training on the new method. In addition, dissolution data should be generated on at least some of the older lots, if available, to establish a data base with the new method. The advantage of having an IVIVR for formulation development and optimization and for regulatory purposes more than compensates for the additional development work.

1.2. Initial Dissolution Assay Development

Developing a dissolution test *a priori* requires a knowledge of the physical/chemical properties of the drug molecule and the dosage form designed to deliver it. Typically, the most important components are drug solubility and stability. The stability issue is relatively straightforward. The drug sample should not decompose before the analytical assay is performed. If some decomposition occurs, the assay should be capable of quantifying not only the parent compound, but also the degradation products in order to get a true measure of drug release rate. This adds a level of complexity to the dissolution assay which may inhibit the ready acceptance of the method in a routine assay environment.

The solubility issue is somewhat more complex, but typically the test medium should be able to completely dissolve the amount of drug added. This is referred to as sink conditions. The USP has recommended sink conditions of 3x or greater[1], i.e., the final drug concentration in the medium should be 1/3 or less of the saturation concentration. Others have recommended up to 10x sink[2], or not more than 1/10 the saturated concentration is obtained if the drug added dissolves completely. Figure 1 shows a hypothetical solubility profile as a function of pH for an ionizable free base. Also plotted is the dose number for a 200 mg formulation. Dose number is the fraction of saturated solubility, so for a dose number of 1.0, the medium would be completely saturated if 100% of the drug added dissolved.[3] To achieve 10x sink conditions, one would be restricted to a pH such that dose number did not exceed 0.1. In this example, it would require using a pH of 3.0 or less. To achieve 3x sink conditions (dose number = 0.33), a pH of about 3.5 would be the maximum allowable. While these sink recommendations may be appropriate for some compounds, one can envision situations where it may be desirable to have concentrations

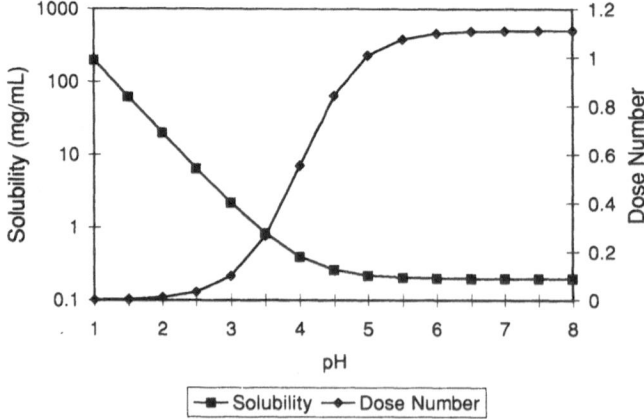

Figure 1. Hypothetical solubility and dose number profile for an ionizable free base with pKa = 4.0, intrinsic solubility = 0.2 mg/mL, and formulated to a 200 mg dose.

closer to saturated solubility in order to obtain an IVIV correlation, for example, a compound where solubility is suspected to be the limiting factor for dissolution *in vivo*.

Other considerations when developing a test are the type of formulation and the relevance of the dissolution medium to physiological conditions where the dosage form is intended to release drug. Medium pH is often considered to be the most important physiologically relevant variable, although perhaps this is because it is the most easily characterized *in vivo* and easy to control *in vitro*. Some standard dissolution media are pH 1.2 simulated gastric fluid without enzymes for immediate release formulations, and pH 6.5–7.5 for sustained release formulations. For modified or sustained release products, the formulation itself is designed to control release of the drug. The medium pH may be important if drug solubility or dissolution of a pH dependent polymer controls drug release. Other release mechanisms, such as matrix erosion or drug diffusion through a membrane/gel matrix may be dependent on hydrodynamics and agitation rate. For the USP 1 (basket) apparatus, the agitation rate is typically 100 rpm, and for the USP 2 (paddle), 50 rpm.[4] Osmotic pump delivery systems utilize osmotic pressure differential to control drug release. In this case, agitation has no effect on drug release rate.[5]

Theoretically, when searching for an appropriate *in vitro* test, if the mechanism which controls *in vivo* drug release operates *in vitro*, one should obtain an IVIVR. However, one should not necessarily expect an IVIVR to hold across formulations if the drug release mechanism changes.

1.3. Dissolution Assay Modification

Dissolution assay modification involves two types of variables, apparatus and medium. The apparatus variables control the hydrodynamics and include type of apparatus (basket, paddle, reciprocating cylinder, or flow through)[6] and agitation or flow rate. From a theoretical viewpoint, apparatus variables should affect formulations where disintegration, deaggregation, erosion, or dispersion plays an important part in release of drug from the dosage form. Medium variables include pH, ionic strength, surfactant type/concentration, and buffer species. These variables control factors like solubility, wettability, osmotic pressure, buffer capacity, etc., and changes in the dissolution medium will affect formulations where these factors have a role in the drug release mechanism.

When searching for a correlation in the universe of possible dissolution variable combinations, response surface methodology or experimental design approaches are an efficient way to screen assay variables and optimize assay conditions.[7] These methods allow many variables to be changed simultaneously, and regression analysis provides the degree of change or response of the dissolution metric to changes in dissolution variables. The following case studies are three examples which illustrate the use of response surface methodology to find appropriate dissolution conditions for *in vitro-in vivo* relationships. The first example is a sustained release tablet formulation, the latter two are immediate release tablet formulations.

2. CASE STUDIES

2.1. Example 1

The first example involves a drug that is very soluble over a wide pH range, and is formulated at low dose in a hydrophilic polymer matrix. Three prototype formulations

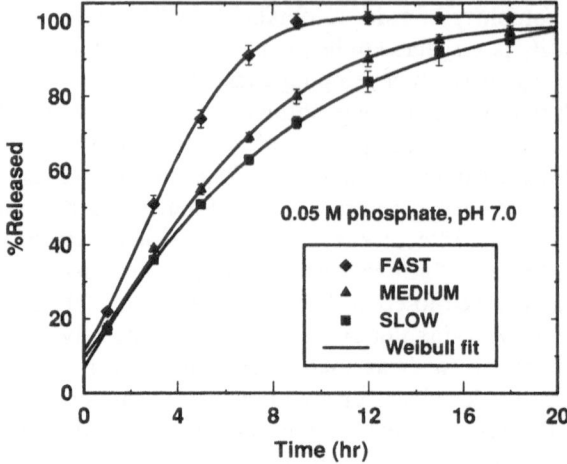

Figure 2. Dissolution profiles for three sustained release formulations in example 1 designed to have slow, medium, and fast release profiles.

were developed with different drug release profiles. The initial dissolution test conditions utilized the USP 1 apparatus (basket) at 100 rpm. The dissolution medium was 500 mL of pH 7.0, 0.05 M phosphate buffer (ionic strength = 0.11 M). Sampling intervals were 1, 3, 5, 9, 12, and 18 hours and HPLC was used to quantify drug concentration in samples.

Dissolution profiles are shown in Figure 2. The solid line is the best fit to the Weibull function shown in equation 1:

$$\%D(t) = \%D_{\infty}\left[1 - \exp\left(\frac{t_0 - t}{t_d}\right)^{\beta}\right] \tag{1}$$

where $\%D_{\infty}$ is the percent dissolved at infinite time, t_0 is the lag time, t_d is a dissolution factor, and β is an empirical exponential factor. Figure 3 contains the corresponding

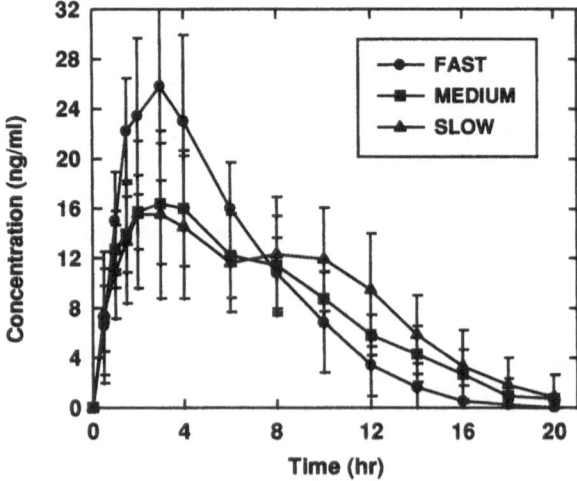

Figure 3. Plasma concentration profiles for slow, medium, and fast sustained release formulations.

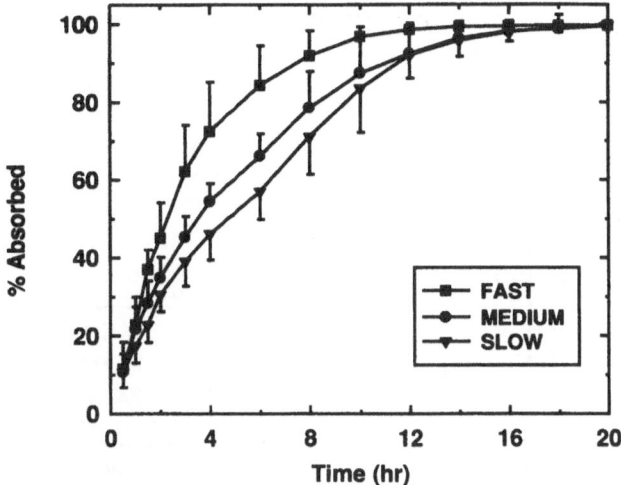

Figure 4. Estimated *in vivo* absorption profiles from plasma concentrations in Figure 3.

plasma profiles. These *in vivo* profiles were transformed to yield the absorption profiles in Figure 4. Using the Weibull function fits to the dissolution profiles, percent dissolved *in vitro* was estimated at times corresponding to those for which *in vivo* data was available. The IVIV relationship is shown in Figure 5 where percent absorbed is plotted against percent dissolved for all three formulations.[8] Also plotted is the best fit quadratic equation. From the plasma profiles, one formulation was chosen for further development.

At this point, the dissolution test was evaluated over a series of different experimental conditions. Even though a good IVIVR had been obtained, there were several reasons for pursuing this evaluation, among them to get information about the ruggedness of the formulation and the drug release mechanism, to maximize discriminating ability of the assay, and to gather data to support test rationale for regulatory filings. A 3-factor 3-level fractional factorial design was used to study the response surface of the dissolution test results with respect to pH, ionic strength, and agitation rate. Table 1 contains the values for the variables studied. Instead of collecting dissolution data at all combinations of the variables, the study design can be envisioned as collecting data only at combinations which form the corners, face centers, and center points of a cube centered at (0,0,0) on a 3D cartesian axis system. This reduces the number of experimental points to collect from 3^3 or 27 for all combinations to 15.

Examples of the response surface can be seen in Figures 6 and 7 where percent dissolved at five hours is plotted against pH and ionic strength at 100 rpm, and pH and rotation speed at I.S. = 0.175 respectively. Although the surfaces look steep because of the expanded z-axis, in reality, the response surfaces are relatively flat. The results from the

Table 1. Example 1 variable values for 3-factor fractional factorial dissolution study

Coded levels	-1	0	1
pH	1.2	4.5	8.0
Ionic strength	0.05	0.175	0.3
Basket rpm	50	100	150

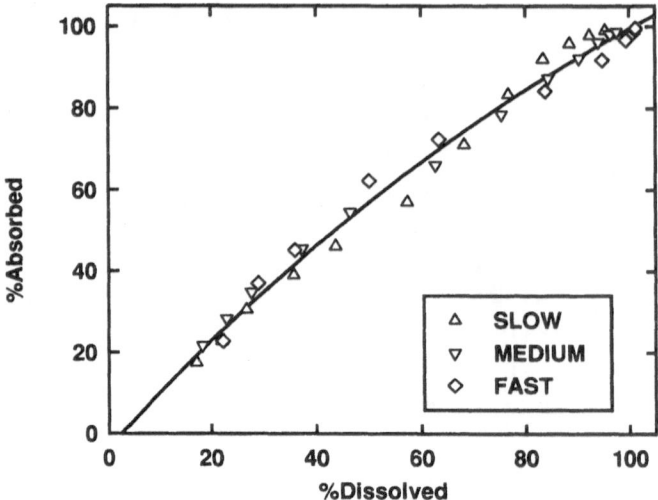

Figure 5. *In vitro-in vivo* correlation for example 1 slow, medium, and fast sustained release formulations.

study led to the conclusions that the matrix type formulations were rugged over a wide variety of conditions. Formulations which varied in polymer content behaved similarly as a function of the test variables so that conditions for a more discriminating test were not identified. Also studied was not just extent of drug release versus time, but also dissolution assay variability. Standard deviations were relatively constant except at low pH and high ionic strength, so there would be no advantage to using other test conditions to reduce assay variability. Figure 8 contains pooled experimental data from four biostudies and the final IVIV correlation curve.

2.2. Example 2

The next example is a compound formulated into an immediate release tablet. The initial dissolution test used a 75 rpm paddle with pH 1.2 simulated gastric fluid without enzyme dissolution medium. Early in development a biostudy was designed to compare a 50 mg tablet, two different formulations of a 100 mg tablet, and a 200 mg tablet at equivalent doses. As shown in Figure 9, a range of about 10% was observed in tablet performance as measured by AUC, but there was no *in vitro* discrimination between tablet lots. One premise we use is that if differences between formulations are observed *in vivo*, there should be conditions that can produce similar differences *in vitro*. The solubility characteristics of the drug made pH the most likely variable to significantly impact the dissolution profiles, so the response of dissolution to pH was measured. Figure 10 contains the results. At higher pH, the dissolution profiles look similar. Based on the dose used for testing, a dose number greater than 1.0 (saturated solubility) is achieved at pH 4 and above. A dissolution medium at pH 3 appeared to be viable since the dose could be completely dissolved, yet the profile was slow enough to perhaps afford some discrimination between formulations. Dissolution of the four tablet lots was performed using the pH 3 medium with a 75 rpm paddle, and, as shown in Figure 11, a good rank order relationship was obtained. This relationship was used to drive further formulation development, and the correlation was verified by later biostudies.

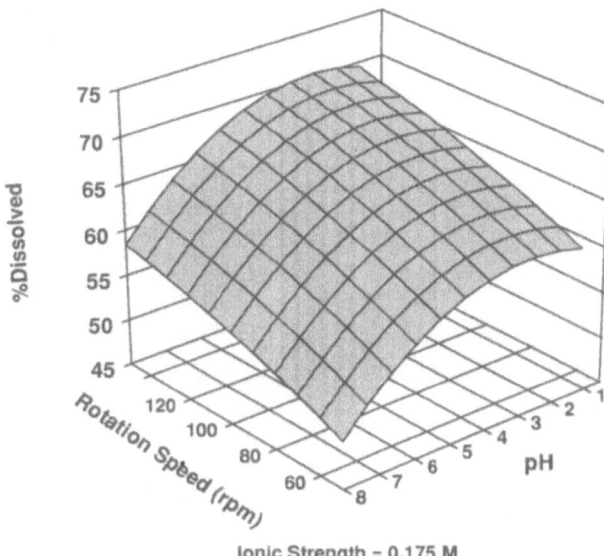

Figure 6. Response surface for pH and rotation rate effect on drug release for 3-factor fractional factorial dissolution study.

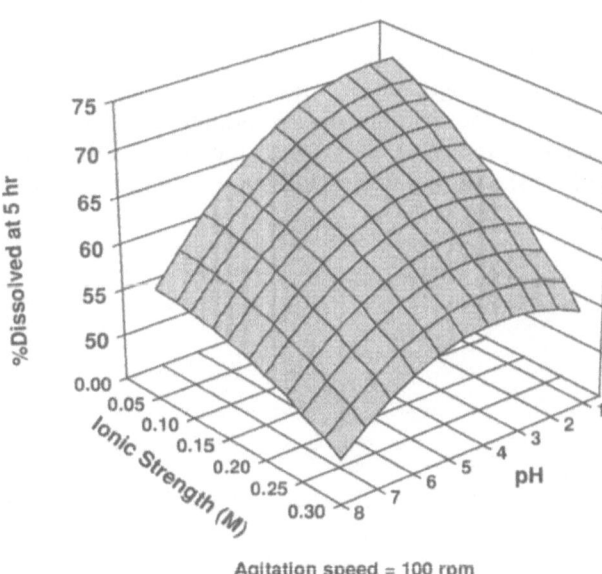

Figure 7. Response surface for pH and ionic strength effect on drug release for 3-factor fractional factorial dissolution study.

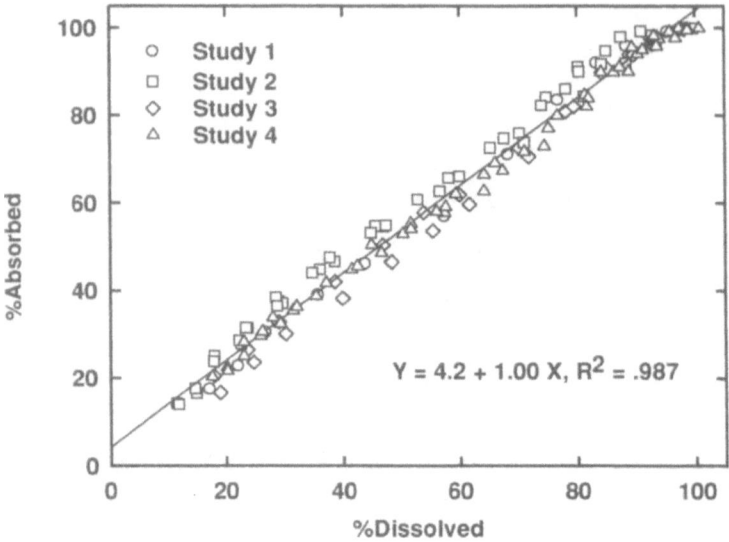

Figure 8. Example 1 *in vitro-in vivo* correlation curve using data pooled from four biostudies.

2.3. Example 3

The final example is a situation where it was not at all obvious which dissolution conditions would lead to a good IVIV relationship. The compound was formulated as the salt of a free base into an immediate release tablet. The initial dissolution test used the paddle apparatus at 50 rpm, and pH 1.2 simulated gastric fluid without enzymes (SGF) as the medium. A high drug load formulation was being pursued, and a formulation finding biostudy was run. Figure 12 shows the relationship between AUC and percent dissolved at 60 minutes for five tablet formulations. It is readily apparent that the dissolution test does not discriminate between formulations in a meaningful way.

Usually, one bases modifications to the dissolution test on an understanding of the bulk drug or formulation properties. In this case, the drug is formulated as a highly soluble salt, so drug solubility should not control dissolution behavior. The formulation was de-

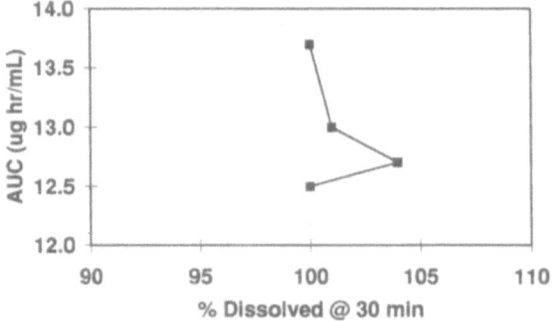

Figure 9. *In vitro-in vivo* relationship for four formulations in example 2 using pH 1.2 simulated gastric fluid without enzymes as the dissolution medium.

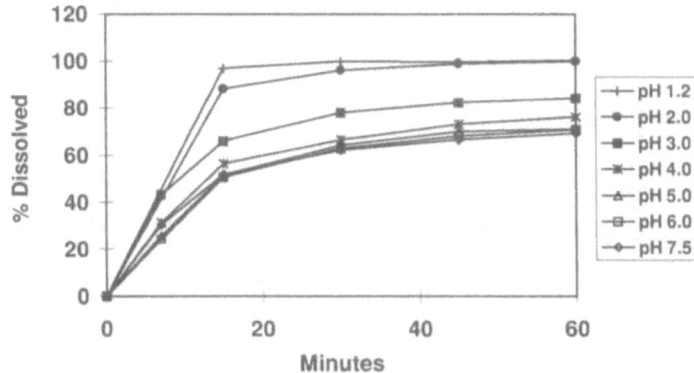

Figure 10. Dissolution profiles in various pH media for example 2,200 mg formulation.

signed to be immediate release, so it should not control release of drug. In fact, the only difference between tablet lots C and D was the lot of bulk drug used for the formulation. This points to some property of the bulk drug as the controlling factor for determining AUC. One hypothesis was that precipitation of a less soluble free base form occurs *in vivo*, and the extent of that precipitation controls the relative bioavailability. Even with that working hypothesis, it was not evident which test variables would affect dissolution.

An initial variable screening study was designed to look at the effect of variations in pH, ionic strength, surfactant concentration, and paddle agitation rate. To choose a surfactant, two were screened for their effect on the extent of dissolution for lots A and D. In order to get a rank order IVIV relationship, the extent of dissolution for lot A needs to increase, whereas that for lot D needs to decrease (see Figure 12). An anionic surfactant, sodium dodecylsulfate (SDS), and a non-ionic surfactant, Triton-X-100, were tested at 0% and 2% in SGF. Figure 13 shows that dissolution of lot A increased slightly with SDS, but decreased with Triton-X-100. For lot D, SDS reduced the extent of dissolution whereas Triton-X-100 had no effect. SDS was therefore selected as the surfactant for further studies. Additional screening showed that pH and SDS concentration have a significant effect on the extent of dissolution, but ionic strength and agitation rate have limited or no effect.

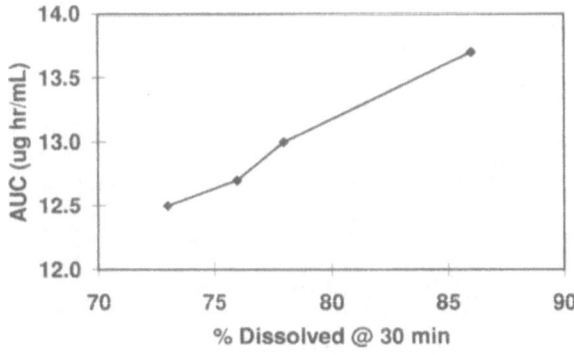

Figure 11. *In vitro–in vivo* relationship for example 2 formulations using pH 3.0 dissolution medium.

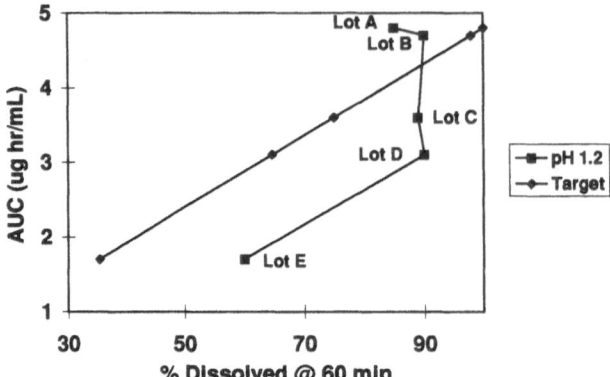

Figure 12. IVIVR for formulations in example 3 using pH 1.2 SGF and 50 rpm paddle apparatus. Also shown are target dissolution values for a linear correlation between AUC and extent of dissolution.

A study was designed to map out the response surface of dissolution versus SDS concentration and pH. The design used was a central composite for two variables, which may be envisioned as the corners and center point of a square centered at coded value (0,0) with data also collected at points ± √2 along the cartesian axes. This type of design defines the curvature of the surface within the square to a reasonable degree. The coded and actual variable values are listed in Table 2 under Study 1.

The pH range was limited to below 5.8 since based on experience, the drug precipitates out of solution above pH 3. The surfactant concentrations spanned the generally used range for SDS in dissolution medium. Lots A and D were run, again with the intent of finding conditions which raised the dissolution extent of A and lowered that of D. To get the appropriate rank order, the difference between dissolution of lots A and D needed to be large and positive. As shown in Figure 14, the difference over most of the response surface was small. Only at two conditions tested was the magnitude of the difference large, but those conditions were at the edge of the variable space. The points were pH 0.17, 1.5% SDS and pH 5, 0.43% SDS. The pH of the first point was considered too low for an acceptable dissolution test, so a second 2-factor central composite study was initiated with pH 5, 0.4% SDS located near the center point. One lesson learned from the first study was

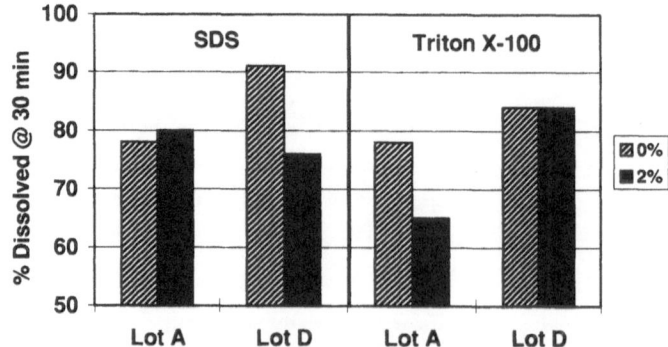

Figure 13. Effect of surfactant type and concentration on extent of dissolution of Lots A and D for example 3.

Table 2. pH and sodium dodecylsulfate concentration (in wt.%) values for 2-factor central composite design studies in example 3

Study number	Coded levels	$-\sqrt{2}$	-1	0	1	$\sqrt{2}$
Study #1	pH	0.2	1.0	3.0	5.0	5.8
	% SDS	0	0.4	1.5	2.6	3.0
Study #2	pH	2.0	2.7	4.5	6.3	7.0
	% SDS	0.01	0.2	0.65	1.1	1.3

that drug precipitation is limited at high pH when surfactant is present, so the pH range could be extended higher. Another conclusion was that the surfactant range was too wide to get a smooth response surface. This information was used to establish the variable ranges, listed in Table 2, for the second study.

Response surfaces were obtained for tablet lots A, D, and E. Figures 15, 16, and 17 show the results and the best fit to the quadratic equation 2:

$$Z = a + b*X + c*Y + d*X^2 + e*Y^2 + f*X*Y \qquad (2)$$

where X and Y are the coded values for pH and SDS concentration respectively. The response surface for lot A is relatively flat, but drops off at the lowest surfactant value, presumably because it is below the critical micelle concentration for SDS. The surface for lot D shows pH having a large effect with a small contribution from SDS. Lot E's surface shows a large influence on dissolution from both pH and SDS.

To find conditions which yield a good IVIV relationship, target values for dissolution were estimated by assuming a 1:1 correlation with AUC could be found. Figure 12 shows the data from the pH 1.2 test and the target dissolution values. The surface equations provide a means to predict dissolution values for any combination of pH and SDS concentration for lots A, D, and E. Using the SOLVER tool within EXCEL, the sum of the square of the differences between predicted dissolution and the target values was minimized. The predicted best fit was pH 6, 0.6% SDS, so dissolution on all five lots was performed using those conditions. The results, plotted in Figure 18, show a very good IVIV relationship between AUC and percent dissolved at 60 minutes. In this example, response

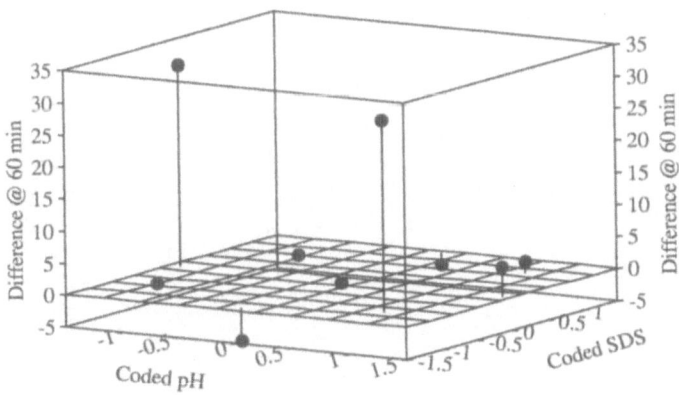

Figure 14. Difference between extent of dissolution at 60 minutes for lots A and D for example 3, first 2-factor CCD study. A large positive difference is desired.

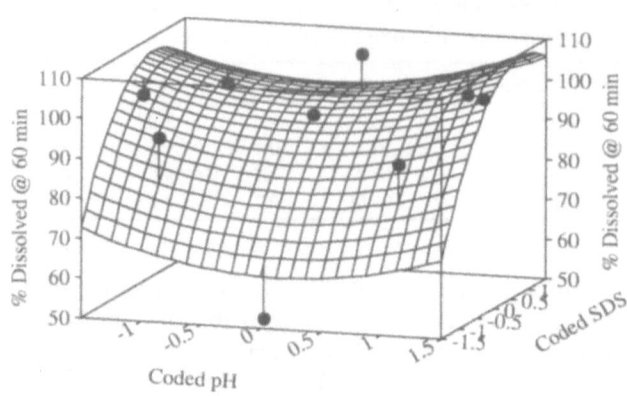

Figure 15. Lot A tablet dissolution response surface, quadratic model Eq. 2.

surface methodology provided a means to find appropriate dissolution conditions where it was not obvious which dissolution test variables or combination of variables to modify.

3. PRACTICAL CONSIDERATIONS

Listed here are some suggestions for developing *in vitro-in vivo* relationships based on our experience.

1. Obtain *in vivo* data on modified formulations as early as possible in development. The sooner an IVIVR can be established, the sooner it can be used to direct formulation development. In addition, changes in the dissolution method are easier to implement early on in drug development before a large historical and stability data base has been established.
2. Reserve samples (50 to 200+ units) from formulation lots for later *in vitro* test development. If test modification needs to occur because *in vivo* differences

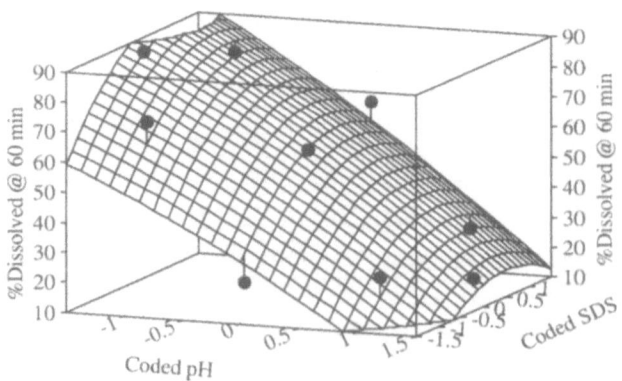

Figure 16. Lot D tablet dissolution response surface, quadratic model Eq. 2.

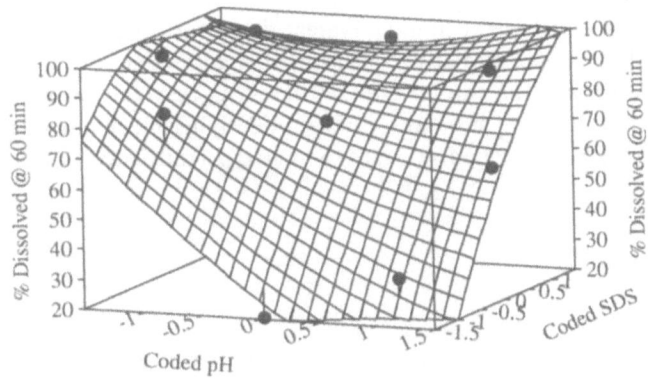

Figure 17. Lot E tablet dissolution response surface, quadratic model Eq. 2.

were not observed *in vitro*, it may be difficult and sometimes impossible to track down enough samples of old clinical lots.

3. Make use of DOE (design of experiment) approaches when searching for appropriate *in vitro* test conditions. The ability to obtain more information from fewer experiments can significantly reduce the time and resource necessary to obtain IVIVRs.

4. When screening test variable effects on dissolution, run three dosage units at each experimental condition rather than the traditional six. Three units give a good estimate of the mean, and two test conditions may be screened within one 'run'. Screening or optimization studies can then occur with about half the number of runs. Of course, study trials may not be completely randomized, since test runs must be grouped by agitation rate.

5. Use differences in dissolution profiles to help direct the *in vitro* test development. Often when developing a correlation, it is not the absolute value of the dissolution profile, but the ability to discriminate between formulations or dosage forms that makes the test conditions appropriate.

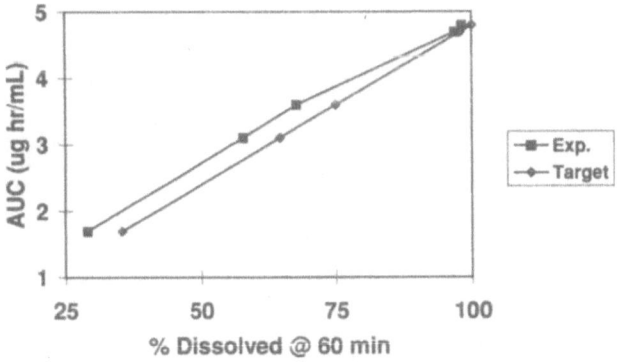

Figure 18. *In vitro-in vivo* relationship for example 3 formulations using pH 6, 0.6% sodium dodecylsulfate dissolution medium.

6. If the dose number is 0.33 (1/3 of saturated solubility) or greater, correlations should be developed using equivalent doses for different formulation strengths. Typically, *in vivo* studies are performed at equivalent doses. If solubility begins to limit dissolution rate *in vitro*, differences between formulations may be observed for different strengths, but the underlying mechanism causing those *in vitro* differences may be irrelevant to *in vivo* performance.

REFERENCES

1. Pharmacopeial Forum, July-Aug 1981, p1226.
2. Abdou HM. Dissolution, bioavailability and bioequivalence. Easton, PA: Mack Publishing Company, 1989, p165.
3. Amidon GL, Lennernas H, Shah VP, Crison JR. A theoretical basis for a biopharmaceutic classification: the correlation of *in vitro* drug product dissolution and *in vivo* bioavailability. Pharm Res 1995;12:413–20.
4. Skoug JW, Halstead GW, Theis DL, Freeman JE, Fagan DT, Rohrs BR. Strategy for the development and validation of dissolution tests for solid oral dosage forms. Pharm Tech 1996; 20:5:58–72.
5. Civiale C, Ritschel WA, Shiu GK, Aiache JM, Beyssac E. *In vivo-in vitro* correlation of salbutamol release from a controlled release osmotic pump delivery system. Meth Find Exp Clin Pharmacol 199; 13(7):491–498.
6. U.S. Pharmacopeia 1995 Chapter 711, Dissolution.
7. Box GEP, Hunter WG, Hunter JS. Statistics for experimenters, an introduction to design, data analysis, and model building. New York: John Wiley & Sons, 1978.
8. Skoug JW, Borin MT, Fleishaker JC, Cooper AM. *In vitro* and *in vivo* evaluation of whole and half tablets of sustained-release adinazolam mesylate. Pharm Res 1991;8(12):1482–87.

3

IN VITRO DISSOLUTION PROFILE COMPARISON AND IVIVR

Carbamazepine Case

Pradeep Sathe, Yi Tsong, and Vinod P. Shah

Office of Pharmaceutical Sciences
Center for Drug Evaluation and Research
U.S. Food and Drug Administration
Metropark North 2
7500 Standish Pl.
Rockville, Maryland 20855

1. INTRODUCTION

USP[1] describes the *in-vivo/in-vitro* correlation as the "establishment of a rational relationship between a biological property, or a parameter derived from a biological property produced by a dosage form and a physicochemical property or characteristic of the same dosage form". In relation to a formulation, the most commonly used biological properties are the pharmacokinetic parameters such as Cmax or AUC, obtained following the administration of the dosage form while the physicochemical property is the dosage form's *in-vitro* dissolution performance such as percent of drug released under a given set of conditions. The relationship between these two biological and physicochemical properties, is then expressed quantitatively. For rapidly dissolving (less than 15 minutes) immediate release formulations, an *in-vitro* characterization could be made using a single dissolution point. This however cannot be said for immediate release formulations consisting either low solubility actives or slowly dissolving matrix or for modified release formulations. In these cases, a dissolution profile and not a single dissolution specification point, is more appropriate. If an adequate *in-vitro/in-vivo* relationship is established using a suitable technique such as statistical moments or deconvolution, an insight into the pharmacokinetics could be obtained from the formulation's *in-vitro* dissolution performance. More importantly, in case of an alteration in the dissolution profile characteristics, it allows one to raise a flag about the possible alteration in the *in-vivo* performance and pharmacokinetics.

In Vitro–in Vivo Correlations, edited by Young *et al.*
Plenum Press, New York, 1997

1.1. IVIVR Classification and Study

The *in-vitro* and *in-vivo* correlation levels are traditionally defined as either A, B or C based on the descending order of correlation and usefulness. Level A correlation represents a point to point relationship between *in-vitro* dissolution and *in-vivo* input rate of the dosage form. Level B correlation compares the mean *in-vitro* dissolution time to either mean residence time or the mean *in-vivo* dissolution time. Level C, which essentially represents a single point correlation, relates a percent dissolved point to a pharmacokinetic parameter.

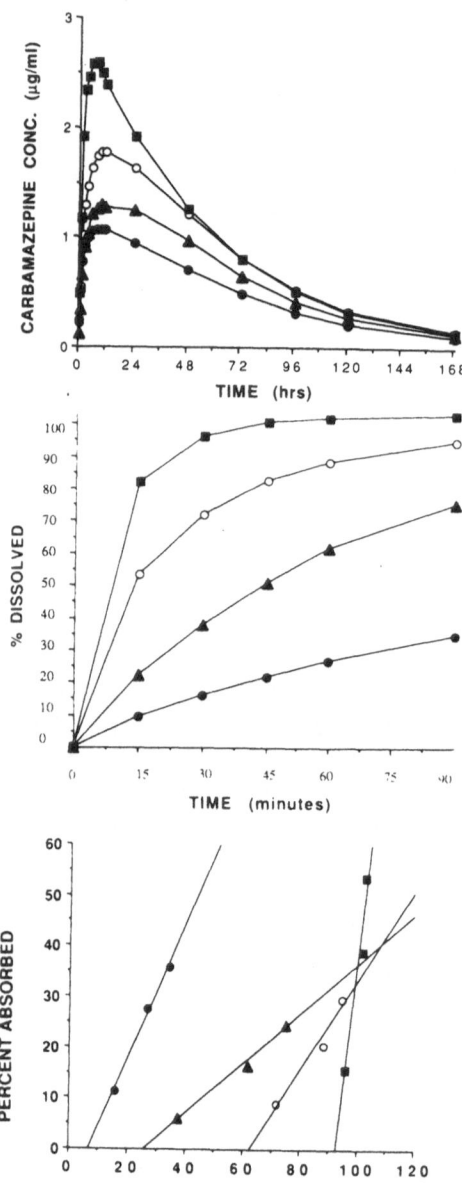

Figure 1. IVIVR of carbamazepine tablet formulations. Reproduced with permission from Meyer et al., Pharmaceutical Research 9(12): 1612–1616 (1992).

Meyer[2] et al. studied *in-vitro/in-vivo* relations (IVIVR) of the carbamazepine immediate release tablet formulations with bioavailability problems. A formulation specific *in-
vitro/in-vivo* relation, Figure I (a-c), was established using four carbamazepine tablet lots
with different dissolution characteristics. In case, such a relationship is established on a
formulation, the dissolution profiles which are predictive of the pharmacokinetics can be
put to effective use. In case the firm desires to make a change in the bio-study formulation
(scale-up and post-approval change), which includes change in the manufacturing site or
equipment, change in additives or increase the batch-size (which is invariably done post-
approval), the dissolution profiles could be statistically compared to ascertain the changes
in the formulation release characteristics and therefore the pharmacokinetic or pharmacodynamic endpoints. Recently, the agency has published scale up and post approval
change (SUPAC) guidances[3a,3b] which state the use of dissolution profiles to compare the
pre-modified and post-modified lots.

2. *IN-VITRO* DISSOLUTION PROFILE COMPARISON METHODS

The similarity of the two dissolution profiles can be considered as either local or
global. The local similarity refers to similarity around a dissolution measurement point
and this in turn gets reflected into the global or overall similarity. The global or overall
similarity on the other hand refers to the similarity on the whole and may or may not get
translated into the local similarity. Currently, we are evaluating the following dissolution
profile comparison approaches a] "Model Dependent" approach, b] "Model Independent"
approach and c] "Principal Component" analysis approach. The "Model Dependent" approach, as the name suggests, does need specification of a mathematical function or
"model" to describe the dissolution data under consideration. We recommend this approach for a dissolution "data rich" scenario, hitherto defined as the dissolution data consisting of four or more data points. "Model Independent" approach does not require
specification of a mathematical model to describe the dissolution characteristics and may
be used with three or less than three dissolution data points. This approach can be further
subgrouped into the "Index" approach and "Multivariate" approach. In "Index" approach[4,5] a mathematical index is calculated using an apriory defined "acceptable" dissolution difference (say 10%). With respect to that difference, and the corresponding
acceptable index value, a new index value calculated from the test and reference mean dissolution difference, is compared and similarity or dissimilarity is declared. In "Model Independent Multivariate" approach, in addition to the mean dissolution value, the variance
covariance of the data is also accounted[6]. The principal component analysis approach is
recommended for unique situations consisting of dissolution "data rich" scenario where an
adequate mathematical model may not be fitted to describe the dissolution data. This approach compares the principal components generated by slicing the dissolution profiles
into different sections. The principal components obtained by combining percent dissolved at various time points, may then be compared using a "Bonferoni"[7] type test.

Each of these dissolution profile comparison approaches have certain pros- and
cons. The f1 and f2 indexes proposed by Moore and Flanner[4] as well as Resigno[5] index are
relatively simple to calculate and are not model dependent. They however do not account
for the variance covariance structure of the data and can be effectively used only for the
data with small variance. Resigno index also has some other problems such as bias[4] towards curves at different locations. The "Model Independent Multivariate" approach of
Tsong[6] et al. may be used effectively with a few data points. It may however become com-

plicated with non-identical covariance structures of the similarity and confidence regions, especially when the number of dissolution data points is large. It also cannot be used with non-identical sample schemes. A detailed comparative discussion of all the dissolution profile comparison approaches is beyond the scope of this presentation and is done elsewhere[8].

2.1. "Model Dependent" *in-Vitro* Dissolution Profile Comparison

The following discussion pertains to "Model Dependent" profile comparison approach and uses a realistic example of carbamazepine tablet dissolution data. The steps involved in the approach are as follows:

1. Define a suitable and best fitting mathematical model function to describe the dissolution data, which is generated from many production size standardized batches or lots.
2. Construct a similarity region (SR) based on the parameter variances of the standard batches. Subsequently, compared to the SR, the similarity or dissimilarity of the dissolution profiles is ascertained.
3. Fit the individual unit dissolution data coming from the pre-modified or reference (on which IVIVR has been established) and post-modified or test lots using the model specified in Step 1, to generate model parameters.
4. Calculate a statistical distance between the parameter means, which incorporates the variance and covariance structures of the parameters in addition to the mean differences.
5. Normalize the statistical distance calculated in step 4, so that it can be compared and analyzed using a statistical distribution.
6. Compute a confidence region (CR) of certain percentile (say 90%), around the normalized statistical distance between the lot means.
7. Compare the CR with SR, calculated in step 2, to declare the similarity or dissimilarity of the dissolution profiles of the two lots.

Now, consider the bio-study lot on which IVIVR has been established (say dissolution profile of product "1" in Figure 1) and also consider that many more similar production size lots have been manufactured by the firm. The mean dissolution profiles of four such standardized carbamazepine lots, besides the reference on which IVIVR has been established, are given in Figure 3 (simulated data).

2.1.1. Step Details. Step 1 involves defining the appropriate and best fitting mathematical function to describe the dissolution data. Various empirical mathematical functions such as "Probit", "Logistic", "Weibull", "Exponential", "Quadratic" etc.(Table I), could be utilized to ascertain the best suitable function to describe the dissolution data coming from different standard lots. The mathematical function selection can be done either by a two-stage or in case all data is not available at identical time points, by a one-stage approach. In this case, we utilized a two-stage approach. The mathematical functions were fitted to individual unit dissolution data and then to the mean dissolution data to ascertain the best fitting function. A standard goodness of fit criteria such as least sum of squares, residual mean square error and "Akaike" information criterion[9] could be used to ascertain the best fitting function. Ideally, a parsimony of model parameters and a small percent coefficient of variation (say <3%) of the parameter estimates, indicating adequate precision, are prerequisites for the proper use of the approach. Upon evaluation of differ-

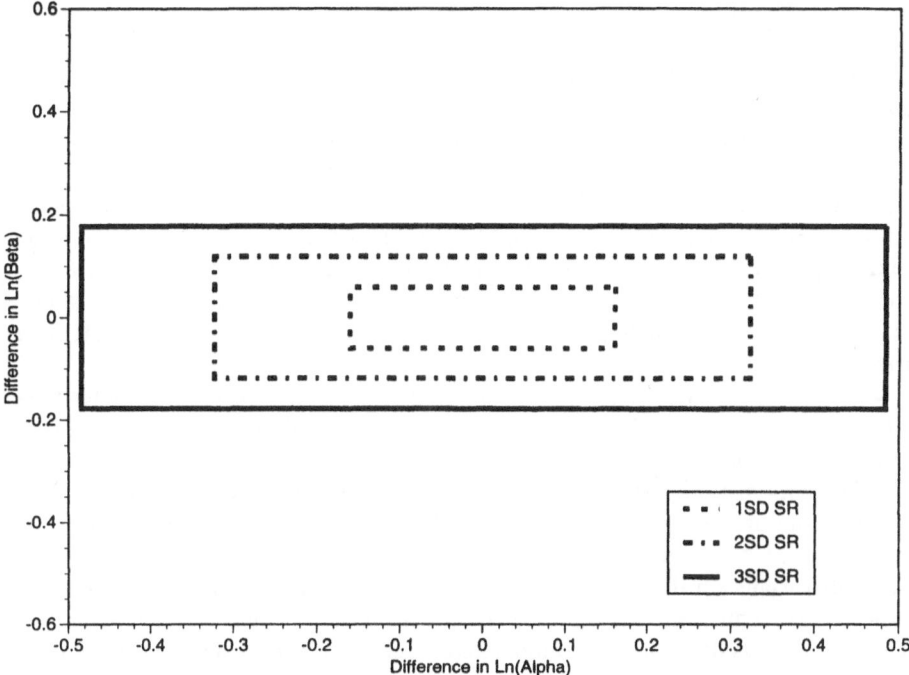

Figure 2. Carbamazepine standard lot mean dissolution profiles; Reference and four other (simulated) profiles.

ent model functions (Table I) , it was found that "Weibull" was the best fitting and most robust function to describe the dissolution data. "Weibull" function and its utility for describing dissolution data has been discussed previously[10].

In Step 2, the similarity region or the criterion for the similarity or dissimilarity of the dissolution profiles is defined. This is done by taking into consideration the intra- and inter-lot variances of the model parameter estimates of the standard lots. The parameter variances are pooled using the following formula and pooled standard deviation is calculated.

$$\text{Pooled SD} = \text{Sqrt}[\{(\text{Var}_1 + \text{Var}_2 + \text{Var}_3 + \text{Var}_4)/4\} + \text{Var}(\bar{x})] \tag{1}$$

where,

$\text{Var}_1, \text{Var}_2, \text{Var}_3, \text{Var}_4$ are the intra—lot variances of the ln parameters,

$\text{Var}(\bar{x})$ is the inter-lot variance of the ln-parameter means.

Table I. Selected mathematical functions describing percent dissolved 'X' with respect to time 't' and model parameters 'α' and 'β'

Function	Form
1. Probit	$X/100 = \Phi(\alpha + \beta*(\log(t)))$ where Φ = Standard normal distribution.
2. Logistic	$X/100 = e^{[\alpha+\beta*\log(t)]}/\{1+e^{[\alpha+\beta*\log(t)]}\}$
3. Weibull	$X/100 = 1-e^{-\alpha*(t)**\beta}$
4. Quadratic	$X/100 = \alpha + \beta 1*(t-t\#) + \beta 2*(t-t\#)^2$ where t# = average of all sampled time values.
5. Exponential	$X/100 = 1-e^{-\alpha*(t)}$

In case the profiles are superimposable, the differences of the model parameter estimates and the pooled standard deviations would be zero. Boxes corresponding to either one two or three pooled standard deviations of the parameter estimates are therefore constructed around zero axes. This results in the construction of a univariate similarity region which accounts for the variance of the model parameters. In our example, the similarity regions corresponding to either one, two or three pooled standard deviation estimates, covering approximately 67%, 95% and 99% cases, were constructed based on the data from four standard lots, each consisting of 12 tablet dissolution data. Each individual unit dissolution profile was fitted using linearized "Weibull" function[10], which was the function of choice as ascertained in Step 1, parameter variances were pooled and standard deviation boxes were constructed around zero to generate similarity region (SR) as seen in Figure 2.

Ideally, a multivariate similarity region which accounts for the parameter covariance in addition to the variance, should be calculated. This calculation is accurate when one has data from many more standard batches. The theoretical discussion of the calculation of the multivariate similarity region is beyond the scope of this article and is done by the authors elsewhere[11]. In our opinion, calculation of the univariate similarity region is more convenient and liberal. On the other hand, if one has sufficient dissolution data on many standard lots, a multivariate similarity region, which is more correct and appropriate, should be constructed. To reduce the skewness of the distribution, a logarithmic conversion of the model parameters is done in this example and is recommended where appropriate.

Step 3 deals with the comparison of the test (post-modified) and reference (pre-modified) data. Figure 4 shows the hypothetical mean test dissolution profile of car-

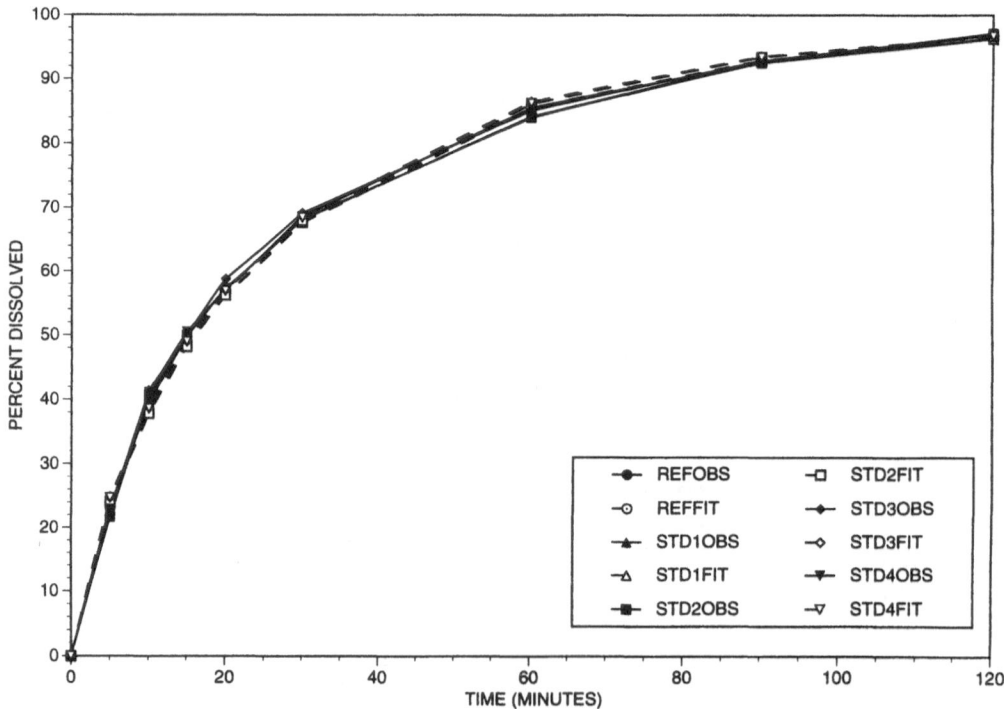

Figure 3. The similarity region boxes corresponding to one, two or three pooled standard deviations.

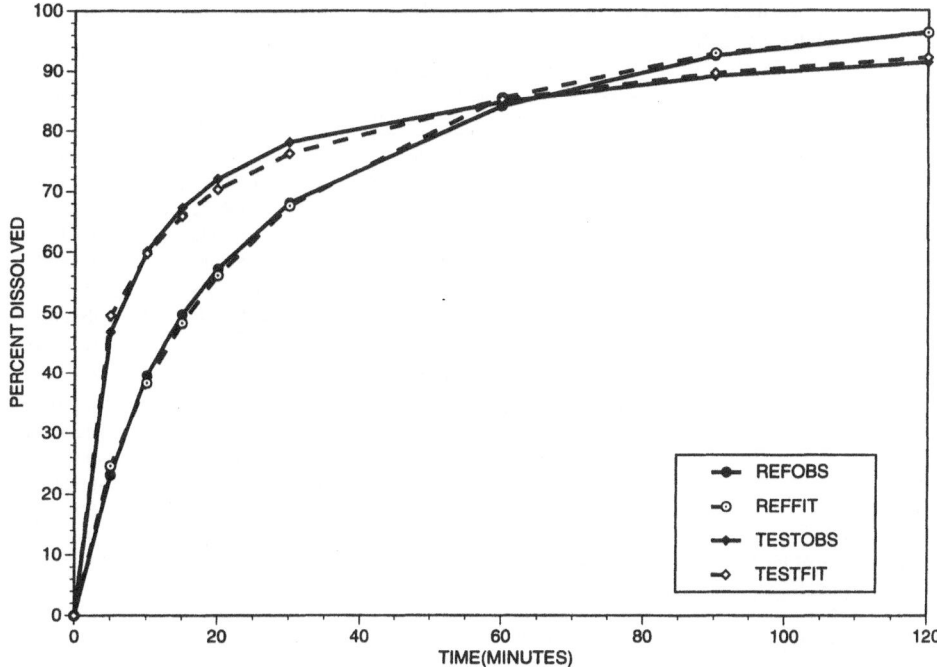

Figure 4. Mean test (post-modified) and reference (pre-modified) carbamazepine profiles with "Weibull" function fits.

bamazepine tablets along with the mean reference profile. The individual unit dissolution data from these two lots, pre-modified and post-modified are fitted using the model function (in this case Weibull), model parameter estimates corresponding to each curve are generated and log-transformed. The individual unit parameters with mean and standard deviations of the two lots are given in Table II. In this example, data consisting of 6 tablet units are used, however, ideally 12 unit dissolution data should be used for better variance assessment.

A statistical distance between the parameter means is calculated in step 4 using the following formula.

$$D^2 = [(X_T - X_R)' \, S_{pooled}^{-1} \, (X_T - X_R)] \qquad (2)$$

where,

D^2 = Squared statistical or "Mahalanobis" or "M" distance,,
X_T, X_R = Vectors corresponding to the sample means of test and reference lot $\ln(\alpha)$'s and $\ln(\beta)$'s
S_{pooled} = Pooled sample variance/covariance matrix, equal to $(S_{test} + S_{ref})/2$,

where S_{test} and S_{ref} are the variance/covariance matrix of $\ln(\alpha)$ and $\ln(\beta)$ in the test and reference batch respectively. For matrix arragement, please refer to "Appendix".

Table II. Test and reference lot, 'Weibull' parameters and subsequent statistics such as mean, standard deviation, Mahalanobis distance, Hotelling T^2 and confidence intervals

Tablet number	Alpha (ref.)	Beta (ref.)	Alpha (test)	Beta (test)
1	1.88062	0.83238	2.08515	0.51335
2	1.8122	0.76114	1.81377	0.41265
3	1.80093	0.77569	1.80972	0.40596
4	1.95862	0.78835	1.95002	0.41668
5	2.05513	0.7596	1.95311	0.38791
6	2.10544	0.78971	1.90724	0.37503
	LnAlpha(ref.)	LnBeta(ref.)	LnAlpha(test)	LnBeta(test)
1	0.63447	-0.18347	0.73484	-0.66679
2	0.59454	-0.27293	0.59541	-0.88516
3	0.5863	-0.254	0.59317	-0.90151
4	0.67224	-0.23781	0.66784	-0.87544
5	0.72034	-0.27497	0.66942	-0.94698
6	0.74452	-0.23609	0.64566	-0.98075
Mean	0.65907	-0.24321	0.6510567	-0.8761038
Std.	0.064817	0.0336369	0.0531327	0.110013
Mean diff.(test-ref.)	-0.008012	-0.63289		
'M' distance, D^2	70.355			
Sc. 'M' distance	211.065			
Con. interval for the diff. in	-0.096452,	-0.754287,		
$\ln(\alpha)$ and $\ln(\beta)$	0.0804285	-0.5115		

It is evident from the formula that the statistical distance, referred as the "Mahalanobis" or "M'[12] distance in the statistical literature, is different than the 'Euclidean" distance between the two means. In statistical distance calculation, in addition to the differences between the parameter means, the variance and co-variance structures of the parameters are also accounted.

The "Mahalanobis" distance which is a composite number of the mean parameter differences, and the parameter pooled variance, covariance structures, is then normalized or scaled to a statistic, known as "Hotelling T^2"[12], which accounts for the number of units in each data set and number of parameters in the model. Step 6 is thus essentially a standardization step to get a mathematical entity, which can be compared using a standard statistical distribution. In this case, it is the "F'-distribution.

Scaling factor:

$$K' = \{(N_1+N_2-P-1)/[(N_1+N_2-2)P]\} * [(N_1 N_2)/(N_1+N_2-P)] \qquad (3)$$

where,

N_1, N_2 = Number of units in the lot,
P = Number of parameters

$$\text{Scaled 'M' distance, } T^2 = K' * D^2 \qquad (4)$$

The normalized "Mahalanobis" distance which follows the statistical "F'-distribution can be utilized in the next step to generate a confidence region. In Step 7, a confidence region (CR) with a particular percentile, in this case we selected 90th percentile, is constructed around the normalized statistical distance. This is done by the following equation.

$$CR = \{K^{'}[(Y-(X_T-X_R))^{'}\ S_{pooled}^{-1}\ (Y-(X_T-X_R))]<=F_{P,N1+N2-P-1,\ .90}\} \tag{5}$$

The response value "Y", when satisfying the equality of the expression, generates a geometric shape which depends on the difference in the mean parameters, their variance/covariance structures as well as number of units per lot and number of model parameters. The values satisfying the inequality fall inside this shape. In this instance, since the used "Weibull" function has two parameters "Alpha" and "Beta", and since the variances of the differences between the "Alpha" and "Beta" parameters are not exactly identical, this results into an elliptical shape. Had the variances of the differences between Alpha's and Beta's been exactly identical, this would result in a circular shape. If the model function had three parameters, the shape would have been spherical. As the model parameters increase to more than three, the shape of the confidence region also becomes increasingly complex. The elliptical 90% confidence region corresponding to the difference in profiles of the two dissolution lots is seen in Figure 5 contrasted against the precalculated similarity region (SR) boxes. It is evident that the ellipse clearly falls outside the three standard deviation box of the similarity region, implying the dissimilarity of the dissolution profiles of the two lots. For the comparison, we also calculated an ellipse for the data from the two standard pre-modified lots (as seen in Figure 3 and mentioned as REF and STD in Figure 5). As expected, this ellipse clearly falls within the three standard

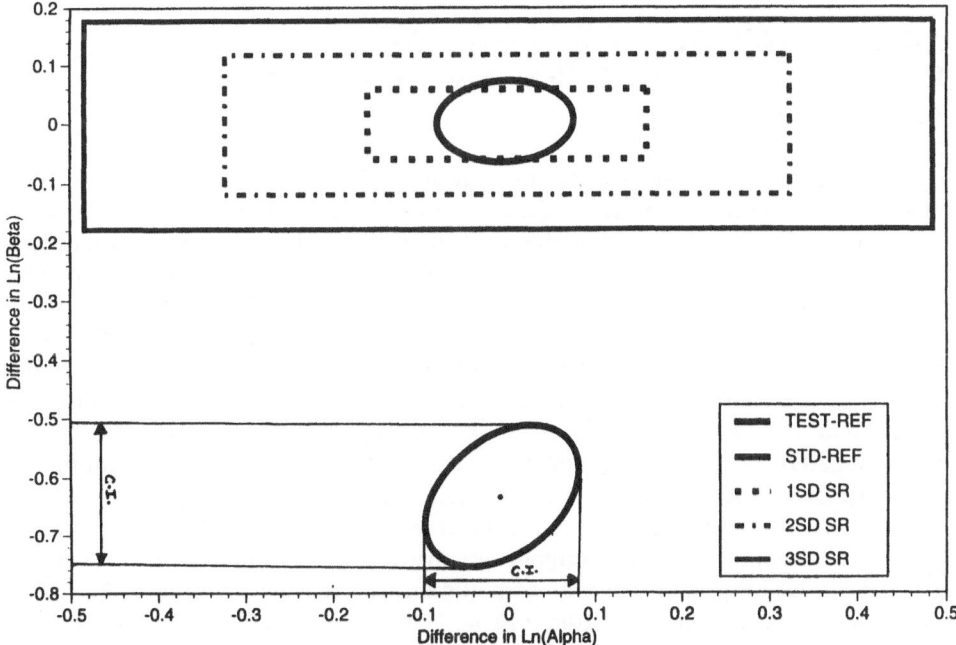

Figure 5. Ninety percent Confidence Regions (CR) of the mean difference in the Ln "Weibull" parameters contrasted with the Similarity Region (SR) boxes. Please note the center of the ellipse and the confidence intervals.

deviation box of the similarity region, suggesting similarity of the dissolution profiles of the two standard lots.

2.2. Meaning of the Ellipse and Its Components

From the study of the ellipse, at one glance, a lot of useful information about the dissolution profiles comparison, could be gathered. The center of the ellipse represents the mean difference of the model parameters, in this case "Ln-Alpha" and "Ln-Beta". Invariably, as could be seen from the test (post-modified) vs. reference (pre-modified) lot comparison, the center of the ellipse dose not include both zeros of the X and Y axes, meaning the differences in mean ln-parameters are statistically significant ($\alpha=0.1$). If on the other hand, the ellipse includes both zeros, as evident from the two standard lot comparison, this means that the difference in the mean ln-parameters are not statistically significant. The direction of the ellipse indicates the positive or negative correlation of the parameter differences. In this example it could be seen that the test and reference lot dissolution profile parameter differences in Alpha's and Beta's (on the natural log scale) are negatively correlated. For the dissolution profile comparison of the two standard lots, the correlation of the dissolution profile parameter differences is almost negligible. The reader is cautioned that the correlations are for the mean ln-parameter differences and not for the actual parameters. If one draws two tangents from the ellipse on the X and Y axes, this defines the confidence intervals for the difference in respective parameters. In short, the ellipse from its position relative to axes, indicates where the mean parameter difference would lie with a certain statistical precision.

3. ADVANTAGES/DISADVANTAGES OF THE MODEL DEPENDENT DISSOLUTION PROFILE COMPARISON APPROACH

By using the model dependent dissolution profile comparison approach, the dissimilarity of the dissolution profiles of the two lots if any, could be detected not only due to the differences in means but also due to differences in the variance covariance structures of the data. The principal advantage of the approach is reduction in dimensions. The dissolution data consisting of four or more dissolution points are reduced to a few dimensions leading to a simpler data handling. The approach accounts for the variance covariance structure of the data. Since dissolution data points are not independent, this aspect also could be considered a major advantage. Due to empirical model fitting, data from different sampling schemes could be accommodated and used. Finally, "Weibull" model parameters "Alpha" and "Beta" may be considered as meaningful from the dissolution perspective, "Alpha" being the scale or extent factor and "Beta" being the shape factor. The *in-vitro* model parameters, as documented earlier[13], could be correlated with the *in-vivo* parameters. The disadvantages of the approach include a need for the "data rich" scenario limiting its use for the formulations not dissolving rapidly. Also, in case the approach is used with a few (three or less) data points, model mis-specification is likely, leading to erroneous results and conclusions. A word of caution about the data collection. The approach may work adequately only if the dissolution data are collected at the meaningfully spaced intervals. If the collected data points are clustered near the origin (0% dissolved) or the end of the dissolution (theoretically 100% dissolved), it may lead to erroneous conclusions.

4. SUMMARY

Dissolution data for the immediate or modified release drug products are usually collected as percent dissolved at multiple time points. Once an *in-vitro/in-vivo* relationship is established on a drug product, the dissolution profile becomes meaningful and important. In that context, if a firm desires to modify its formulation on which the *in-vitro/in-vivo* association has been established, a meaningful insight into the pharmacokinetics may be obtained by comparing the dissolution profiles of the two lots. In this presentation, we demonstrated a model dependent dissolution profile comparison approach using example of carbamazepine tablet dissolution data. Once a mathematical function was selected to describe the dissolution data coming from various standard lots, a similarity region could be constructed using the model parameter variances. To compare the test and reference lot dissolution profiles, a statistical distance was calculated between the mean parameters. A confidence region generated around the normalized mean statistical distance could then be compared with the similarity region to assess the similarity or dissimilarity of the dissolution profiles.

5. APPENDIX

For "Mahalanobis" distance, the arrangement of vector matrix is as follows :
$(X_T - X_R)$ is arranged as a 2*1 and 1*2 matrix with the following form....

$$\text{Spooled} = \begin{Bmatrix} 1 & 2 \\ 3 & 4 \end{Bmatrix}$$

$$(X_T - X_R) = (\ln(\alpha_T)-\ln(\alpha_R)), (\ln(\beta_T)-\ln(\beta_R)) \quad \text{and} \quad \begin{Bmatrix} (\ln(\alpha_T)-\ln(\alpha_R)) \\ (\ln(\beta_T)-\ln(\beta_R)) \end{Bmatrix}$$

$$\text{where } 1 = [\text{Var}(\ln(\alpha_T))+\text{Var}(\ln(\alpha_R))]/2$$
$$2,3 = [\text{Cov}(\ln(\alpha_T), \ln(\beta_T)) + \text{Cov}(\ln(\alpha_R), \ln(\beta_R))]/2$$
$$4 = [\text{Var}(\ln(\beta_T))+\text{Var}(\ln(\beta_R))]/2$$

REFERENCES

1. U.S. Pharmacopeia 23 and National Formulary 18, 1995.
2. M.C.Meyer et al. The bioequivalence of carbamazepine tablets with a history of clinical failures. Pharmaceutical Research 9(12): 1612–1616 (1992).
3a. Guidance for Industry. Immediate release solid oral dosage forms; Scale-up and post-approval changes. FDA/CDER, November 1995.
3b. Guidance for Industry. Extended release solid oral dosage forms; Scale-up and post-approval changes. FDA/CDER, July 1996
4. J.W.Moore and H.H.Flanner. Mathematical Comparison of Dissolution Profiles Pharmaceutical Technology 20(6):64–74 (1996).
5. A.Resigno. Bioequvalence Pharmaceutical Research 9/7:925–928 (1992).

6. Y.Tsong, T.Hammerstrom, P.Sathe and V.P.Shah. Statistical assessment of mean differences between two dissolution data sets. Drug Information Journal 30(4):1105–1112 (1996).

7. Fred M. Hoppe. ed., Multiple Comparisons, Selection and Applications in Biometry, Marcel Dekker Inc., New-York, 1993.

8. Y.Tsong, P.Sathe and V.P.Shah, American Statistical Association Proceedings of the Biopharmaceutical Section (in print) 1996.

9. H.Akaike. IEEE Trans.Automat.Contr. 19:716–723 (1974).

10. F.Langenbucher. Linearization of dissolution rate curves by Weibull distribution. J.Pharm.Pharmac., 24:979–981 (1972).

11. P.M.Sathe, Y.Tsong and V.P.Shah. *in-vitro* dissolution profile comparison: statistics and analysis; model dependent approach. Pharmaceutical Research, 13(12):1798–1802 (1996).

12. R.A.Johnson and D.W.Wichern, Applied Multivariate Analysis, Prentice-Hall Inc, New Jersey, 1989.

13. A.Kayali. Bioequivalency evaluation by comparison of *in vitro* dissolution and *in vivo* absorption using reference equations. Eur. J. Drug Met. and Pharmacok. 3:271–277 (1994).

4

A GENERAL FRAMEWORK FOR NON-PARAMETRIC SUBJECT-SPECIFIC AND POPULATION DECONVOLUTION METHODS FOR *IN VIVO–IN VITRO* CORRELATION

Davide Verotta[1,2]

[1]Department of Biopharmaceutical Sciences and Pharmaceutical Chemistry
[2]Department of Epidemiology and Biostatistics
University of California San Francisco
San Francisco, California 94143

1. INTRODUCTION

Suppose that a drug is given in some formulation to a system resulting in an (unknown) input $A(t)$, and n observations are collected at different times following the input. The i-th observation takes the form:

$$y_i = \int_0^{t_i} A(\tau)K(t-\tau)d\tau + \varepsilon_i \tag{1}$$

where ε_i indicates a random measurement error.

The general input identification problem is to estimate $A(t)$ given $K(t)$ (see [5]). In an *in vivo in vitro* correlation situation there is (partial) knowledge about the input $A(t)$ which is obtained from *in vitro* release experiments. However such knowledge is not complete because of the changing conditions between *in-vitro* and *in-vivo*. Given an *in-vitro* release profile and observations from an individual receiving a similar formulation of drug, the main problem facing the data analyst is to determine how different is the *in-vitro* release with respect to the *in vivo* one. An additional aspect of the problem is that the fonts of variability present in the *in-vivo* and *in-vitro* release are different. The *in-vitro* release is generally observed under strictly controlled experimental situations, while the *in-vivo* release is observed indirectly in an experiment involving different humans.

In this paper we describe a general modeling framework to determine the degree of difference between the *in-vivo* and *in-vitro* release which also describes and allows to estimate interindividual differences. The paper is divided in three sections: (I) we describe the

In Vitro–in Vivo Correlations, edited by Young *et al.*
Plenum Press, New York, 1997

general framework, (ii) we report results from a set of simulations demonstrating the approach, and (iii) close the paper with a final discussion.

2. GENERAL FRAMEWORK

2.1. Mathematical Model

We propose an approach which partions the input, $A(t)$, into the observation site (typically plasma) as follows:

$$A_{vivo}(t) = A_{change}(t)\hat{A}_{vitro}(t) \tag{2a}$$

$$A(t) = \int_0^t A_{vivo}(\tau)K_{absorption}(t-\tau)d\tau \tag{2b}$$

In this formulation the function $A_{vivo}(t)$ represents the actual profile of drug released (for example in the gastro intestinal tract). This in turn is the product of $\hat{A}_{vivo}(t)$, which indicates the known (estimated) *in-vitro* release input function, and $A_{change}(t)$ which indicates an unknown component which modifies $\hat{A}_{vitro}(t)$ to account for the changed *in-vivo* conditions.

Finally the convolution of $A_{vivo}(t)$ with $K_{absorption}(t)$ represents the absorption process following the *in-vivo* release. The (*in-vivo*) observations are given (for a system with linear pharmacokinetics) by:

$$y(t) = \int_0^t A(\tau)K_{disposition}(t-\tau)d\tau \tag{2c}$$

where $K_{disposition}(t)$ indicates the disposition function of the drug. Figure (1) depicts the model just described. A few examples of particular instances of the previous general model will help clarify the situation.

Example 1: Identical *in-vivo in-vitro* release; instantaneous absorption In this case

$$A_{vivo}(t) = \hat{A}_{vitro}(t) \tag{3a}$$

and

$$A(t) = A_{vivo}(t) = \hat{A}_{vitro}(t) \tag{3b}$$

which implies $A_{change}(t) = 1$ and $K_{absorption}(t)$ approaching an impulse (delta Dirac, $\delta(t)$) function. This example represents the luckiest situation: a complete equality between *in-vitro* and *in-vivo* release in the observation site.

Example 2: Identical *in-vivo in-vitro* release. First order absorption In this case $A_{vivo}(t) = \hat{A}_{vitro}(t)$, and $K_{absorption}(t) = e^{-k_a t}$ where k_a is the rate of absorption of released drug into the plasma site. The example represents a somewhat more complicated situation:

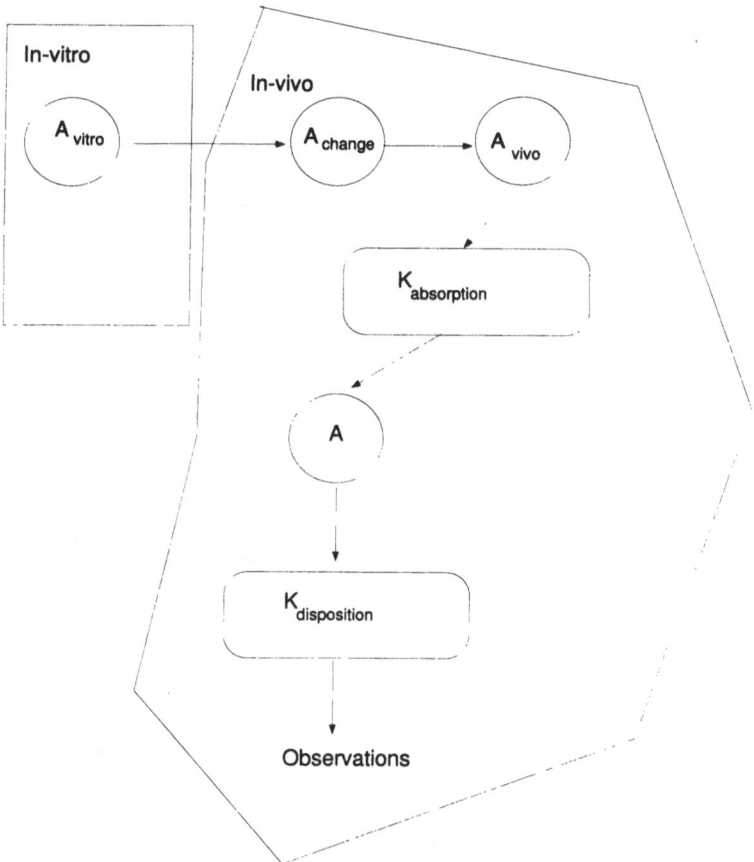

Figure 1. The general model for *in vivo–in vitro* correlation. The input into the observation site, $A(t)$, depends on the in vitro release A_{vitro}, the change of release between in vivo and in vitro $A_{change}(t)$, the absorption process $K_{absorption}$, and the disposition function of the drug $K_{disposition}$.

there is a complete correspondence between *in vitro* and *in vivo* release, but a slow absorption process delays the input into the plasma resulting in an apparent discrepancy between *in vitro* release and the observed release. The model is:

$$A_{vivo}(t) = \hat{A}_{vitro}(t) \tag{4a}$$

$$A(t) = \int_0^t A_{vivo}(\tau) e^{-k_a(t-\tau)} d\tau \tag{4b}$$

Example 3. Different *in-vivo-in-vitro* release. First order absorption An example of this situation is obtained if we substitute $A_{change}(t)$ with e^{-kt} obtaining:

$$A_{vivo}(t) = e^{-kt} \hat{A}_{vitro}(t) \tag{5a}$$

$$A(t) = \int_0^t \hat{A}_{vitro}(\tau) e^{-k\tau} e^{-k_a(t-\tau)} d\tau \qquad (5b)$$

where now e^{-kt} represents a time dependent attenuation of the *in-vitro* release profile, a continuous approximation to, e.g., the elimination of the formulation due to gastric emptying.

The three examples show the flexibility of the general model described above. In general it is easy to see how the approach can describe a variety of situations, and how in particular non-parametric functions (as opposed to the parametric exponentials used in the examples above) can further improve the flexibility of the approach (see the simulations below). We now show how to incorporate inter-individual variability in the *in-vivo in-vitro* correlation model just described.

2.2. Inter-Individual Variability

Inter-individual variability can be present in both the processes altering the *in-vivo/in-vitro* correlation. Individuals can differ in the process changing the *in-vitro* release and in the absorption process. Accordingly, the model for the *in-vivo* release rate from the *j*-th individual ($A_{vivo, j}$) takes the form:

$$A_{vivo,j}(t) = A_{change,j}(t)\hat{A}_{vitro}(t) \qquad (6a)$$

$$A_j(t) = \int_0^t A_{vivo,j}(\tau) K_{absorption,j}(t-\tau) d\tau \qquad (6b)$$

An example of incorporating inter-individual variability in the mathematical framework is provided by the third example above, now:

$$A_{vivo,j}(t) = e^{-k_j t} \hat{A}_{vitro}(t) \qquad (7a)$$

$$A_j(t) = \int_0^t \hat{A}_{vitro}(\tau) e^{-k_j \tau} e^{-k_{a,j}(t-\tau)} d\tau \qquad (7b)$$

where k_j and $k_{a,j}$ represent the *j*-th individual attenuation and absorption constants.

2.3. Semi-Parametric Representation

A semi-parametric representation of the general model just described is given by:

$$A_{change}(t) = Spline_1(t) \qquad (8a)$$

$$K_{absorption} = e^{-k_a t} + Spline_2(t) \qquad (8b)$$

where *Spline*$_1$ and *Spline*$_2$ indicate spline functions. *Spline*$_1$ is always constrained to be between zero and one and to be monotonic non-increasing. *Spline*$_2$ is always constrained to be greater than -e^{-kat}. [Splines are the sum of *ni* polynomials of a certain degree (cubic degree, for a cubc spline), which satisfy continuity conditions of all the derivative up to the degree minus at the 2nd, ..., *ni*-th point of an increasing sequence of points called breakpoints. For example a linear spline which goes through a set of points results in the familiar broken line connecting the points.] The semi-parametric representation allows to investigate the *in-vivo/in-vitro* correlation without having to assume a priori a particular model.

3. SIMULATED DATA

To test the different methods we simulated three data sets. In all of them there are 20 subjects with 20 serum samples taken in each person at log-equispaced times between two minutes and 20 hours after dose administration. Each subject has a biexponential disposition function:

$$K_j(t_{ij}) = \delta_{1j}\, e^{-\delta_{2j}\, t_{ij}} + \delta_{3j}\, e^{-\delta_{4j}\, t_{ij}} \tag{9}$$

with δ_{1j}, δ_{2j}, δ_{3j}, and δ_{4j} log-normally distributed with mean $2_1 = 0.04$, $2_2 = 15.0$, $2_3 = 0.02$ and $2_4 = 0.4$, respectively, and covariance matrix:

$$\begin{pmatrix} 0.04 & & & \\ 0.02 & 0.04 & & \\ 0.0001 & 0.0001 & 0.04 & \\ 0.0001 & 0.00001 & 0.02 & 0.04 \end{pmatrix}$$

In all the simulated data sets the *in vitro* release rate function is given by the (unimodal, increasing and than decreasing) biexponential:

$$\hat{A}_{vitro}(t) = \frac{\theta_7}{\theta_6 - \theta_5}\left[e^{-\theta_5 t} - e^{-\theta_6 t}\right] \tag{10}$$

with $2_5 = 2.0$, $2_6 = 2.6$ and $2_7 = 10^4$, respectively.

3.1. First Simulated Data Set

Identical *in-vivo in-vitro* release; instantaneous absorption. In this data set

$$A_{vivo}(t) = 1, K_{absorption\, j}(t) = \delta(t)\ and\ A_j(t) = \hat{A}_{vitro}(t) \tag{11a}$$

3.2. Second Simulated Data Set. Identical *in-Vivo in-Vitro* Release; First Order Absorption

In this simulation $\hat{A}_{vitro}(t) =$ is the same of the first simulation, $K_{absorption,j}(t) = e^{-k_a j t}$ where k_a is log normally distributed with mean .5 and approximately 25% variance. The resulting individual input function is:

$$A_j(t)=\int_0^t A_{vivo}(\tau)e^{-k_{a\,j}(t-\tau)}d\tau \qquad (11b)$$

3.3. Third Simulated Data Set

Different *in-vivo in-vitro* release; first order absorption

In this simulation $A_{change,\,j}(t)(t)=e^{-kjt}$ where k_j is log-normally distributed with mean 0.05 and variance 0.0625, and $K_{absorption,\,j}$ is the same of simulation 2.

$$A_{vivo\,j}(t)=e^{-k_j t}\hat{A}_{vitro}(t) \qquad (12a)$$

$$A_j(t)=\int_0^t \hat{A}_{vitro}(\tau)e^{-k_j\tau}e^{-k_{a\,j}(t-\tau)}d\tau \qquad (12b)$$

In all the simulations observations are obtained by calculating the convolution of $A_j(t)$ with $K_j(t)$ and adding a random proportional error of 10%. The data corresponding to the three simulations are shown in figure (2).

In the analysis of the simulated data, the subject disposition function is assumed to be known exactly, i.e. the correct subject disposition functions are used during the analysis. To simulate a situation where the data analyst does not know the correct model underlying the data, six kinds of *in-vivo/in-vitro* correlation models are fitted to each of the different data sets. The most general model fit to the data (M1) is as follows:

$$A_j(t)=\eta_{1j}\int_0^t \hat{A}_{vitro}(\tau)Spline_1(\tau)\left[e^{-k_{a\,j}\eta_{2j}(t-\tau)}+\eta_{3j}Spline_2(t-\tau)\right]d\tau \qquad (13)$$

where η_{1j}, η_{2j}, and η_{3j} are fixed to 1, for method 1, and are log normally distributed for method 2. The other five models (M2 to M6) are characterized as follows:

M2: $Spline_2(t)=0$ (*in-vivo* change; first order absorption)
M3: $Spline_1(t)=1$ (no *in-vivo* change; complex absorption)
M4: $Spline_2(t)=0$, $e^{-k_{a,\,j}\eta_{2j}(t-\tau)}=1$ (*in-vivo* change; fast absorption)
M5: $Spline_1(t)=1$, $Spline_2(t)=0$ (no *in-vivo* change; first order absorption)
M6: $Spline_1(t)=1$, $Spline_2(t)=0$, $e^{-k_{a,\,j}\eta_{2j(t-\tau)}}=1$ (no *in-vivo* change; fast absorption)

3.4. Estimation Methods

We use two different estimation methods to obtain the desired estimates. Method 1 is a subject-specific two-stages method which takes advantage of the richness of the data available in a in- vitro/*in-vivo* correlation study. Method 2 is a mixed effect population method which in the present context assumes the parameters 0_j to be log-normally distributed and uses maximum-likelihood to obtain the estimates of the parameters in the input model, and the variance covariance matrix of the random effects. Individual estimates are in this case empirical Bayes ones. We used the computer program [2], to obtain the desired estimates in both cases. To select between the different models we used the methods reported in [3] for method 2, and the Akaike criterion [1] for method 1.

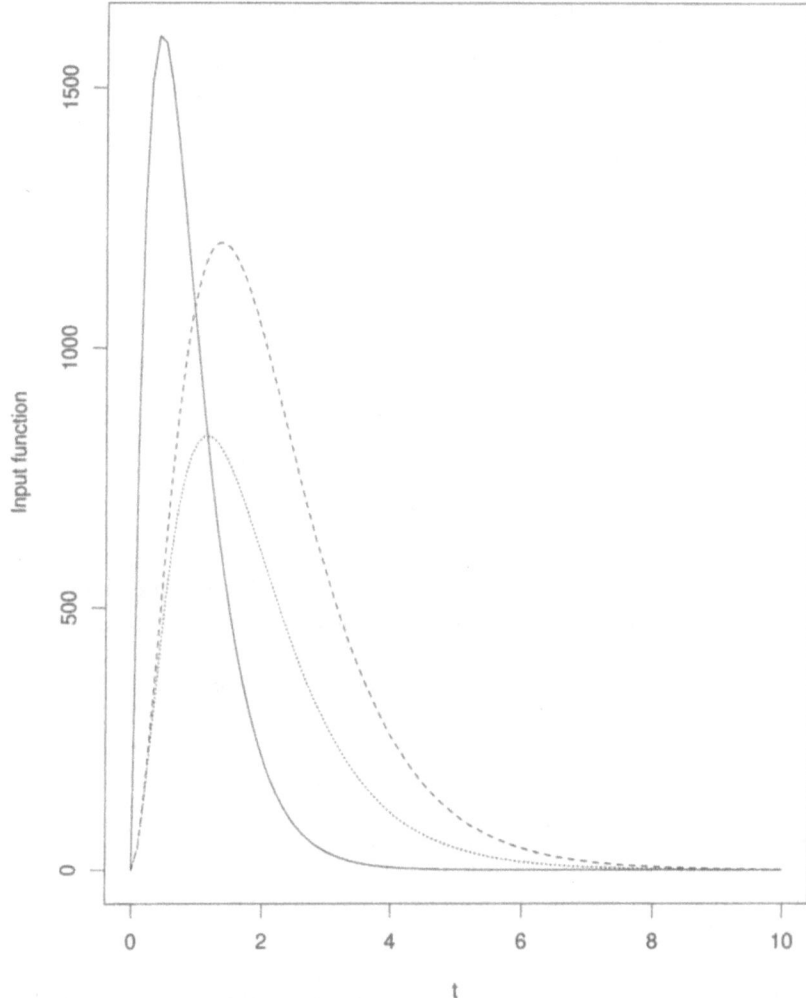

Figure 2. Simulated $A(t)$. Solid line, first simulation (identical *in vivo–in vitro* release; instantaneous absorption); widely dashed line, third simulation (different *in vivo–in vitro* release; first order absorption).

3.5. Measures of Performance

To compare across methods we used the following measures of performance: RMSE, the square root of the mean of the squared differences between true and the estimated individual input rate function. ME, the mean of the differences between true and estimated individual input rate function. RMSE and ME are evaluated at the sampling points, and divided by the standard deviation of the true input to obtain normalized RMSE (NRMSE) and ME (NME), this simply allows to compare better across simulations. The expected value NME if the estimates of close on average to the true model tends to zero.

Table 1. First simulated data set individuals' input functions

	TRUE	M_1	M_2	M_3	M_4	M_5	M_6
NRMSE							
Method 1	0.112	0.241	0.228	0.226	0.202	0.192	0.185
Method 2	0.132	0.241	0.227	0.234	0.22	0.191	0.203
NME							
Method 1	0.0235	0.021	0.0153	0.025	0.0136	0.0235	0.0151
Method 2	0.0223	0.0193	0.0147	0.00793	0.00939	0.0173	0.0222

TRUE indicates the model used to generate the data. M1-M6 the different models described in the text.

4. RESULTS

For all the simulations we will report the results obtained by fitting the "true" model (the same used to generate the data) and models M1-M6 to the data.

4.1. Simulated Data: First Data Set

Table 1 reports the simulations results corresponding to the first data set. In this simulation model M6 has the same functional shape of the model used to simulate the data (the "true" model), while for all the other models are overspecified in respect to the true model. For example model M2 approximates the true model if $Spline_1(t) \approx 1$, $(Spline_2(t) \approx 0)$, and $e^{-ka, j \, \eta 2j(t-\tau)} \approx 1$. By a relatively small margin model M6 obtains the best results for both methods of estimation. However all models obtain really similar performance, as it should be expected since model M6 is always a special case of models M1-M5.

4.2. Simulated Data: Second Data Set

Table 2 reports the simulations results corresponding to the second data set. In this simulation model M5 has a similar functional shape of the model used to simulate the data, while models M1-M3 are overspecified in respect to the true model, and models M4 and M6 are misspecified. For example model M1 approximates the true model $Spline_1(t) \approx e^{-kt}$ and $Spline_2(t) \approx 0$. Models M4 and M6 on the other end cannot approximate the true model because they do not incorporate an absorption function. Models M1-M3 and M5 obtain similar performance in respect to both NRMSE and NME. Models M4 and M6 obtain almost twice the size of NRMSE.

Table 2. First simulated data set individuals' input functions

	TRUE	M_1	M_2	M_3	M_4	M_5	M_6
NRMSE							
Method 1	0.192	0.261	0.258	0.266	0.582	0.252	0.605
Method 2	0.212	0.261	0.257	0.274	0.6	0.251	0.623
NME							
Method 1	0.0264	0.023	0.0183	0.029	0.619	0.0295	0.672
Method 2	0.0252	0.0213	0.0177	0.0119	0.614	0.0233	0.679

See legend to table I

Table 3. Third simulated data set indiviuals' input functions

	TRUE	M_1	M_2	M_3	M_4	M_5	M_6
NRMSE							
Method 1	0.179	0.261	0.258	0.576	0.602	0.572	0.555
Method 2	0.199	0.261	0.257	0.584	0.62	0.571	0.573
NME							
Method 1	0.0267	0.023	0.0183	0.579	0.619	0.599	0.582
Method 2	0.0255	0.0213	0.0177	0.562	0.614	0.593	0.589

See legend to table I

4.3. Simulated Data: Third Data Set

Table 3 reports the simulations results corresponding to the third data set. In this simulation model M1 and M2 have a functional shape similar to the model used to simulate the data (model M2 is closer), models M3-M6 are misspecified in respect to the true model. (they cannot approximate the true model). Model M1 approximates the true model if $Spline_1(t) \approx e^{-kt}$ and $Spline_2(t) \approx 0$. Model M2 approximates the true model if $Spline_2(t) \approx 0$. Models M1-M2 obtain similar performance in respect to both NRMSE and NME while models M3-M6 obtain higher values for NRMSE.

Table 4 reports the fraction of times a particular model (M1-M6) is selected by the model selection criterion, for all the simulations and estimation methods. Overall the correct functional shape is selected in all the simulations, although sometimes an overspecified model is selected. Importantly misspecified models are almost never selected.

5. DISCUSSION

In this paper we describe a general framework for *in-vivo/in-vitro* correlation modeling. We report a general model to describe possible changes between *in-vivo* and *in-vitro*, provide a semi-parametric implementation of the methodology and describe a way to incorporate random effects in the model, thus providing a way to devise population models for *in-vivo/in-vitro* correlation. The general model we propose allows the partition of the difference between *in-vivo in-vitro* situations into physiological events: difference due to a change in release of drug, and difference due to the (*in-vivo*) absorption of drug. The semi-parametric implementation of the model allows the investigation of different modes

Table 4. Percent times a model is selected

	M_1	M_2	M_3	M_4	M_5	M_6
Method 1						
First simulation	< 0.05	< 0.05	< 0.05	< .05	< .05	0.98
Second simulation	< 0.05	< 0.05	0.22	0	0.76	0
Third simulation	.21	0.76	< 0.1	0	0	0
Method 2						
First simulation	< 0.05	< 0.05	< 0.05	< .05	< .05	0.99
Second simulation	< 0.05	< 0.05	0.32	0	0.64	0
Third simulation	.31	0.66	< 0.1	0	0	0

of change by making limited assumptions on the functional shape induced by the changes in the *in-vivo* input function.

In a limited and preliminary set of simulations we test two different estimation methods to obtain the desired estimates: Method 1 is a standard subject-specific method which directly estimates the individuals input functions using least squares. Method 2 is a mixed effect population method which obtains individual estimates as empirical Bayes ones. In respect to the individual estimates the two methods perform similarly and they also obtain similar performance in respect to mean and variance population functions (not shown). These results shows clear evidence that the extra complication of method 2 (which includes random effects and requires a much more elaborate model selection strategy) is not necessary in a data rich situation like *in-vivo/in-vitro* correlation studies. Moreover, in other data rich situations [4], method 2 can obtain biased estimates of population mean input function, and individual estimates which are too close to the population mean, and therefore biased in respect to the individual true input function.

The same simulations indicate that available model selection criteria allow to correctly select (on average) the correct model for change between *in-vivo* and *in-vitro* situations. Overall misspecified models are never selected, but sometimes overspecified models are. The selection of an overspecified model has practical consequences if the model for the a particular individual is important because it will increase the variance of the estimated input function. However if the target is the estimation of a population mean function, and its inter-individual variability, a two-stage estimate (based on individual estimates) will obtain unbiased mean input function, while the inter-individual variability will be in general more influenced by the intrinsic inter-individual variability than the (comparatively little) extra variability induced by using a overspecified model.

The author is currently investigating further generalizations of the methodology, devising ways to carry over measures of precision of the *in-vitro* release estimates to obtain overall measures of precision for the estimated *in-vivo* input function, and assessing the performance of the different methods in respect to the estimation of mean population estimates.

REFERENCES

1. Akaike, H., "A New Look at the Statistical Model Identification Problem," *IEEE Trans Automat Contr*, 19:716–723, 1974.
2. Boeckmann, A.J., S.L. Beal, and L.B. Sheiner, *NONMEM Users Guides*, 1989.
3. Fattinger, K.E. and D. Verotta, "A Non-Parametric Subject-Specific Population Methods for Deconvolution. I. Description, Internal Validation and Real Data Examples," *J Pharmacokin Biopharm*, 23:1996.
4. Fattinger, K.E. and D. Verotta, "A Non-Parametric Subject-Specific Population Methods for Deconvolution. II. External Validation," *J Pharmacokin Biopharm*, 23:1996.
5. Verotta, D., "Concepts, Properties, and Applications of Linear Systems to Describe the Distribution, Identify Input, And Control Endogenous Substances and Drugs in Biological Systems," *Critical Review Bioengineers*, 1996. In Press.

5

CONVOLUTION-BASED APPROACHES FOR *IN VIVO–IN VITRO* CORRELATION MODELING

William R. Gillespie

Pharmacometrics Staff
Office of Clinical Pharmacology and Biopharmaceutics
Center for Drug Evaluation and Research
U.S. Food and Drug Administration
5600 Fishers Lane (HFD-855)
Rockville, Maryland 20857

ABSTRACT

One approach to *in vivo-in vitro* correlation (IVIVC) for extended release (ER) oral dosage forms is to directly model the relationship between the time courses of *in vitro* release and plasma drug concentrations. For drugs that exhibit linear, time-invariant disposition this can be done using models based on the convolution integral. Advantages of this approach relative to deconvolution-based IVIVC approaches include the following:

- The relationship between measured quantities (*in vitro* release and plasma drug concentrations) is modeled directly in a single stage rather than via an indirect two stage approach.
- The model directly predicts the plasma concentration time course. As a result:
  - The modeling focuses on the ability to predict measured quantities (not indirectly calculated quantities such as the cumulative amount absorbed).
  - The results are more readily interpreted in terms of the effect of *in vitro* release on conventional bioequivalence metrics.
- It is easier to construct methods that do not require the administration of an IV, oral solution, or IR reference dose.

A variety of convolution-based IVIVC models and modeling strategies are possible depending on the relationship between *in vivo* and *in vitro* release, the existence of nonlinear absorption or presystemic biotransformation, and the *in vivo* study design. The simplest approach is applicable to the case where the *in vitro* release rate equals the *in vivo* release (or absorption) rate and the study design includes the administration of an IV, oral solution, or IR dose. That basic convolution-based method can be extended to adjust for differences between the *in vitro* and *in vivo* release rates. This is accomplished by formally modeling those differences. Potential models include time-scaling and convolution. The extent of drug absorption may sometimes depend upon the release rate. This may be due to phenomena such as saturable presystemic biotransformation or truncated absorption due to intestinal transit past the sites of absorption. The relationship between the *in vitro* release rate and extent of absorption may be modeled empirically or mechanistically. Such models may be coupled with convolution to construct an overall IVIVC model for the relationship between *in vitro* release and plasma drug concentrations. It is also possible to apply convolution-based IVIVC models to study designs in which no IV, oral solution, or IR dose has been administered. Details of the various modeling approaches listed above are presented. Selected approaches are illustrated by examples of their application to real data.

1. INTRODUCTION

1.1. Background

The USP definition of a Level A *in vivo-in vitro* correlation (IVIVC) states that such a correlation "represents a point-to-point relationship between *in vitro* dissolution and the *in vivo* input rate of the drug from the dosage form" [1]. This is most consistent with deconvolution-based approaches to IVIVC—the class of methods that has dominated Level A IVIVC applications in the literature and in submissions to the FDA. The USP also briefly acknowledges the use of convolution but only in the context of establishing dissolution specifications. The report of a 1992 AAPS/FDA workshop on "Scaleup of Extended-Release Dosage Forms" makes indirect reference to convolution modeling by including in its description of Level A methods the use of "an appropriately validated plasma level simulation based on sound pharmacokinetic principles" [2].

Smolen and Erb described a strategy for IVIVC development that was based on convolution methods [3]. Langenbucher described a conventional convolution-based IVIVC method—essentially the same approach as that termed the "basic method" in this article [4].

1.2. Level A IVIVC—Alternative to USP Definition

It is the author's opinion that the USP definition of a Level A IVIVC overemphasizes the notion of "point-to-point" models. The objective of a Level A IVIVC is to predict the time course of some *in vivo* quantity, e.g., cumulative amount absorbed or plasma concentration, based on the *in vitro* release time course. This suggests the following alternative to the USP definition:

Level A IVIVC—A predictive mathematical model for the relationship between the *entire in vitro* release time course and the *entire in vivo* response time course.

For example, this would include models that describe the relationship between the cumulative amount released *in vitro* and the cumulative amount absorbed, or the relationship between the cumulative amount released *in vitro* and the plasma drug concentration. It would certainly include "point-to-point" models such as models relating the amount released at a specific time to the amount absorbed at a specific time (usually the same time). However, it also includes more complex relationships such as models relating the entire *in vitro* release and plasma concentration time courses via convolution or a compartmental model. The fundamental requirement is that the model should predict the entire *in vivo* time course from the *in vitro* data.

Two important categories of Level A IVIVC approaches are the deconvolution-based and the convolution-based methods. A deconvolution-based IVIVC method is a two-stage modeling procedure. In the first stage a deconvolution method, such as the Wagner-Nelson, Loo-Riegelman or a general noncompartmental method, is used to estimate the time course of *in vivo* absorption or release. In the second stage, an IVIVC model is constructed that relates the *in vitro* release profile to the time course of *in vivo* absorption or release (rate or cumulative amount). The most commonly used model is a point-to-point one that relates the *in vitro* and *in vivo* amounts at the same time according to a straight line. However, the IVIVC model need not be restricted to that simple case.

A convolution-based IVIVC method is a one-stage modeling approach in which the IVIVC model directly relates the *in vitro* release profile to the plasma drug concentration time course. Strictly speaking, it refers to models in which a convolution integral is used to describe the *in vivo-in vitro* relationship. However, the term might also be loosely applied to a wider range of models, e.g., nonlinear pharmacokinetic models, that predict plasma concentrations based on *in vitro* release data.

The objectives of this article are:

- To describe the standard convolution approach for Level A IVIVC—a method that assumes the *in vivo* release or absorption rate equals the *in vitro* release rate.
- To describe extensions to the basic convolution approach for cases where the *in vivo* absorption or release is not equal to the *in vitro* release.
- To describe a convolution-based approach that does not require an intravenous (IV) or immediate release (IR) reference dose.

The methods are briefly illustrated by application to real examples. In particular, the extended approach is illustrated using two nonlinear IVIVC models that are potentially applicable to cases where the extent of absorption is affected by the release profile.

2. CONVOLUTION-BASED IVIVC METHODS

2.1. The Basic Method

This section describes the simplest convolution-based IVIVC method, termed the "basic method" for the remainder of the article. In addition to being a useful method in its own right, it also establishes the framework from which more flexible and complex models are built.

The basic method requires plasma concentration data resulting from an IV dose or from the administration of an IR dosage form, preferably an aqueous oral solution. More correctly, it requires the results of a dose where the *in vivo* absorption or release time course is known, at least approximately. If an IV reference dose is used then the basic method follows from the following assumptions:

- The *in vitro* release rate is (approximately) equal to the *in vivo* absorption rate.
- The kinetic relationship between the absorption and plasma concentration time courses has the properties of linearity and time-invariance.
- The pharmacokinetics of intravenously administered or absorbed drug are indistinguishable, i.e., once an orally administered drug molecule reaches the systemic circulation, it behaves just like an intravenously administered one.

Similarly, the following assumptions apply to the case when an IR reference dose is used:

- The *in vitro* release rate is (approximately) equal to the *in vivo* release rate.
- The kinetic relationship between the *in vivo* release and plasma concentration time courses has the properties of linearity and time-invariance.
- The pharmacokinetics of drug administered as the IR reference dosage form or drug released from an ER dosage form are indistinguishable, i.e., once an orally administered drug molecule is released into the GI tract, it behaves just like a drug molecule administered via the IR reference dose.

Under those assumptions, the IVIVC model takes the following form,

$$c(t) = \int_0^t c_\delta(t-u) \, x'_{rel,vitro}(u) \, du \qquad (1)$$

where c = plasma drug concentration, $x_{rel,vitro}$ = cumulative amount released *in vitro*, $x'_{rel,vitro}$ = *in vitro* release rate (i.e., the first derivative of $x_{rel,vitro}$), and c_δ = the unit impulse response, i.e., the plasma concentration time course resulting from the instantaneous *in vivo* release (or absorption) of a unit amount of drug.

An IVIVC is constructed and evaluated according to the following procedure.

1. If the reference dose is an IV bolus, oral solution or IR dosage form, then the unit impulse response function (c_δ) may be estimated as the dose normalized plasma concentration time course resulting from reference dose, i.e., $c_\delta = c_{ref}/D_{ref}$. More generally, c_δ may be estimated from the plasma concentrations resulting from more complex inputs, e.g., IV infusion, via deconvolution if the input time course is known.
2. Calculate the predicted plasma concentrations resulting from the ER dose(s) by convolution of the *in vitro* release rate time course ($x'_{rel,vitro}$) with c_δ.
3. Assess the predictive performance of the IVIVC by comparing the predicted and observed plasma concentrations.

Table 1. List of symbols used in this article

$a_i, \hat{\alpha}_i$	=parameters of c_δ
b, β	=parameters of function used to extrapolate the cumulative amount released *in vitro*
c	=plasma drug concentration
c_δ	=unit impulse response
c_{ref}	=plasma concentration resulting from a reference dose (usually IV or IR).
c_{sol}	=plasma concentration resulting from an oral solution.
D_{ref}	=reference dose
D_{sol}	=oral solution dose
f	=generic symbol for a function or operator
$F(t)$	=fraction absorbed for drug released at time t
t	=elapsed time
t_0	=lag time
u	=variable of integration or "dummy" variable
$x_{rel,vitro}$	=cumulative amount released *in vitro*
$x'_{rel,vitro}$	=*in vitro* release rate
x_{vivo}	=cumulative amount released or absorbed *in vivo*
x'_{vivo}	=*in vivo* release or absorption rate

Example 1—Basic Convolution-Based IVIVC Approach: The basic convolution-based IVIVC approach is illustrated using the mean *in vitro* release and plasma concentration data resulting from three diltiazem ER oral dosage forms with clearly different *in vitro* release time courses (see Figure 1). A polyexponential function of the form,

$$c_{sol}(t) = D_{sol} c_{\delta}(t) = \begin{cases} 0, & t \le t_0 \\ D_{sol} \sum_{i=1}^{n} a_i e^{-\alpha_i(t-t_0)}, & t > t_0 \end{cases} \tag{2a}$$

where

$$c_{sol}(t_0) = D_{sol} \sum_{i=1}^{n} a_i = 0 \tag{2b}$$

is fit to the mean plasma concentration data resulting from the oral solution (see Figure 2). The unit impulse response function is obtained according to,

$$c_{\delta}(t) = \frac{c_{sol}}{D_{sol}} = \begin{cases} 0, & t \le t_0 \\ \sum_{i=1}^{n} a_i e^{-\alpha_i(t-t_0)}, & t > t_0 \end{cases} \tag{3}$$

The plasma concentrations corresponding to each ER formulation are predicted according to eq. (1) where $x_{rel,vitro}$ (expressed as percent dissolved) is estimated by linear interpolation of the mean *in vitro* release data. For the "slow" and "medium" release formulations, the function is extrapolated beyond 12 hours by fitting the terminal portion

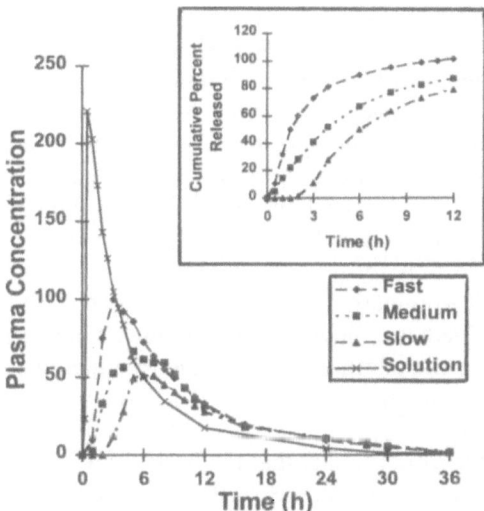

Figure 1. Mean *in vitro* release and plasma concentrations resulting from three diltiazem ER formulations and an oral solution.

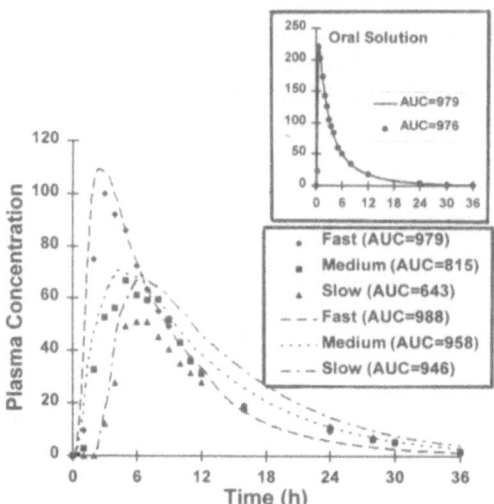

Figure 2. Application of the basic convolution-based IVIVC method to the diltiazem data (see Figure 1). The large plot shows the mean observed and predicted plasma concentrations for the 3 ER formulations. The inset graph shows the mean observed plasma concentrations resulting from the oral solution and the fitted equation used to estimate c_δ.

of the data (6 to 12 hours) to a function of the form $x_{rel,vitro}(t) = 100 - be^{-\beta t}$. The "fast" release formulation had already reached 100% release by 12 hours. Thus, $x'_{rel,vitro}$ is approximated as a piecewise constant function up to 12 hours and as an exponential function $(b\beta e^{-\beta t})$ (or 0 for the "fast" formulation) thereafter. The approach could be somewhat refined by the use of a smooth interpolant such as a cubic spline.

The data analyses for all of the examples in this article are perofrmed with the ADAPT II program for simulation and nonlinear regression [5]. The user-written ADAPT II model uses a numerical convolution method based on trapezoidal rule.

This example illustrates a situation where the basic convolution method clearly fails (see Figure 2). The convolution model predicts the mean plasma concentrations following the "fast" formulation with reasonable accuracy but poorly predicts the "medium" and "slow" formulation results. A couple lessons can be learned from this example. One is the desirability of evaluating an IVIVC model using multiple formulations with different release profiles.

A second lesson is that the basic convolution model cannot accurately predict plasma concentrations of drugs for which the area under the curve is affected by the formulation. This not only precludes its use with drugs that exhibit nonlinear disposition (esp. elimination); it also fails for drugs or formulations that exhibit nonlinear absorption or nonlinear presystemic elimination. Thus, saturable first pass elimination is a contraindication for the method. The basic model will overpredict concentrations during the post-absorption phase following the administration of ER formulations for which drug release is so slow that a significant fraction of the dose is not released in a portion of the GI tract where absorption can occur.

Such failures result from the fact that the basic model is based on the assumption of a linear, time-invariant pharmacokinetic relationship between drug release and plasma concentrations. A consequence is that the model predicts that the ER formulations result in the same dose-normalized AUC as the reference dose used to estimate c_δ. This is a particularly severe restriction when c_δ is estimated based on an IV dose since it is equivalent to assuming 100% absorption. Thus, an oral solution or IR dosage form is generally a more appropriate reference dose with this method.

When an IV reference dose is used and absolute bioavailability of an oral dose is less than 100%, a relatively simple extension of the basic convolution approach would be to multiply the *in vitro* release rate by an estimate of the absolute bioavailability. In fact this is the simplest of the extensions to the basic method discussed in the next section.

2.2. Extensions to the Basic Method

The *in vitro* release profile may not be equal to the *in vivo* release or absorption profile for a variety of reasons. To address such discrepancies the convolution-based IVIVC model may be extended by incorporating a model for the relationship between the cumulative amount released *in vitro* ($x_{rel,vitro}$) and the cumulative amount released or absorbed *in vivo* (x_{vivo}), or between the *in vitro* release rate ($x'_{rel,vitro}$) and the *in vivo* release or absorption rate (x'_{vivo}), i.e.,

$$x_{vivo} = f(x_{rel,vitro}) \tag{4a}$$

or

$$x'_{vivo} = f(x'_{rel,vitro}) \tag{4b}$$

where f is a function or operator relating the two functions. The overall IVIVC model is constructed by combining eq. (4) with the following convolution equation,

$$c(t) = \int_0^t c_\delta(t-u)\, x'_{vivo}(u)\, du \tag{5}$$

Some plausible models for eq. (4) are the following.

- Linear function
  - $x_{vivo}(t) = a + b\, x_{rel,vitro}(t)$ or $x'_{vivo}(t) = a + b\, x'_{rel,vitro}(t)$
- Nonlinear or time-variant functions
  - $x_{vivo}(t) = f(x_{rel,vitro}(t))$ or $x'_{vivo}(t) = f(x'_{rel,vitro}(t))$
  - $x_{vivo}(t) = f(x_{rel,vitro}(t),t)$, e.g., $x_{vivo}(t) = f(t)\, x_{rel,vitro}(t)$

or

$x'_{vivo}(t) = f(x'_{rel,vitro}(t),t)$, e.g., $x'_{vivo}(t) = f(t)\, x'_{rel,vitro}(t)$
- Time scaling
  - Linear: $x_{vivo}(t) = x_{rel,vitro}(a+bt)$ or $x'_{vivo}(t) = x'_{rel,vitro}(a+bt)$
  - Nonlinear: $x_{vivo}(t) = x_{rel,vitro}(f(t))$ or $x'_{vivo}(t) = x'_{rel,vitro}(f(t))$
- Convolution
  - $x'_{vivo}(t) = \int_0^\infty f_{vitro \to vivo}(t-u)\, x'_{rel,vitro}(u)\, du$

The specific functional form of such models may be selected empirically based on success with respect to model fitting and prediction. Ideally, model selection should also be supported by a mechanistic understanding of the *in vitro-in vivo* relationship. In fact, a formal mechanistic model could be used for eq. (4) that explicitly considers factors that influence release and absorption such as GI transit, and changes in pH and permeability along the GI tract.

The extended convolution-based IVIVC models are estimated and evaluated according to the following procedure that is significantly modified from that used for the basic method.

1. Estimate the unit impulse response function (c_δ) based on the IV or IR reference dose results as described previously for the basic method (Section 2.1).
2. Substitute the chosen model for the x_{vivo} vs $x_{rel,vitro}$ relationship into the convolution equation.
3. Estimate the model parameters by fitting the overall convolution model to the plasma concentrations resulting from the ER dosage forms.
4. Assess the predictive performance of the IVIVC by comparing the predicted and observed plasma concentrations. At least 2 ER's with different release profiles are required for a convincing demonstration of the model's predictive ability.

The requirement of at least 2 ER's with different release profiles follows from the fact that it is always possible to pick a model for eq. (4) that will exactly fit the results of any single ER formulation. Such a curve fitting procedure provides no assurance that the proposed IVIVC model is capable of predicting the consequences of a change in the release profile. Evidence of the successful prediction of at least one more formulation with a different release time course is needed for such assurance.

The extended convolution-based methods are illustrated by applying two different IVIVC models to the same diltiazem data set previously used in Example 1. Both models have the property that the extent of absorption depends on the *in vitro* release profile, but they are consistent with very different mechanisms, e.g., truncated absorption due to intestinal transit past the sites of absorption versus saturable presystemic elimination.

Both models can be described in terms of the following general form for rate- or time-dependent extent of absorption,

$$x'_{vivo}(t) = F(t) x'_{rel,vitro}(t) \qquad (6)$$

where $F(t)$ is the fraction absorbed (relative to the reference dose) for drug released at time t.

Example 2—Simple Truncated Absorption: For this model the fraction absorbed ($F(t)$) has the simple time-dependent form,

$$F(t) = \begin{cases} 1, & 0 \le t \le T \\ 0, & otherwise \end{cases} \qquad (7)$$

that is also depicted graphically in Figure 3. This model describes a case where the fraction absorbed is equal to the reference dose for drug released up to time T. Drug released after that time is not absorbed at all. A possible mechanism for this would be intestinal transit of the entire unreleased dose past the sites of absorption at time T. This could certainly occur with a monolithic ER dosage form that remains largely intact in the GI tract, e.g., osmotic pump devices, since the entire unreleased dose would pass out the GI tract at one time.

Application of this model to the example data set produces the results shown in Figure 3. The model fits the data much better than the basic convolution model. In particular, the model predicts the decrease in *AUC* as the release profile is increasingly prolonged. In fact, the *AUC*'s of the fitted curves are close approximations of the observed *AUC*'s. However, the model overestimates the peak concentrations for the "slow" formulation.

This model describes the intestinal transit time of the drug molecules as a single value. It can readily be generalized to consider a distribution of transit times. For example, if the transit times are distributed according to some cumulative distribution function $G(t)$ = $\Pr(T \le t)$, then the fraction absorbed ($F(t)$) might be described by $F(t) = 1 - G(t)$ where $G(t)$ has the properties $G(0) = 0$ and $\lim_{t \to \infty} G(t) = 1$. This model still assumes that the initial fraction absorbed is equal to that for the reference dose. To allow for a difference in bioavailability between the reference dose and the ER doses, the model can be further generalized to $F(t) = F_{max}[1 - G(t)]$ where $F_{max} = F(0)$ is the maximum value of $F(t)$.

Example 3—Saturable Presystemic Elimination: The following model equation for $F(t)$ is consistent with a saturable presystemic elimination process, e.g., hepatic first pass, that may be described in terms of Michaelis-Menten kinetics,

$$F(t) = \frac{F_{min} x'_{50} + x'_{rel,vitro}(t)}{x'_{50} + x'_{rel,vitro}(t)} \qquad (8)$$

A function of this type is illustrated in Figure 4. When applied to the example data set, the model produces the results shown in Figure 4.

Though the model predicts a decrease in AUC with decreasing release rates, the predicted AUC's do not match the observed AUC's as well as the truncated absorption model.

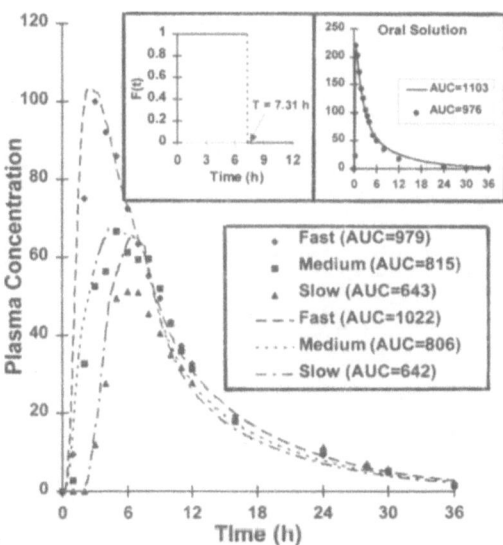

Figure 3. Application of an extended convolution-based IVIVC model for truncated absorption to the diltiazem data (see Figure 1). The large plot shows the mean observed and fitted plasma concentrations for the 3 ER formulations. The inset graphs show the estimated fraction absorbed function ($F(t)$) and the mean observed and fitted plasma concentrations for the oral solution.

On the other hand, both models fit the data more closely than the basic convolution model. Ideally, final model selection should be based on a combination of mechanistic knowledge and empirical success with respect to model fitting and prediction error. Thus, a choice between the "truncated absorption" and the "saturable presystemic elimination" models should be supported by other evidence that the proposed mechanism(s) is relevant.

2.3. Method Not Requiring an IV or IR Reference Dose

The previously described convolution-based IVIVC methods, both basic and extended, utilize data resulting from an IV or IR reference dose. Such data provides an estimate of the unit impulse response (c_δ) that is independent of the ER data. This has the advantage of assuring that the c_δ estimate is not affected by the potentially confounding influence of the release or absorption rate. Though desirable for that reason, an IV or IR reference dose is not a requirement.

Consider the following alternative interpretation of eq. (1) that was used for the basic convolution method:

$$c(t) = \int_0^t c_\delta(t-u)\, x'_{rel,vitro}(u)\, du$$

(9)

where c_δ represents the hypothetical plasma concentration time course that would result from the instantaneous *in vitro* release of a unit amount of drug—a function that is not necessarily equal to the *in vivo* unit impulse response used for the previously described methods. The model may be estimated and evaluated using only ER data according to the following procedure.

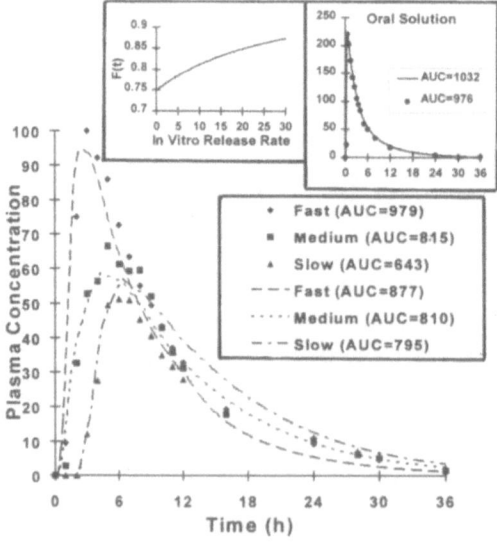

Figure 4. Application of an extended convolution-based IVIVC model for saturable presystemic elimination to the diltiazem data (see Figure 1). The large plot shows the mean observed and fitted plasma concentrations for the 3 ER formulations. The inset graphs show the estimated fraction absorbed function ($F(t)$) as a function of the *in vitro* release rate, and the mean observed and fitted plasma concentrations for the oral solution.

1. Substitute a function describing the *in vitro* release profile ($x_{rel,vitro}$) into the convolution equation.
2. Select a suitable parametric function for the unit impulse response (c_δ), e.g., a polyexponential function, and insert it into the convolution equation.
3. Estimate the parameters of c_δ by fitting the overall convolution model to the plasma concentrations resulting from the ER dosage forms.
4. Assess the predictive performance of the IVIVC by comparing the predicted and observed plasma concentrations. At least 2 ER's with different release profiles are required for a convincing demonstration of the model's predictive ability.

The requirement of two or more ER formulations with different release rates is particularly critical with this method. This is because it is theoretically possible to find a c_δ function such that the model exactly fits the data for any single ER. Thus, successfully fitting the model to the results for one ER formulation provides no assurance that the model can predict the *in vivo* consequences of a change in the release rate. Such assurance is only provided by successfully predicting the *in vivo* results of additional formulations with different release rates, or equivalently, by simultaneously fitting the results for two or more formulations.

Example 4–Convolution-Based IVIVC Without an IV or IR Reference Dose. The proposed IVIVC approach is applied to the mean *in vitro* release and plasma concentration data resulting from three metoprolol ER formulations (see Figure 5). The unit impulse response is approximated as a polyexponential function of the form,

$$c_\delta(t) = \begin{cases} 0, & t \le t_0 \\ \sum_{i=1}^{n} a_i e^{-\alpha_i(t-t_0)}, & t > t_0 \end{cases} \tag{10a}$$

where

$$c_\delta(t_0) = \sum_{i=1}^{n} a_i = 0 \tag{10b}$$

$x_{rel,vitro}$ is obtained by linear interpolation of the cumulative amount releases *in vitro* data, so that $x'_{rel,vitro}$ is a piecewise constant function. The parameters of c_δ are estimated by simultaneous fitting of eq. (9) to the plasma concentration data resulting from all three formulations. The result (see Figure 6) is a reasonably good fit to the data, suggesting a successful IVIVC.

3. DISCUSSION

The methods and examples presented in this article demonstrate the versatility and flexibility of the convolution-based IVIVC approach, particularly when it is extended to include nontrivial submodels of the relationship between *in vitro* release and *in vivo* release or absorption (eqs. (4–5)). However, that versatility and flexibility does not come without a cost. As with any mathematical model estimation problem, the ability to successfully assess the predictive performance of a model is inversely related to its complex-

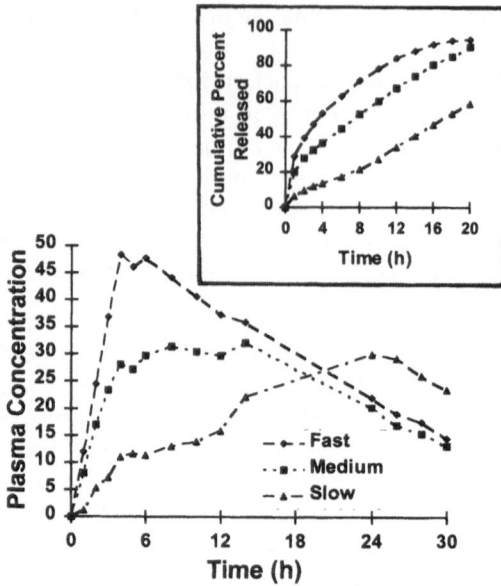

Figure 5. Mean *in vitro* release and plasma concentrations resulting from three metoprolol ER formulations.

ity and flexibility. Therefore, some models may be difficult or even impossible to validate using practical study designs for IVIVC characterization.

Convolution-based IVIVC methods offer a number of advantages over the more commonly used deconvolution-based methods:

- The model directly predicts the plasma concentration time course. As a result:

$$x'_{vivo}(t) = \int_0^\infty f_{vitro-vivo}(t-u) \, x'_{rel,vitro}(u) \, du.$$

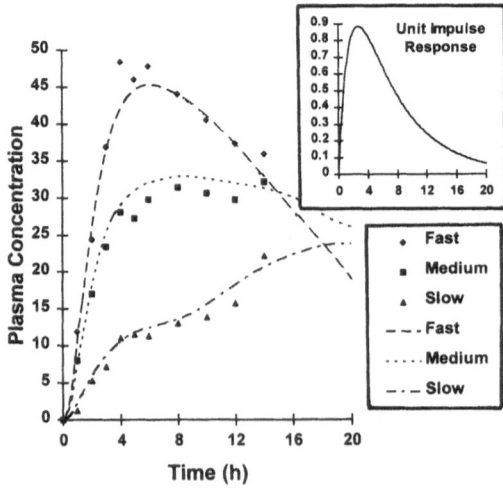

Figure 6. Application of a convolution-based IVIVC method that does not require an IV or IR reference dose to the diltiazem data (see Figure 1). The large plot shows the mean observed and predicted plasma concentrations for the 3 ER formulations. The inset graph shows the estimated unit impulse response function (c_δ).

- The modeling focuses on the ability to predict measured quantities, not indirectly calculated quantities such as the cumulative amount absorbed.
- The results are more readily interpreted in terms of the effect of *in vitro* release on conventional bioequivalence metrics, e.g., AUC and C_{max}.
- The relationship between measured quantities (*in vitro* release and plasma drug concentrations) is modeled directly in a single stage rather than via an indirect two stage approach.
- It is easier to construct methods that do not require the administration of an IV, oral solution, or IR dose.

Such advantages lead the author to advocate the use of convolution-based IVIVC approaches over deconvolution-based methods for most applications. This is not to say that deconvolution should be abandoned. Instead it leads to an alternative paradigm for the roles of deconvolution and convolution in IVIVC. Deconvolution-based methods are extremely useful for exploratory data analysis during the model building process. In particular, deconvolution (for estimation of the time course of *in vivo* drug release or absorption) coupled with graphic presentations (such as plots of the cumulative amount absorbed versus the cumulative amount released *in vitro*) can greatly facilitate identification of an appropriate submodel for eq. (4) in the extended convolution-based method. The selected submodel is then incorporated into the more complete convolution-based model for estimation and evaluation of the final IVIVC model.

This article focuses primarily on the mathematical form of the convolution-based IVIVC models and only briefly addresses the issue of model evaluation or validation. Such model evaluation should focus on predictive performance. The reader is encouraged to read other articles in this compilation that address this issue as well as relevant section of the FDA draft guidance entitled "Extended Release Solid Oral Dosage Forms: Development, Evaluation and Application of In Vitro/In Vivo Correlations".

ACKNOWLEDGMENTS

The author would like to thank Peter Lockwood (Office of Clinical Pharmacology and Biopharmaceutics, CDER, FDA) for his collaboration in the development of the method not requiring an IV or IR reference dose. Thanks also to David Young and Debbie Piscitelli (University of Maryland, College of Pharmacy) for their role in facilitating the development and presentation of the extended convolution-based approach.

REFERENCES

1. *In Vitro* and *in Vivo* Evaluation of Dosage Forms. USP 23 chapter <1088>, United States Pharmacopeial Convention, Inc., 1995, pp. 1927–1929.
2. J.P. Skelly, et al. Workshop Report. Scaleup of oral extended-release dosage forms. Pharm. Res. 10:1800–1805 (1993).
3. V.F. Smolen and R.J. Erb. Predictive conversion of *in vitro* dissolution data into *in vivo* drug response versus time profiles exemplified for plasma levels of warfarin. J. Pharm. Sci. 66:297–304 (1977).
4. F. Langenbucher. Numerical convolution/deconvolution as a tool for correlating *in vitro* with *in vivo* drug availability. Pharm. Ind. 44:1166–1172 (1982).
5. D.Z. D'Argenio and A. Schumitzky. ADAPT II User's Guide. Biomedical Simulations Resource, University of Southern California, Los Angeles, 1992.

6

APPROACHES TO IVIVR MODELLING AND STATISTICAL ANALYSIS

Adrian Dunne,[1] Tom O'Hara,[2] and John Devane[2]

IVIVR Co-operative Working Group
[1]Department of Statistics
University College Dublin
Dublin 4, Ireland
[2]Élan Corporation plc
Monksland, Athlone
Ireland

1. INTRODUCTION

The current general approach to the development of a level A *in vivo-in vitro* correlation (IVIVC) is open to criticism in two major respects. The statistical methodology used is not based on the statistical properties of the data being analysed and consequently parameter estimates may be biased and the analysis may be inefficient. The second criticism is that a linear model is used and this is clearly very restrictive and limits the number of instances where we might expect to find such a relationship. This chapter addresses both of these issues. New statistical methods for the current linear model are proposed and their effectiveness demonstrated by means of a simulation experiment. In addition, new non-linear models which are generalisations of the linear model are proposed together with appropriate statistical methodology for fitting them. These models are shown to have some promise by using them to describe the *in vivo-in vitro* relationship for a number of batches of an extended release drug product.

2. CURRENT MODEL AND METHODS

At present the commonly used approach to developing a level A IVIVC is based on calculating (using deconvolution methods[1]) the fraction dissolved (or absorbed) *in vivo* for a number of dosage unit/subject combinations across a range of time points. These data may be averaged across subjects and plotted as shown in figure 1. The fraction dissolved *in vitro* for a number of dosage units at the same set of time points as used *in vivo* is also measured and may be averaged and plotted as in figure 2. These two plots can be combined into a single

In Vitro–in Vivo Correlations, edited by Young *et al.*
Plenum Press, New York, 1997

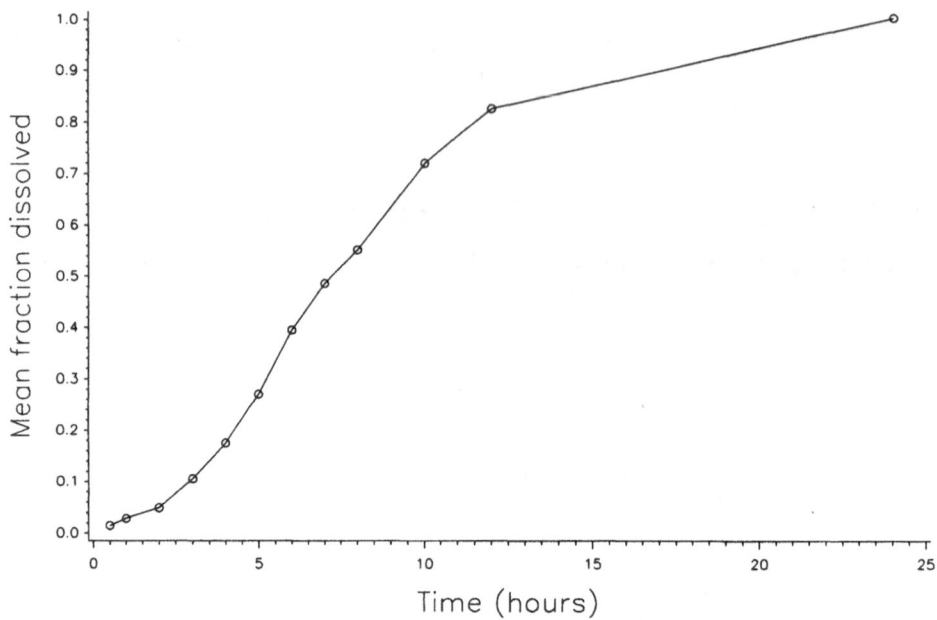

Figure 1. Shows how the mean fraction of drug dose dissolved *in vivo* varies with time.

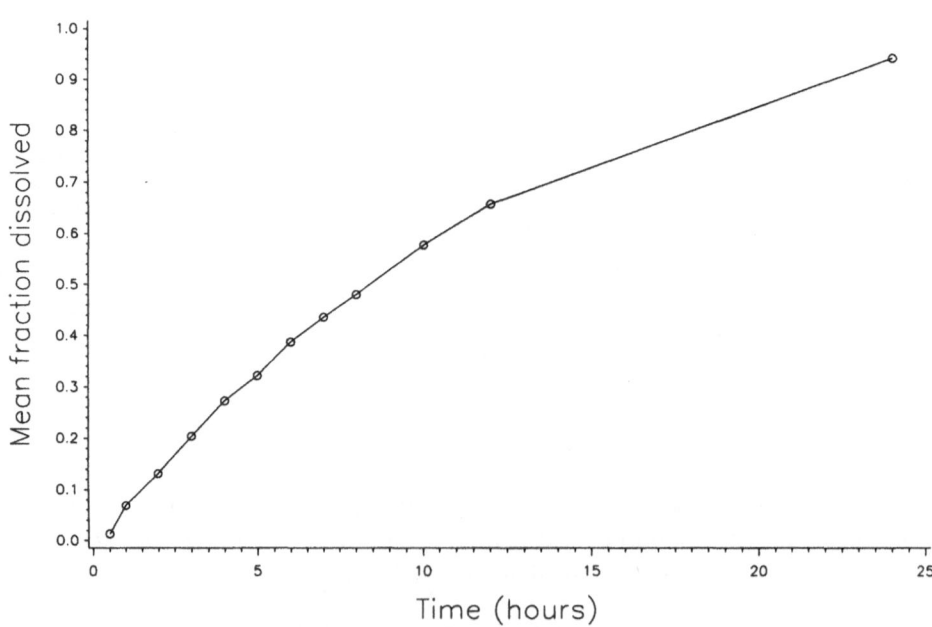

Figure 2. Shows how the mean fraction of drug dose dissolved *in vitro* varies with time.

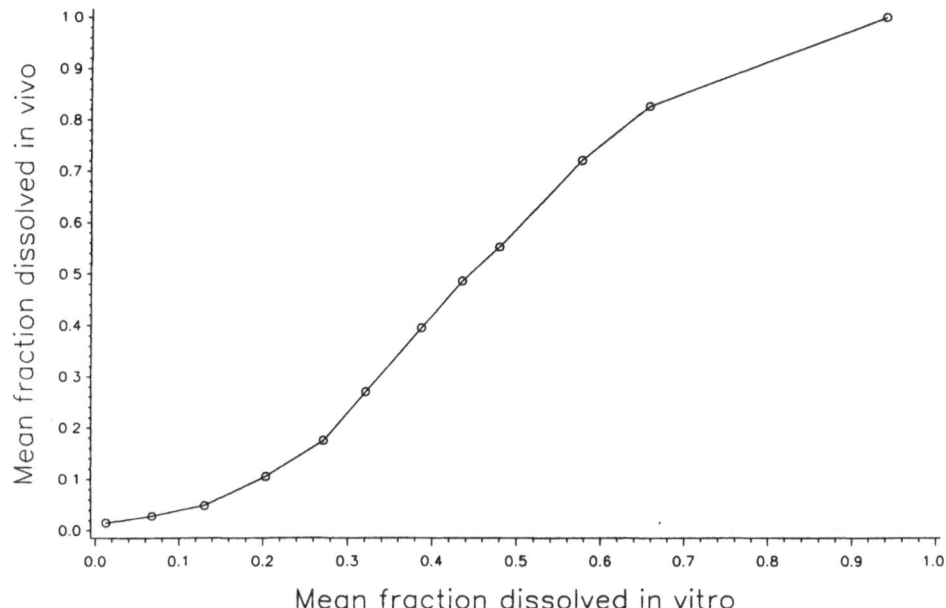

Figure 3. A combination of figures 1 and 2 which eliminates time as a variable and demonstrates the relationship between the mean fraction of drug dissolved *in vivo* and that dissolved *in vitro* at the same time.

graph of mean fraction dissolved *in vivo* versus the mean fraction dissolved *in vitro* at the same time[2–7], as demonstrated in figure 3. In this way, time is eliminated as a factor and the relationship between *in vivo* and *in vitro* dissolution is clearly displayed.

The ensuing data analysis is a linear regression analysis with the mean fraction dissolved *in vivo* as the response variable and the mean fraction dissolved *in vitro* as the independent variable. There are a number of variations on this analysis including;

a. A simple regression analysis and the hypotheses that the intercept is zero and the slope is unity are tested separately. Only if neither of these hypotheses is rejected is a level A IVIVC said to have been established[6].

b. As in (a) above but the intercept is ignored and the only hypothesis tested is that the slope is unity[8].

c. A regression through the origin is performed and the hypothesis that the slope is unity is tested[7].

From a statistical viewpoint these analyses are deficient in a number of respects i.e.

i. The measurement errors in the independent variable are ignored. This problem is minimised (but not eliminated) by using the *in vitro* data as the independent variable because its variability is considerably less than that of the *in vivo* data.

ii. The data consist of repeated observations on the same subject/dosage unit which are therefore correlated. The regression analysis ignores these correlations and treats the observations as being independent.

iii. In method (a) above two hypothesis tests are used (one for the slope and another for the intercept) to answer a single question rather than constructing a single test for both slope and intercept. This is the well known multiple comparisons problem[9].

The linear model with slope of unity and zero intercept corresponds with the mean time profiles for *in vivo* and *in vitro* being coincident or superimposable[10]. This represents a very simple type of *in vivo-in vitro* relationship, consequently one should not expect to find such a relationship very often in practice. More complex relationships are likely to be found and there is therefore a need for more complex models which could describe such relationships. Non-linear models with the linear model as a special case would be attractive in that they represent generalisations of the current linear model.

3. NEW METHODS

Methods for fitting the current linear model taking account of the statistical properties of the data may be developed using standard statistical methodology. The times at which the data are collected will be denoted by t_i where $i = 1, 2, ...,p$ and these times are common to both *in vivo* and *in vitro* data. Let Y_{1ij} represent the measured fraction of dosage unit j dissolved *in vitro* at time t_i and Y_{2ik} the measured fraction for dosage unit/subject k dissolved *in vivo* at time t_i. Consider the data collected *in vitro* from the jth dosage unit to be a p dimensional random vector[11]

$$\mathbf{Y_{1j}} = (Y_{11j}, Y_{12j}, ..., Y_{1pj})'$$

with mean vector μ_1 and covariance matrix Σ_1. Similarly the *in vivo* data for the kth subject may be written as

$$\mathbf{Y_{2k}} = (Y_{21k}, Y_{22k}, ..., Y_{2pk})'$$

with mean vector μ_2 and covariance matrix Σ_2. The *in vitro* and *in vivo* curves being superimposable is equivalent to the hypothesis that $\mu_1 = \mu_2$ i.e. the mean *in vitro* profile across time is coincident with the mean *in vivo* profile. Making the assumption that $\Sigma_1 = \Sigma_2$ and that the data are normally distributed, the hypothesis $\mu_1 = \mu_2$ can be tested using Hotelling's T^2 test[11] which is a special case of MANOVA (multivariate analysis of variance). As stated earlier, the *in vivo* data would be expected to be considerably more variable than the *in vitro* data with consequent differences in the covariance matrices. For this reason and because Hotelling's T^2 test is lacking in power, another test which is based on an assumed structure for the covariance matrices was considered. The assumption is that of compound symmetry which assumes that all observations made on the same dosage unit/subject are equally correlated with one another and corresponds with a mixed effects model[12–13].

4. SIMULATION EXPERIMENT

The methods described above were compared by means of a simulation experiment. Multivariate normal data were generated at observation times 0.5, 1, 2, 3, 4, 5, 6, 7, 8, 10, 12, 24 hours with the vector μ_1 set at (0.087, 0.171, 0.332, 0.494, 0.609, 0.706, 0.783, 0.835, 0.875, 0.925, 0.952, 0.999)' and each element of μ_2 calculated using the formula

$$\mu_{2i} = 1 - (1 - \mu_{1i})^r$$

where r (more correctly $ln(r)$) is a measure of 'distance' between the two vectors. This relationship is illustrated in figure 4 for a range of values of r. When $r=1$ ($ln(r)=0$) the two vectors are equal or coincident.

The *in vitro* data were generated with a random 'between tablet' term which was common to all observations on the same tablet and had a standard deviation of 0.02 and a 'within tablet' error term which was unique to each observation and had a standard deviation of 0.015. Similarly, the *in vivo* data were generated with a random 'between subject' term which was common to all observations on the same subject and had a standard deviation of 0.06 and a 'within subject' error term which was unique to each observation and had a standard deviation of 0.08. There were 12 dosage units *in vitro* and 12 subjects *in vivo*. For each of 31 values of r (equally spaced on a logarithmic scale) in the range 0.165 to 3.32, one thousand sets of data were generated. These values used for the various parameters were in fact estimates derived from a set of real data and were chosen so that the simulated data sets would be as realistic as possible.

Each set of data was analysed (using a nominal 5% type I error rate) by a number of different methods as follows;

    a. Simple linear regression of mean (across subjects) percentage dissolved *in vivo* on mean (across tablets) percentage dissolved *in vitro* followed by separate t tests for slope=1 and intercept=0. The joint hypothesis is not rejected only if neither t test results in rejection.

    b. The same as (a) above except that a single simultaneous test was conducted of the hypothesis that the slope was unity and the intercept was zero.

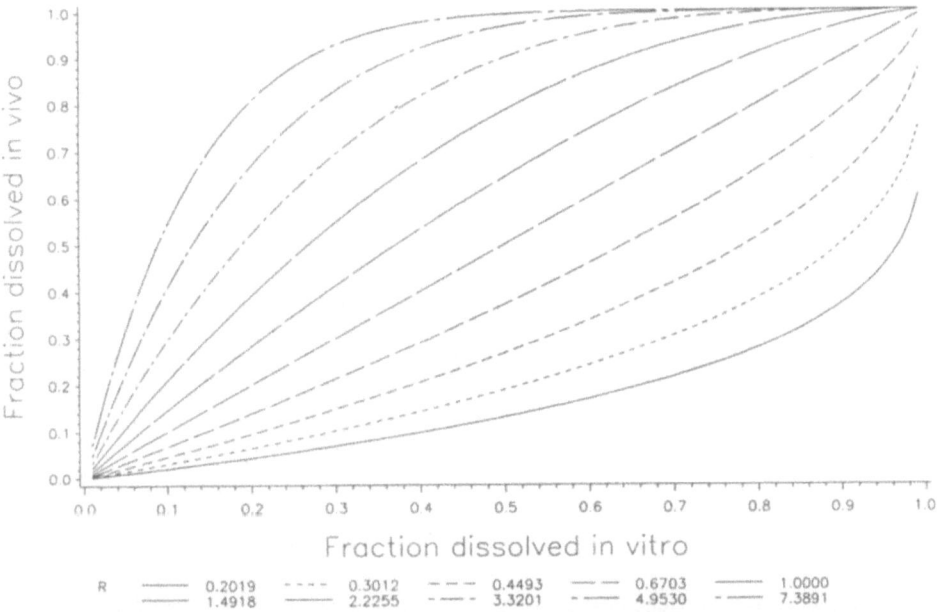

Figure 4. Illustrates how the mean fraction dissolved *in vivo* was computed from that *in vitro* for a number of different values of r (see text).

c. The same as (a) above except that the only hypothesis tested was that the slope was unity.

d. Linear regression through the origin using the same response and independent variables as in (a) above, followed by a t test for slope=1.

e. Hotelling's T^2 test of the hypothesis $\mu_1 = \mu_2$.

f. Mixed effects analysis testing the hypothesis $\mu_1 = \mu_2$.

For each value of r and for each method of analysis the percentage of the 1000 data sets for which an IVIVC was declared to have been established was noted. These percentages are plotted against $ln(r)$ in figure 5.

The data were generated using SAS and the analyses conducted using the GLM and MIXED procedures in SAS.

5. NONLINEAR MODELS

The current level A IVIVC model is a linear model insofar as it is based on a linear relationship between the mean fraction of drug dissolved *in vivo* and the mean fraction dissolved *in vitro*. This model is very restrictive since we have no reason to expect the *in vivo-in vitro* relationship to be linear. Alternative nonlinear models were developed in the following manner.

Consider the time at which a drug molecule goes into solution *in vitro* to be a random variable. Similarly, the time at which a drug molecule is dissolved *in vivo* is also a random variable. The distribution functions for these two random variables may vary from dosage unit to dosage unit (and from subject to subject) and can be written as $F_{ij}(t)$ for the ith unit under condition j ($j=1$ (*in vitro*), $j=2$ (*in vivo*)). These distribution functions are in fact the fractions of drug dissolved from unit i at time t under condition j.

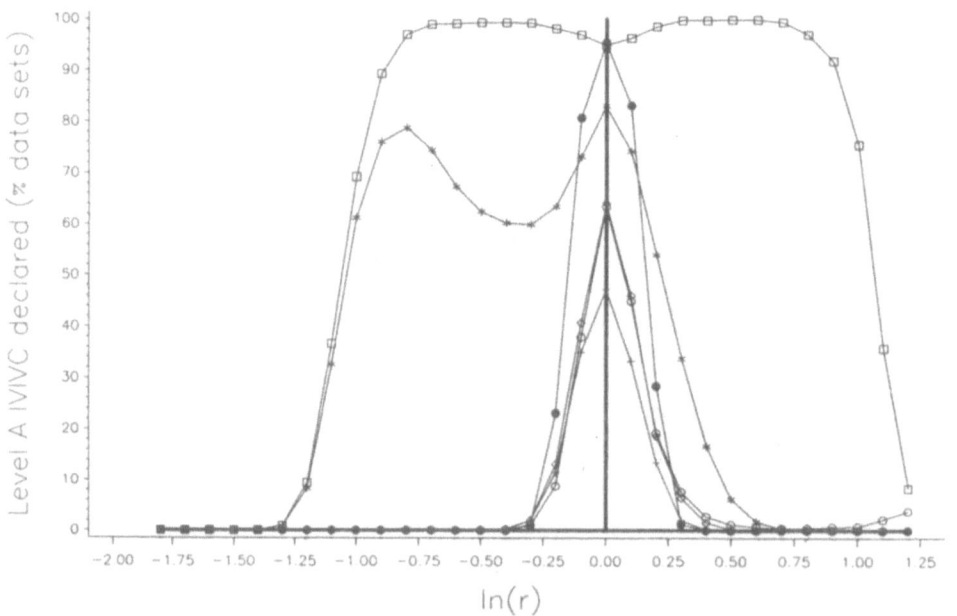

Figure 5. The percentage of simulated data sets for which a level A IVIVC was declared to have been established plotted against $ln(r)$ for each of the methods examined. Methods (a), (b), (c), (d), (e) and (f) of section 4 are depicted by the symbols ∗, ◇, □, ○, ✛, and ● respectively. For reference purposes the behaviour of an ideal method is shown by the thick solid line.

The relationship between the *in vivo* and *in vitro* dissolution may be expressed in terms of a relationship between the distribution functions or between related functions. Consequently we can use some of the following functions (all of which are related to the distribution function $F_{ij}(t)$) to express an *in vivo-in vitro* relationship.

The odds function: $F_{ij}(t)/(1- F_{ij}(t))$ which expresses the ratio of the probability that a molecule will enter solution prior to time t to the probability that it will not.

The hazard function: $h_{ij}(t) = F'_{ij}(t)/(1-F_{ij}(t))$ which is the probability that a molecule is about to enter solution at time t given that it has not done so prior to time t.

The reverse hazard function: $r_{ij}(t) = F'_{ij}(t)/F_{ij}(t)$ which is the probability that a molecule has entered solution just immediately prior to time t given that it has done so at some time between time 0 and time t.

The prime (') denotes differentiation with respect to time.

5.1. The Identity Model

The simplest possible model is the identity model which states that

$$F_{i2}(t) = F_{i1}(t)$$

This is the linear model which corresponds with the mean time profiles for *in vivo* and *in vitro* being coincident or superimposable and is therefore the currently most often used model. The model can be rewritten as

$$F_{ij}(t) = \gamma_i(t) \quad j = 1,2$$

where $\gamma_i(t)$ represents the fraction of drug dissolved from unit i at time t irrespective of whether it is *in vivo* or *in vitro* dissolution. This model is illustrated in figure 6.

5.2. The Proportional Odds Model

The proportional odds model states that at any time the odds function *in vivo* is proportional to the corresponding function *in vitro* i.e.

$$\frac{F_{i2}(t)}{1 - F_{i2}(t)} = \alpha_i \frac{F_{i1}(t)}{1 - F_{i1}(t)}$$

where α_i is the constant of proportionality for the ith unit. This model is illustrated in figure 7 and has the following properties;

Note:
$$F_{i1}(t) = 0 \Rightarrow F_{i2}(t) = 0$$
$$F_{i1}(t) = 1 \Rightarrow F_{i2}(t) = 1$$
$$F_{i1}(t) = F_{i2}(t) \quad \text{When } \alpha_i = 1$$

the last of which shows that the identity model is a special case of the proportional odds model. This model can be rewritten as

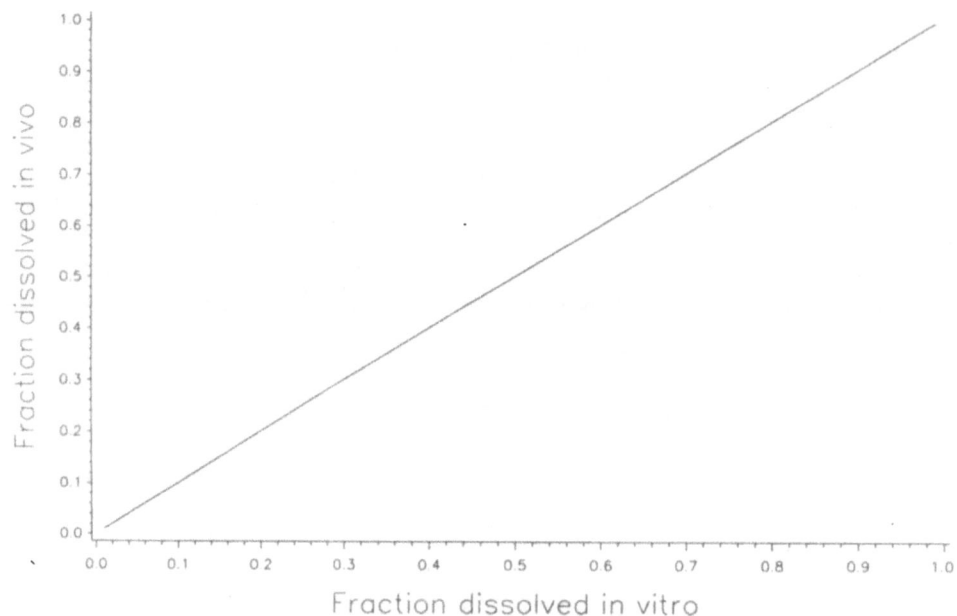

Figure 6. The relationship between the fraction of drug dissolved *in vivo* and that *in vitro* according to the identity model.

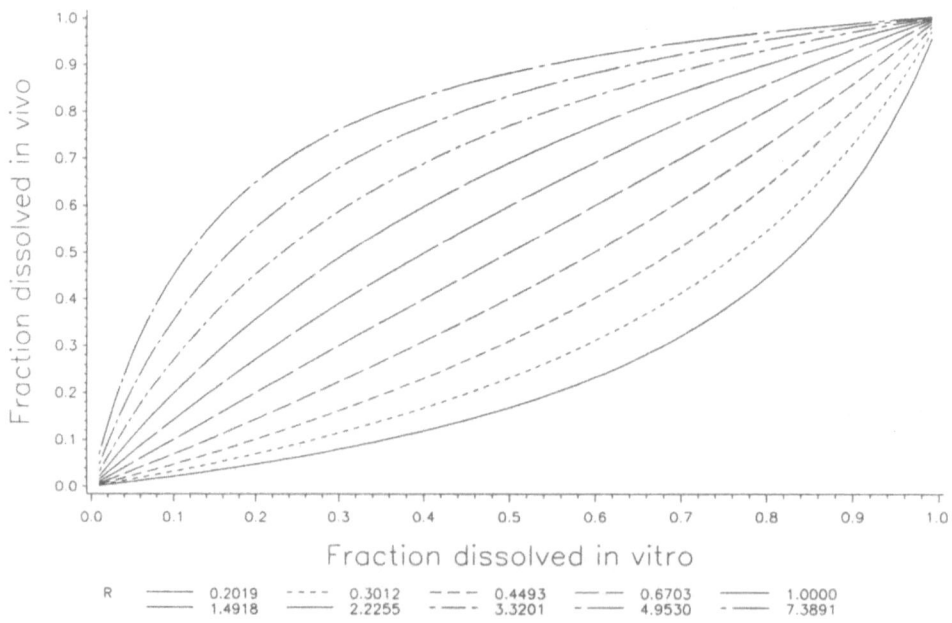

Figure 7. The relationship between the fraction of drug dissolved *in vivo* and that *in vitro* according to the proportional odds model for a range of values of the constant of proportionality (note α_i in the text is denoted by R in the legend).

$$\log\left(\frac{F_{i2}(t)}{1-F_{i2}(t)}\right) = \log(\alpha_i) + \log\left(\frac{F_{i1}(t)}{1-F_{i1}(t)}\right)$$

$$\text{logit}(F_{i2}(t)) = \log(\alpha_i) + \text{logit}(F_{i1}(t))$$

$$\text{logit}(F_{ij}(t)) = \beta_{ij} + \gamma_i(t) \quad \beta_{ij} = \begin{cases} 0 & j=1 \\ \log(\alpha_i) & j=2 \end{cases}$$

where the logit function is defined as

$$\text{logit}(y) = \log\left(\frac{y}{1-y}\right) .$$

and $\gamma_i(t)$ represents the logit of the fraction of drug dissolved from unit i at time t *in vitro*.

5.3. The Proportional Hazards Model

The proportional hazards model states that at any time the hazard function *in vivo* is proportional to the corresponding function *in vitro* i.e.

$$h_{i2}(t) = \alpha_i \, h_{i1}(t)$$

Following integration this gives

$$1 - F_{i2}(t) = (1 - F_{i1}(t))^{\alpha_i}$$

where α_i is the constant of proportionality for the ith unit. This model is illustrated in figure 8 and has the following properties;

Note:
$$F_{i1}(t) = 0 \Rightarrow F_{i2}(t) = 0$$
$$F_{i1}(t) = 1 \Rightarrow F_{i2}(t) = 1$$
$$F_{i1}(t) = F_{i2}(t) \quad \text{When } \alpha_i = 1$$

the last of which shows that the identity model is a special case of the proportional hazards model. This model can be rewritten as

$$\log(1 - F_{i2}(t)) = \alpha_i \log(1 - F_{i1}(t))$$

$$\log(-\log(1 - F_{i2}(t))) = \log(\alpha_i) + \log(-\log(1 - F_{i1}(t)))$$

$$\log(-\log(1 - F_{ij}(t))) = \beta_{ij} + \gamma_i(t) \quad \beta_{ij} = \begin{cases} 0 & j=1 \\ \log(\alpha_i) & j=2 \end{cases}$$

where the function on the left hand side is known as the complementary log-log function and $\gamma_i(t)$ represents the complementary log-log of the fraction of drug dissolved from unit i at time t *in vitro*.

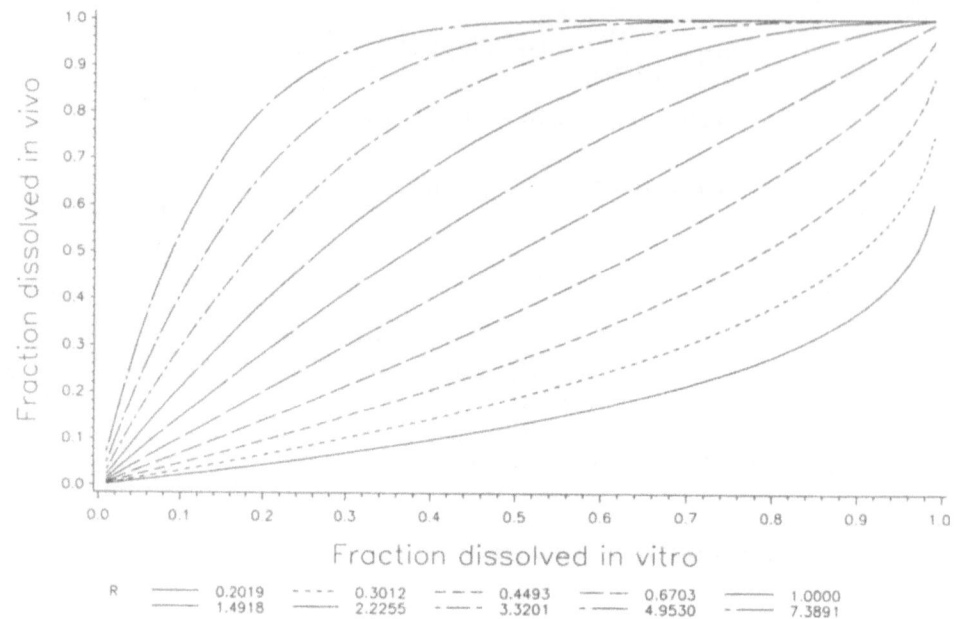

Figure 8. The relationship between the fraction of drug dissolved *in vivo* and that *in vitro* according to the proportional hazards model for a range of values of the constant of proportionality (note α_i in the text is denoted by R in the legend).

5.4. The Proportional Reverse Hazards Model

The proportional reverse hazards model states that at any time the reverse hazard function *in vivo* is proportional to the corresponding function *in vitro* i.e.

$$r_{i2}(t) = \alpha_i \, r_{i1}(t)$$

Following integration this gives

$$F_{i2}(t) = F_{i1}(t)^{\alpha_i}$$

where α_i is the constant of proportionality for the ith unit. This model is illustrated in figure 9 and has the following properties;

Note:
$$F_{i1}(t) = 0 \Rightarrow F_{i2}(t) = 0$$
$$F_{i1}(t) = 1 \Rightarrow F_{i2}(t) = 1$$
$$F_{i1}(t) = F_{i2}(t) \quad \text{When } \alpha_i = 1$$

the last of which shows that the identity model is a special case of the proportional reverse hazards model. This model can be rewritten as

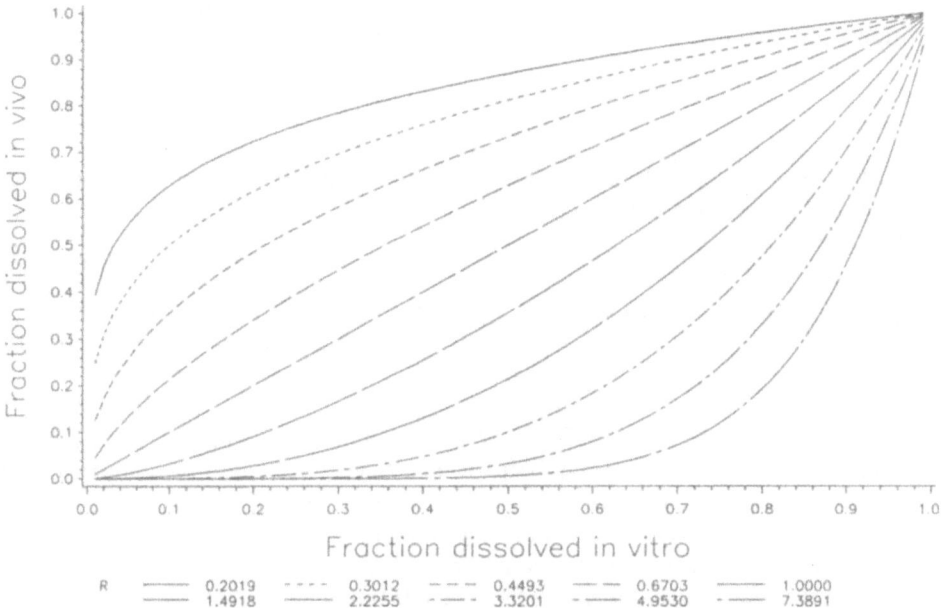

Figure 9. The relationship between the fraction of drug dissolved *in vivo* and that *in vitro* according to the proportional reverse hazards model for a range of values of the constant of proportionality (note α_i in the text is denoted by R in the legend).

$$\log(F_{i2}(t)) = \alpha_i \log(F_{i1}(t))$$

$$\log(-\log(F_{i2}(t))) = \log(\alpha_i) + \log(-\log(F_{i1}(t)))$$

$$\log(-\log(F_{ij}(t))) = \beta_{ij} + \gamma_i(t) \quad \beta_{ij} = \begin{cases} 0 & j = 1 \\ \log(\alpha_i) & j = 2 \end{cases}$$

where $\gamma_i(t)$ represents the log-log of the fraction of drug dissolved from unit i at time t *in vitro*.

6. STATISTICAL MODEL

The models described above define relationships between the fraction of drug dissolved *in vivo* and that *in vitro* but do not make any statement about the statistical properties of the data. These properties are described by a statistical model which influences the methods used for fitting the models described in section 5 to data. A statistical model was constructed by writing

$$Y_{ij}(t) = F_{ij}(t) + \varepsilon_{ij}(t)$$

$$\varepsilon_{ij}(t) \overset{iid}{\sim} N(0, \sigma_j^2)$$

where $Y_{ij}(t)$ represents the measured fraction of drug dissolved from unit i at time t under condition j and $\varepsilon_{ij}(t)$ includes measurement errors. Let

$$g(F_{ij}(t)) = \beta_{ij} + \gamma_i(t)$$

where $g(.)$ is the so-called link function. The form of the link function is dictated by the choice of model as shown in table 1.

Furthermore

$$\text{Let} \quad \beta_{ij} + \gamma_i(t) = \beta_j + \gamma(t) + u_i$$

$$u_i \overset{iid}{\sim} N(0, \sigma_u^2)$$

where β_j is the average value of β_{ij} and $\gamma(t)$ is the average value of $\gamma_i(t)$. The random unit effect u_i describes differences in both β_{ij} and $\gamma_i(t)$ from unit to unit and accounts for the correlation between repeated observations made on the same unit. If a separate parameter is used at each time point to define $\gamma(t)$ the model is described as a generalised linear mixed effects model (GLMM). If a nonlinear function of time is used to define $\gamma(t)$ the model is a nonlinear mixed effects model. The advantage of the former is that no assumption is required about the form of $\gamma(t)$ which means that no assumptions are necessary about the time dependence of either the *in vivo* or *in vitro* dissolution profile, we simply model the relationship between them.

7. EXTENDING THE MODEL

In all the above models β_2 is a constant i.e. the relationship between *in vivo* and *in vitro* dissolution lies on one of the curves shown in figures 7, 8 or 9. Increased flexibility can be attained by allowing β_2 to change with time i.e. the relationship may slide from one curve to the other. The simplest relationship between β_2 and time is a linear one i.e.

$$\beta_2 = \theta_0 + \theta_1 t$$

Table 1. Shows the appropriate link function corresponding to each of the models considered

Model		Link function
Identity	Identity	$F_{ij}(t)$
Proportional odds	Logit	$\log\left(\dfrac{F_{ij}(t)}{1 - F_{ij}(t)}\right)$
Proportional hazards	Complementary log-log	$\log(-\log(1 - F_{ij}(t)))$
Proportional reverse hazards	Log-log	$\log(-\log(F_{ij}(t)))$

Allowing β_2 to change with time could be thought of as accounting for the fact that the relationship between *in vivo* and *in vitro* dissolution is affected by the changing environment as the dosage form proceeds along the gastrointestinal tract.

8. EXAMPLES

The above GLMM's were fitted to data sets for two batches (PD14151 and PD14152) of drug product using an approximate maximum likelihood fitting technique implemented by the SAS macro GLIMMIX[14]. The *in vivo* data were based on a study with 10 subjects and the *in vitro* data were collected from 12 (PD14151) or 24 (PD14152) dosage units. The deconvolution of the *in vivo* data was performed using PCDCON[15] and an immediate release dosage form of the drug as a reference. Both the data and the fitted curves are shown in figures 10–21.

9. DISCUSSION

Figures 1 and 2 illustrate the time dependence of drug dissolution both *in vivo* and *in vitro*. In the context of IVIVC the relationship of interest is that between *in vivo* and *in vitro* drug dissolution and time is a 'nuisance' variable which can be eliminated by plotting *in vivo* dissolution versus *in vitro* dissolution at the same time as shown in figure 3. However, basing the data analysis on this graphical device rather than on the statistical properties of the data is not to be recommended. The curves displayed in figure 5 are described as operating characteristic (OC) curves because they illustrate the essential characteristics of the methods of analysis examined in the simulation experiment. For a set of data any

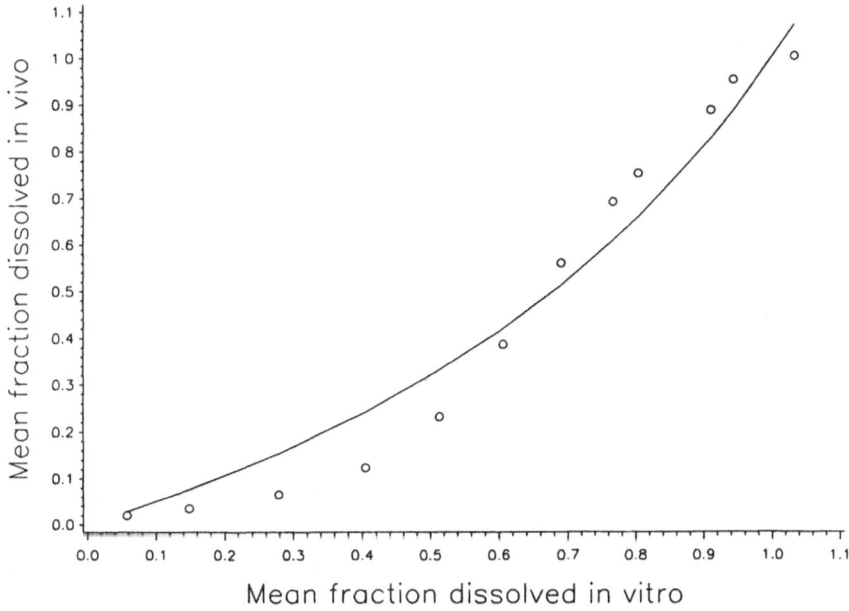

Figure 10. The mean data (open circles) and fitted curve (solid line) when the proportional odds model with β_2 constant was fitted to the data for PD14151 as described in section 6 of the text.

A. **Dunne** *et al.*

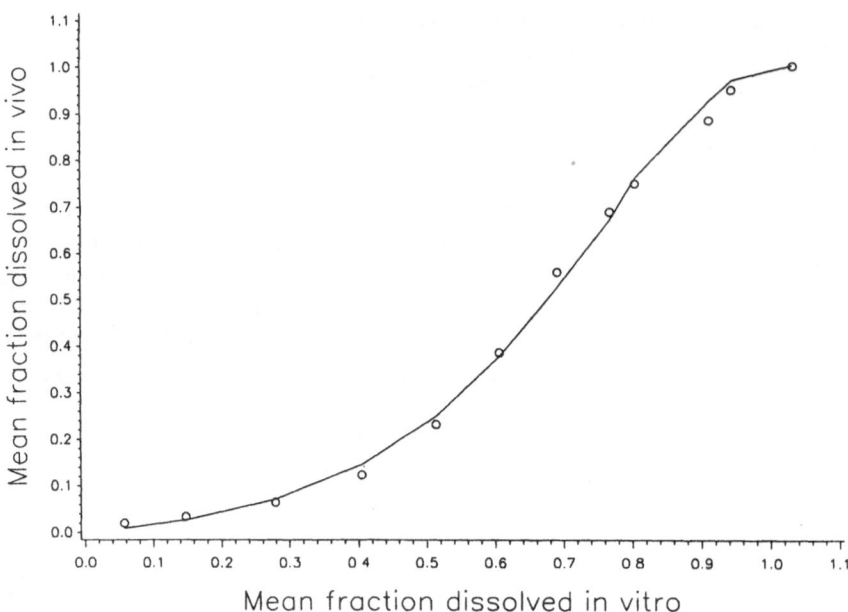

Figure 11. The mean data (open circles) and fitted curve (solid line) when the proportional odds model with β_2 as a linear function of time was fitted to the data for PD14151 as described in section 6 of the text.

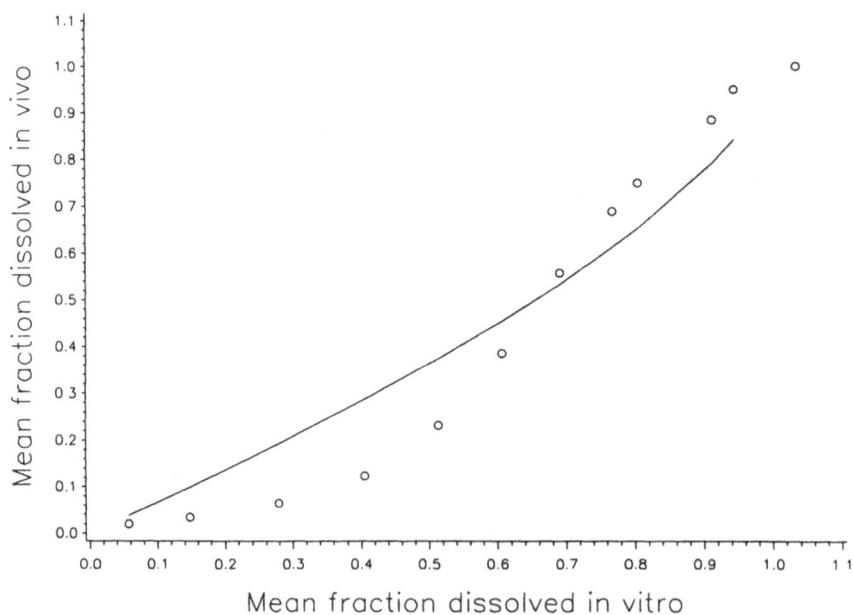

Figure 12. The mean data (open circles) and fitted curve (solid line) when the proportional hazards model with β_2 constant was fitted to the data for PD14151 as described in section 6 of the text.

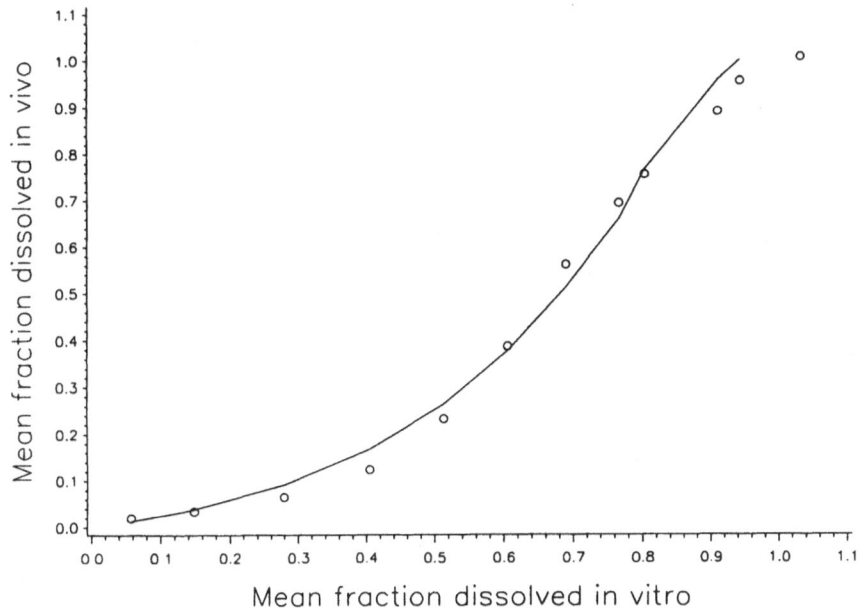

Figure 13. The mean data (open circles) and fitted curve (solid line) when the proportional hazards model with β_2 as a linear function of time was fitted to the data for PD14151 as described in section 6 of the text.

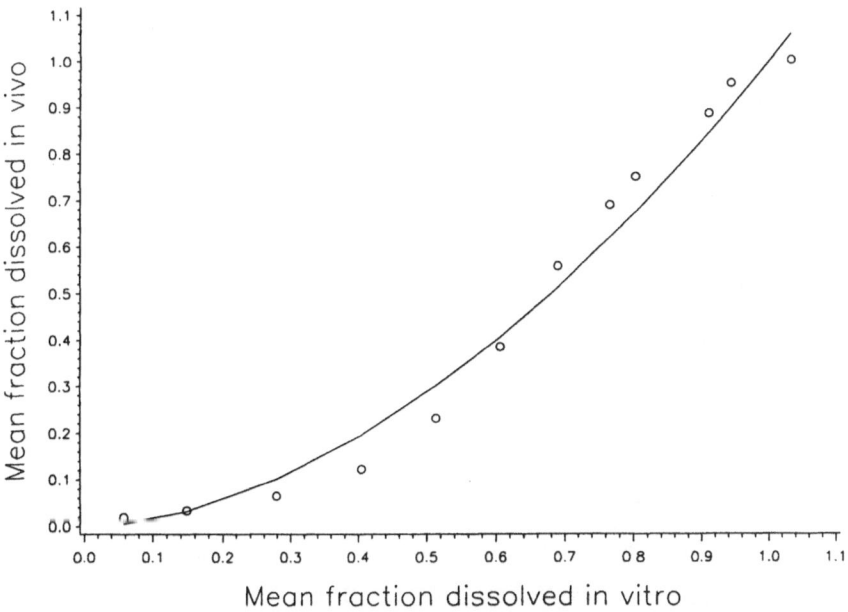

Figure 14. The mean data (open circles) and fitted curve (solid line) when the proportional reverse hazards model with β_2 constant was fitted to the data for PD14151 as described in section 6 of the text.

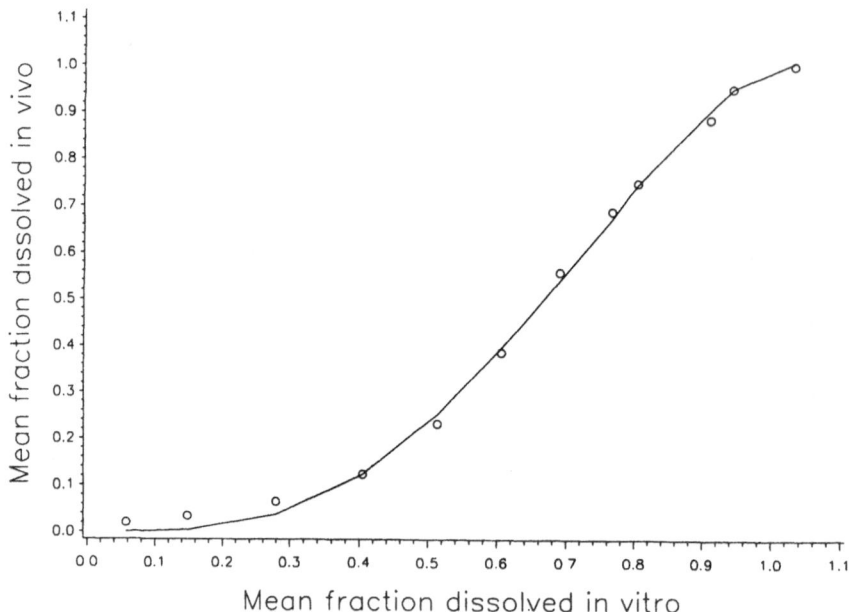

Figure 15. The mean data (open circles) and fitted curve (solid line) when the proportional reverse hazards model with β_2 as a linear function of time was fitted to the data for PD14151 as described in section 6 of the text.

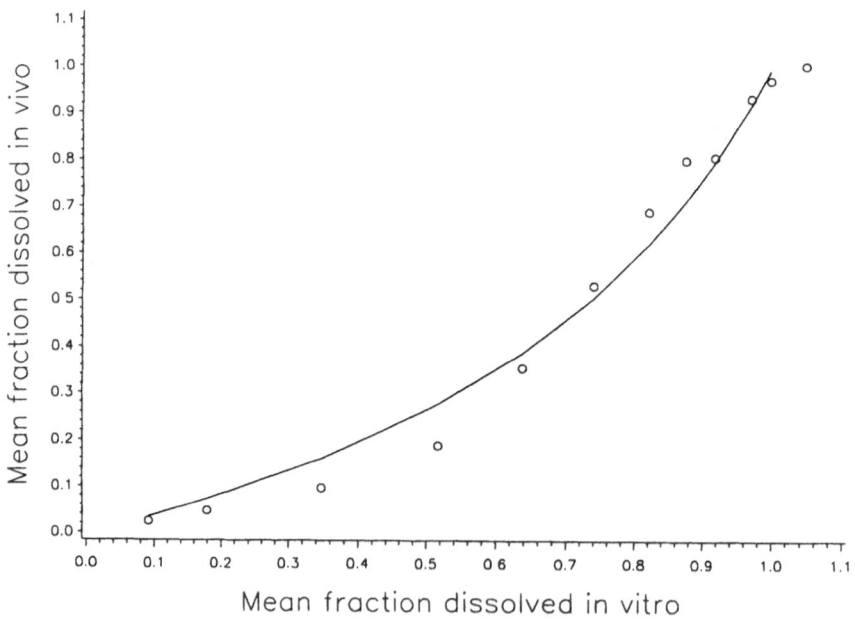

Figure 16. The mean data (open circles) and fitted curve (solid line) when the proportional odds model with β_2 constant was fitted to the data for PD14152 as described in section 6 of the text.

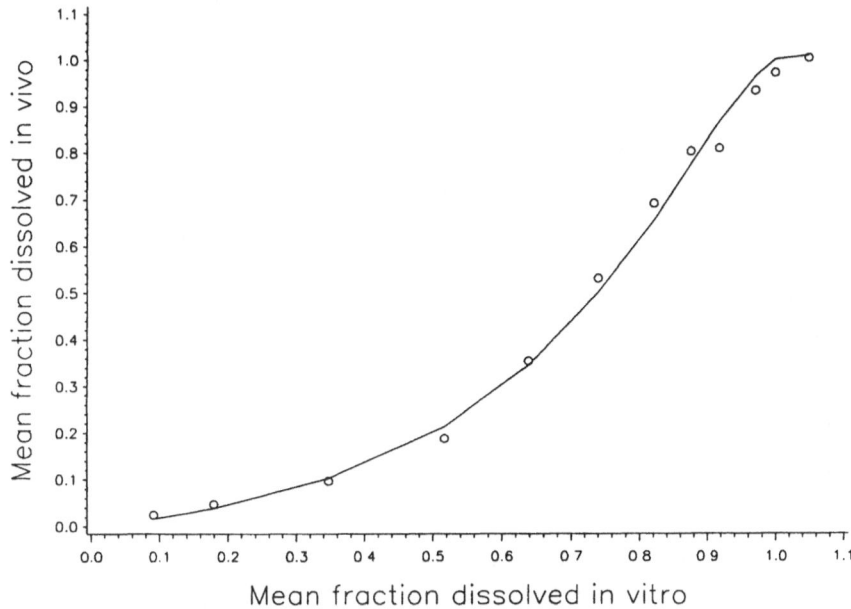

Figure 17. The mean data (open circles) and fitted curve (solid line) when the proportional odds model with β_2 as a linear function of time was fitted to the data for PD14152 as described in section 6 of the text.

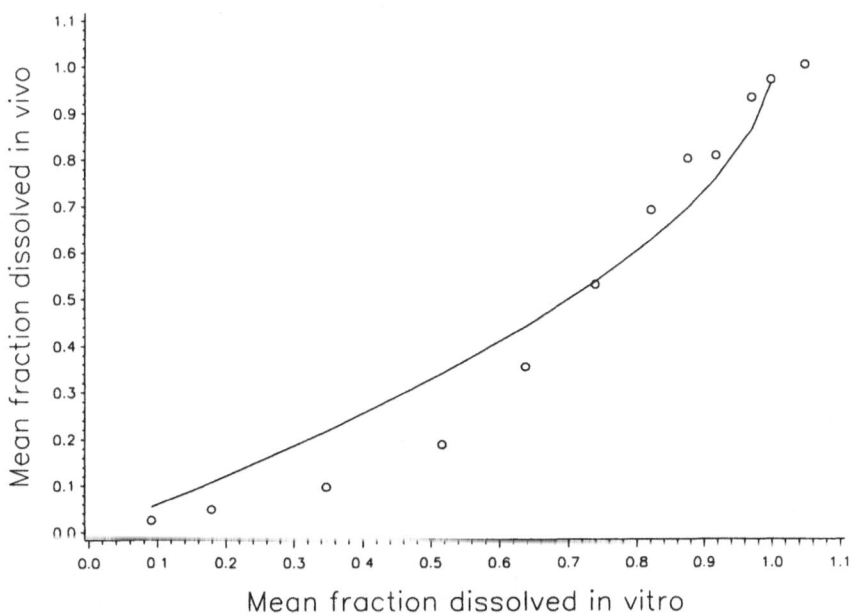

Figure 18. The mean data (open circles) and fitted curve (solid line) when the proportional hazards model with β_2 constant was fitted to the data for PD14152 as described in section 6 of the text.

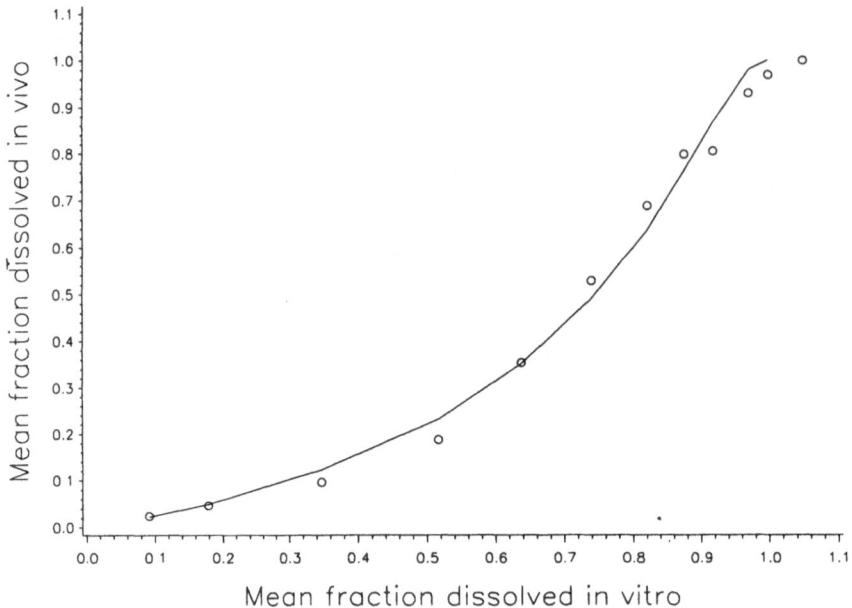

Figure 19. The mean data (open circles) and fitted curve (solid line) when the proportional hazards model with β_2 as a linear function of time was fitted to the data for PD14152 as described in section 6 of the text.

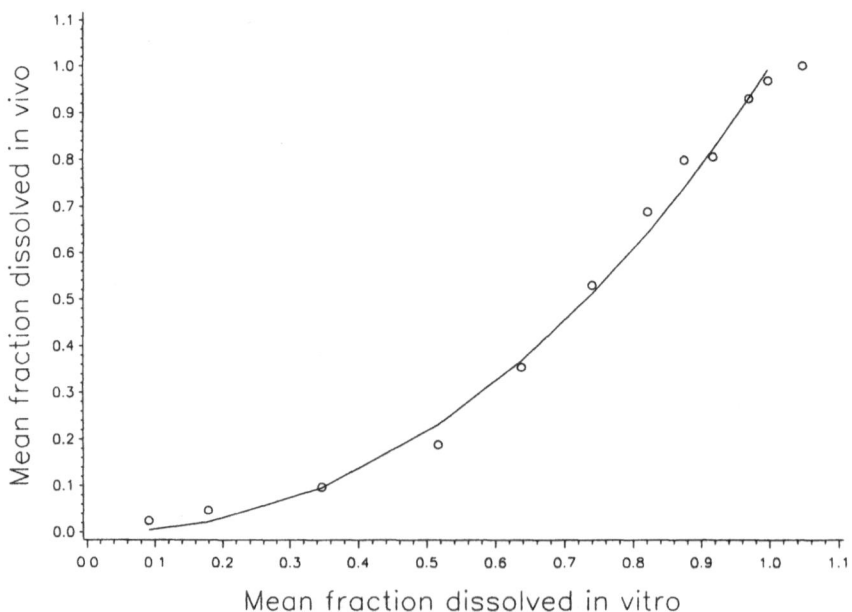

Figure 20. The mean data (open circles) and fitted curve (solid line) when the proportional reverse hazards model with β_2 constant was fitted to the data for PD14152 as described in section 6 of the text.

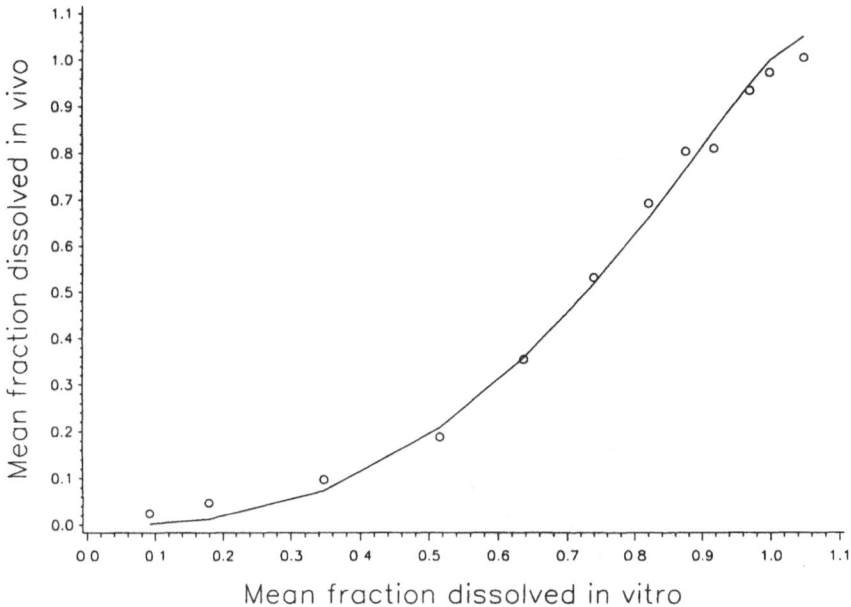

Figure 21. The mean data (open circles) and fitted curve (solid line) when the proportional reverse hazards model with β_2 as a linear function of time was fitted to the data for PD14152 as described in section 6 of the text.

particular method of analysis leads us to conclude that a level A IVIVC either does or does not exist. There are two possible incorrect conclusions i.e. declaring that an IVIVC does not exist when in actual fact it does (known as a type I error) and declaring that an IVIVC exists when in fact it does not (known as a type II error). The probabilities of each of these errors occurring depend on the method of analysis used and may be used to decide which methodology is optimal. Instead of talking about the probability of a type II error we frequently refer to the 'power' which is the probability of not making a type II error. Figure 5 shows the probability (in terms of a percentage) of declaring a level A IVIVC as a function of $ln(r)$. When $ln(r)$ is nonzero an IVIVC (in the sense discussed in section 2) does not exist and this probability is therefore that of a type II error and when $ln(r)$ is zero an IVIVC does exist and this probability is 1-probability of a type I error. The thick solid line in figure 5 is the OC curve for the ideal method (it does not actually exist because of random variation) which never makes type I or type II errors i.e. always declares an IVIVC when $ln(r)$ is zero and never declares an IVIVC otherwise. Many of the methods of analysis whose OC curves are displayed in figure 5 have type I error rates far greater than the nominal 5% and are lacking in power (have high type II error rates). Simple regression analysis followed by separate tests for a slope of unity and an intercept of zero performs badly with a high percentage of failures to reject the null hypothesis across a wide range of values of r for which the null hypothesis is false. Simple regression with a simultaneous test of slope and intercept performs better but has a type I error rate of approximately 36%. Regression through the origin with a test for a slope of unity behaved in a similar fashion. Simple regression with a test of slope equals unity is almost totally lacking in power over a wide range of values of r. Hotelling's T^2 had a type I error rate of more than 50% and a separate simulation study showed that this was due to the inequality of the *in vivo* and *in vitro* covariance matrices. The linear mixed effects analysis of variance is obviously the best of the methods studied with highest power and the correct type I error rate. Consequently it can be concluded from the simulation study that of the methods considered the linear mixed effects analysis is the method of choice when the covariance structure is compound symmetric. In the case of nonlinear models it is to be expected that similar considerations would apply and that nonlinear regression analysis[7] is not to be recommended. For the nonlinear models considered in section 5 above the GLMM outlined in section 6 corresponds with the linear mixed effects model in the linear case.

The proportional odds, proportional hazards and proportional reverse hazards models are generalisations of the identity (linear) model because it is a special case of each of them. Hence we have nothing to lose by using such models because when the identity model is appropriate it is automatically included. On the other hand we have much to gain because figures 7–9 show that these models describe a broad range of nonlinear relationships which cannot be described using a linear model. Furthermore, the model extension described in section 7 above serves to further generalise and enhance the scope of these models. Figures 10–21 show that with the exception of the proportional reverse hazards model fitted to batch PD14152, the models did not fit the data very well without the extension. However, with the extension the fits were quite good and this indicates that these models may prove to be useful tools for modelling level A IVIVC in the future.

10. REFERENCES

1. Gibaldi, M. and Perrier, D. (1982) *Pharmacokinetics*. Marcel Dekker, New York.
2. Dietrich, R., Brausse, R., Benedikt, G. and Steinijans, V.W. (1988) Feasibility of *in-vitro/in-vivo* correlation in the case of a new sustained-release theophylline pellet formulation. *Arzneim. Forsch. Drug Res.*, 38, 1229–1237.
3. Hussein, Z. and Friedman, M. (1990) Release and absorption characteristics of novel theophylline sustained-release formulations: *In vitro-in vivo* correlation. *Pharm Res.*, 7, 1167–1171.
4. Mojaverian, P., Radwanski, E., Lin, C.C., Cho, P., Vadino, W.A. and Rosen, J.M. (1992) Correlation of *in vitro* release rate and *in vivo* absorption characteristics of four chlorpheniramine maleate extended-release formulations. *Pharm. Res.*, 9, 450–456.
5. Humbert, H., Cabiac, M.D. and Bosshardt, H. (1994) *In vitro-in vivo* correlation of a modified-release oral form of ketotifen: In vitro dissolution rate specification. *J. Pharm. Sci.*, 83, 131–136.
6. Hwang, S.S., Gorsline, J., Louie, J., Dye, D., Guinta, D. and Hamel, L. (1995) *In vitro* and *in vivo* evaluation of a once-daily controlled-release pseudoephedrine product. *J. Clin. Pharmacol.*, 35, 259–267.
7. Polli, J.E., Crison, J.R. and Amidon, G.L. (1996) Novel approach to the analysis of *in vitro-in vivo* relationships. *J. Pharm. Sci.*, 85, 753–759.
8. Leeson, L.J. (1995) *In vitro/in vivo* correlations. *Drug Information Journal*, 29, 903–915.
9. Snedecor, G.W. and Cochran, W.G. (1989) *Statistical Methods*. Iowa State University Press, Ames, Iowa.
10. USP (1988) *In-vitro/in-vivo* correlation for extended-release oral dosage forms. *Pharmacopeial Forum*, 14, 4160–4161.
11. Morrison , D.F. (1978) *Multivariate Statistical Methods*, McGraw-Hill, Singapore.
12. Crowder, M.J. and Hand, D.J. (1990) *Analysis of Repeated Measures*. Chapman and Hall, London.
13. Searle, S.R., Casella, G. and McCulloch, C.E. (1992) *Variance Components*. Wiley, New York.
14. Wolfinger, R. and O'Connell, M. (1993) Generalized linear mixed models: a pseudo-likelihood approach. *J. Statist. Comput. Simul.* 48, 233–243.
15. Gillespie, W.R. (1992) PCDCON: Deconvolution for Pharmacokinetic Applications. Documentation for PCDCON computer program, University of Texas at Austin.

7

VALIDATION OF *IN VITRO–IN VIVO* CORRELATION MODELS

David Young,[1] James A. Dowell,[1] Deborah A. Piscitelli,[1] and John Devane[2]

[1]IVIVR Cooperative Working Group
Pharmacokinetics-Biopharmaceutics lab
School of Pharmacy
University of Maryland
100 Penn Street Room 540
Baltimore, Maryland 21201
[2]IVIVR Cooperative Working Group
Elan Corporation
Athlone Ireland

I. INTRODUCTION

A valid model can be defined as one that is a) well-grounded on principles or evidence, b) able to withstand criticism or objection, and c) capable of serving the purpose of the model. In the area of mathematical modelling, there are two criteria in the validation process: internal criteria and external criteria (1). The internal criteria establish the validity of a model within the conditions of the model itself (e.g., the model's consistency and algorithmic integrity). The external criteria, on the other hand, are based on the model's purpose, theory, and data, considering those conditions that are not internal properties of the model.

The FDA has appropriately realized that, although *in vitro-in vivo* correlation (IVIVC) models can be developed, investigators need to evaluate the validity of the models (2). The draft guidance on the development, evaluation and application of IVIVC models states that the "validation is a broad term encompassing experimental and statistical techniques used during development and evaluation of a correlation which aid in deciding whether the correlation is both predictive and of good quality" (2). Although the emphasis in the Draft Guidance is on the predictive performance of the models (a part of the external criteria), the other components of the validation process cannot be ignored given the definition of validation within the guidance. In order to thoroughly understand the validation of an IVIVC model, this chapter will briefly describe 1) other components of the validation process that one should consider, 2) the methods that are discussed in the FDA Draft Guidance, and 3) alternate approaches to the validation of IVIVC models. The intention of this chapter is not to be an exhaustive review of the issues associated with IVIVC model validation (e.g., experimental design, number and type of formulations). Instead, the purpose of this chapter is to discuss some approaches that can be used in the validation of Level A IVIVC models.

II. INTERNAL CRITERIA FOR VALIDATION

The internal criteria evaluate the appropriateness of the internal properties of the model (e.g., model consistency, algorithm) (1). A model's consistency refers to the suitability of the model and mathematical structure. The model should be mathematically logical and have no mathematical or conceptual misgivings. A very simple example of this is the ability to correctly set up the set of first order differential equations that describe a two compartment pharmacokinetic model with first order elimination. If an error is made setting up these equations, the results from this model would not be reliable. Another consideration regarding model structure, especially for complicated compartmental pharmacokinetic models, is that a unique solution for the equations must exist, or the results obtained would be meaningless.

The algorithm or method by which we obtain estimates of the model parameters must also be evaluated. The algorithm must be numerically and statistically appropriate given the model and the type of data. Usually established software is used for these analysis and the integrity of the algorithm has already been tested. Newer methodologies and software, however, must be validated and cross-validated with proven methods. Errors due to the implementation of a less than robust algorithm, soft termination criteria, or large round off errors are examples of a solution method or software that does not meet internal criteria.

III. EXTERNAL CRITERIA FOR VALIDATION

There are four external criteria for model validation: theoretical, heuristic, empirical, and pragmatic validity (1). The theoretical validity must be continually applied to a model at all levels of development. It is a criteria which states that the model must be consistent with accepted theories or previously validated models. Parameter values, model identities and model structure should not conflict with the proven and validated information available. Also, any expressions and relationships derived from the model must be able to exist with the proven science. For instance, a drug is modeled by a linear two compartment model, yet is known to be eliminated by means of a saturable enzymatic conversion. A linear model would probably be inappropriate in this situation.

The potential value of a model must also be considered as a practical question before a model is even developed. A model may be used to identify important relationships, explain an observation, to test a possible hypothesis, or discover an alternative hypothesis. The failure of a model to pass tests of heuristic validity indicates a model is not useful. Heuristic validity is a qualitative question asked at the onset of modeling and when a model is established. It questions the potential value of the developed model.

Empirical validity involves looking at the overall goodness-of-fit of a model. There are many proven tests available to determine empirical validity of a model, all of which examine the data and quality of the fit. The majority of these methods examine the residuals in some manner.

The last part of the external criteria is an evaluation of the pragmatic validity. The pragmatic validity determines to what extent the model satisfies the purpose of the model. This purpose often includes either a desire to describe the biological system or to predict what occurs under new conditions. For example, if the purpose of the model is to predict the *in vivo* response from the *in vitro* dissolution, then the predicted *in vivo* response (based on the *in vitro* data from a given formulation) must be statistically compared to the observed *in vivo* response when the formulation is administered to human volunteers. Regardless of the specific purpose, pragmatic validity often requires some evaluation of the model's ability to represent the actual data. Most tests for this include an evaluation of a models ability to predict the observed data.

IV. FDA DRAFT GUIDANCE ON INTERNAL/EXTERNAL PREDICTABILITY

Since the purpose of the IVIVC model is to predict the *in vivo* response from an *in vitro* dissolution profile, the evaluation of these models in the FDA Draft Guidance has emphasized their predictive performance. Two approaches to estimate this performance are internal and external predictability.

Internal predictability evaluates how well the IVIVC model describes the data used to develop the model. This approach is recommended for all IVIVC analysis. The Draft Guidance states that the internal predictability can be assessed using the method of cross-validation, as long as three or more formulations are used in the model development. Cross-validation is widely used in regression analysis to validate models but, in the context of IVIVC, the term cross-validation is slightly different. The procedure of cross-validation in IVIVC is:

1. leave the data from one formulation out and develop the model with the remaining formulations
2. predict the *in vivo* response of the formulation left out from the *in vitro* dissolution of the formulation and the IVIVC model
3. calculate the prediction error by comparing the observed *in vivo* response to the predicted response
4. repeat steps 1–3 for each formulation

Other measures of the internal predictability which are not discussed in the draft guidance are some of the measures used in evaluating the goodness-of-fit of the model (i.e., empirical validation). These measures are the weighted residual sum of squares and the weighted residual plots. These measures should be evaluated when all the formulations are used in developing the model and when comparing alternate models.

External predictability is the second method of evaluating the IVIVC model. This approach is a more comprehensive evaluation of the ability for IVIVC model to predict and should be used when a) internal predictability is not conclusive, b) only two formulations with different release rates are available to develop the IVIVC model, c) the correlation is being developed for a drug with narrow therapeutic index, and d) whenever one desires a more comprehensive validation of the model. This approach determines how well the IVIVC model describes data which are not used in the model development. The ideal situation is that the formulations not used in the model development should be different enough in release rate to appropriately evaluate the predictability of the model. The procedure of calculating the external predictability is:

1) develop the IVIVC model with a given set of formulations
2) predict the *in vivo* response of a formulation not used in the model development based on the *in vitro* dissolution of that formulation and the IVIVC model
3) calculate the prediction error by comparing the observed *in vivo* response to the predicted response

V. PREDICTABILITY METRICS

Given that the validity of the IVIVC model depends on the internal and/or external predictability, their are two other features of the validation that are important: predictability metrics and statistical criteria for defining "good" predictability or a valid model. In the previous description of predictability, the metric used to evaluate the predictability has been described simply as the prediction error. However, the most appropriate metrics to measure bias, precision, or both in this application are still under investigation. Some of the prediction metrics being evaluated are prediction error, absolute value of the prediction error, squared prediction error, and root mean squared prediction error.

VI. STATISTICAL CRITERIA FOR A VALID MODEL

The statistical approaches to determine the validity of an IVIVC model is another area of active research. There are a number of approaches that one can use to determine how well an IVIVC model predicts *in vivo* response for both internal or external predictability.

The first statistical approach is to evaluate the accuracy or bias and the precision of the predictions by statistically comparing the mean prediction error (bias) and root mean squared error to zero (e.g., H0: Mean Prediction Error = 0, Ha: mean prediction error not equal to zero). A similar approach has been more fully described by Sheiner and Beal in 1977 (3).

A second approach statistically determines if the IVIVC model predicts better than a naïve model. One naïve model that has been discussed by the FDA (and presented in the Draft Guidance)is the naïve averaged model. This model actually represents the mean *in vivo* response at each time for the formulations used in the IVIVC model. This mean *in vivo* response is then assumed to have either the *in vitro* dissolution of the formulation left out in cross-validation or the *in vitro* dissolution of the formulation not used in external validation. The naïve averaged *in vivo* response at each *in vitro* value can then be compared to the observed *in vivo* response for the formulation left out or not used in the model development. The prediction error can then be calculated. The predictability of the naïve averaged model can then be compared statistically to the IVIVC model.

To evaluate the relative accuracy (e.g., prediction error) and/or precision (e.g., absolute prediction error) of the IVIVC model with the accuracy and precision of a naïve model, the null hypothesis for this approach would be defined as: the prediction metrics for the IVIVC model and the naïve model are statistically equivalent at a $p<0.05$. The alternative hypothesis would be that the prediction metrics for the IVIVC model is statistically less than the naïve model. The IVIVC model is then considered valid only when both the accuracy and precision are significantly smaller in the IVIVC model. This approach can also be modified to compare the IVIVC model to another IVIVC model (e.g., linear model) instead of the naïve model.

A third approach uses the principles of resampling statistics to determine the probability that the predictability from the IVIVC model may occur just "by chance" (4). There are a number of ways to approach this problem and one is presented below:

1) calculate the change in the *in vivo* response at each time point for each formulation used in developing the model

2) randomly choose from any of the formulations the first *in vivo* response value

3) randomly choose the change in the *in vivo* (calculated from step 1) between the first time point and second time point for any formulation and add it to the value obtained in step 2.

4) repeat the random choice at each time point and add it to the *in vivo* response calculated at the previous time point. The final curve represent an *in vivo* response that may occur "by chance".

5) the "by chance" *in vivo* response is then assumed to have either the *in vitro* dissolution of the formulation left out in cross-validation or the *in vitro* dissolution of the formulation not used in external validation. For cross-validation, the "by chance" model will be repeated for each formulation left out.

6) the ability of the IVIVC model to predict the formulation not used in developing the IVIVC model can then be compared to the predictability of the "by chance" model.

7) this is repeated 100–1000 times and one counts how many times the IVIVC model prediction metrics is less than the "by chance" metrics. If the IVIVC model predictability is less than or equal to the "by chance" model more than 95% of the time, than the IVIVC model predictability could not have occurred by chance and the model is valid.

VII. CONCLUSION

Although the methods used to develop an IVIVC model are critical, the validation of the IVIVC model is equally important. This chapter has attempted to describe some of the principles of IVIVC model validation. Additional research is presently underway to evaluate the different approaches to validation, to determine which metric is most important, and to investigate the use of more complex resampling statistics to develop and validate IVIVC models.

REFERENCES

1. "The Validation of Models of Metabolic and Endocrine Systems", Chapter 9, *The Mathematical Modeling of Metabolic and Endocrine Systems: Model Formulation, Identification, and Validation*. E.Carson, C. Cobelli, L. Finkelstein, 1983, John Wiley & Sons, Inc., Toronto, Canada
2. Guidance for Industry: Extended Release Solid Oral Dosage Forms. Development, Evaluation and Application of *In Vitro/In Vivo* Correlations. Center for Drug Evaluation and Research (CDER), July 1, 1996, FDA Draft Guidelines or Chapter 25 of this publication.
3. Scientific Commentary: Some Suggestions for Measuring Predictive Performance. Sheiner, L.B., Beal, S. *Journal of Pharmacokinetics and Biopharmaceutics*, Vol. 9, No. 4, 1981.
4. *Resampling Stats: User's Guide* by Julian Simon, 9/1995, Resampling Stats, Inc., Arlington, Virginia.

8

EXAMPLES OF DEVELOPING *IN VITRO–IN VIVO* RELATIONSHIPS

Jackie Butler

IVIVR Cooperative Working Group
Elan Corporation
Athlone, Ireland

This manuscript focuses on developing IVIVR for two quite distinctive formulation approaches to drugs with different physicochemical properties. Each of these formulations is designed for once-daily dosing (See Table 1).

Formulation 1 is a multiparticulate formulation of an acidic drug, poorly soluble at low pH's and reasonably soluble at higher pH's. This compound has a low molecular weight and is permeable throughout the gi tract. The P-effective for these compounds were determined using an *in-vitro* rat gut model, with the drug in solution. A more detailed description of this method is outlined in a paper by Araz Raoof .

Formulation 2 is a matrix tablet formulation of a drug which is highly soluble, independent of pH and non-ionisable. This drug has a low molecular weight and is permeable at all sites of the gi tract.

For Formulation 1, drug release is controlled by the level of rate controlling polymer applied and for Formulation 2, release is controlled by the methocel content in the formulation.

Each of the formulations was initially evaluated *in-vitro* using a 'standard' dissolution test (Table 2).

For formulation 1, the multiparticulate formulation of an acidic drug, the standard test was performed in the USP II apparatus at 50rpm, based on an existing monograph for the immediately releasing form of this compound. 0.05M phosphate buffer at pH 7.2 was chosen for the dissolution medium to ensure optimal solubility of the drug.

For formulation 2, the matrix tablet formulation of a highly soluble drug, the USP II apparatus at 100rpm was chosen as the standard method, based on early *in-vitro* development work with this formulation. This test was performed using water as the medium due to the high pH-independent solubility of this compound.

Early in the development of these products, a prototype biostudy evaluating a range of qualitatively and quantitatively different formulations was performed. The choice of products for this study was based on an assumed IVIVR, that is *in-vivo* = *in-vitro*. Based on the results of this biostudy, the non-standard dissolution methods were developed retrospectively to mimic as closely as possible the *in-vivo* performance of these products.

Table 1. Formulation and Physicochemical Characteritics

	Formulation 1			Formulation 2		
Active						
pKa	5 - 6			Non-ionized		
MW	254			191		
Solubility (mg/ml)	pH 1 -	pH 6.5 2.07	pH 7.4 10.7	pH 1 >236.7	pH 6.5 >145.1	pH 7.4 >162
Permeability (P_{EFF}) (10^{-6} cm.sec^{-1})	Upper SI 49.99	Lower SI 18.39	Colon 12.20	Upper SI 22.64	Lower SI 18.24	Colon 10.40
Dosage Form	Multiparticulate			Matrix Tablet		
Dosage Regimen	Once-a-day			Once-a-day		

For Formulation 1, the *in-vivo* absorption profiles show a lag in absorption for the first 2 hours after dosing (Figure 1). This is most likely due to the poor solubility of this drug in the lower pH of the stomach. This lag in release is not apparent using the standard dissolution system, as this was performed at optimal pH for solubility of the drug.

If we assume that minimal dissolution occurs in the first two hours and timeshift the dissolution curve, we see that the dissolution is considerably slower than the absorption for these batches (Figure 2). We performed some experiments evaluating the effect of pre-incubation of the product for two hours in pH 1 (Figure 3). The results show that there is little effect on the subsequent dissolution in pH 7.2. Therefore optimisation of the dissolution system to match *in-vivo* absorption concentrated on the dissolution at the higher pH's. The dissolution was then compared to the absorption post-lag, that is from 2 hours in order to chose the most suitable dissolution method.

The non-standard method developed for formulation 1 uses the Biodis (USP III) system at 10 dips per minute (Table 2). This method was chosen to increase the agitation of the beads compared to the USP II method and hence to speed up the dissolution. 0.2M phosphate buffer at pH 6.5 was chosen as the medium for the test, due to the higher ionic strength of this buffer compared to the 0.05M buffer used in the standard test and the fact that ionic strength had also been shown to speed up the dissolution of this product. The non-standard dissolution system in this case is not ideal in that the dissolution profile needs to be timeshifted to incorporate the lag in the absorption profile. Although the lag in

Table 2. Dissolution methods

	Formulation 1	Formulation 2
Standard Method	USP II (Paddle)	USP II (Paddle)
	50 rpm	100 rpm
	900 ml phosphate buffer (0.05M) pH 7.2	900 ml distilled water
	37^{0}C	37^{0}C
Non-Standard Method	USP III (Biodis)	USP III (Biodis)
	10 dpm	30 dpm
	250 ml phosphate buffer (0.2 M) pH 6.5	250 ml 0.32 M phosphate buffer
	37^{0}C	37^{0}C

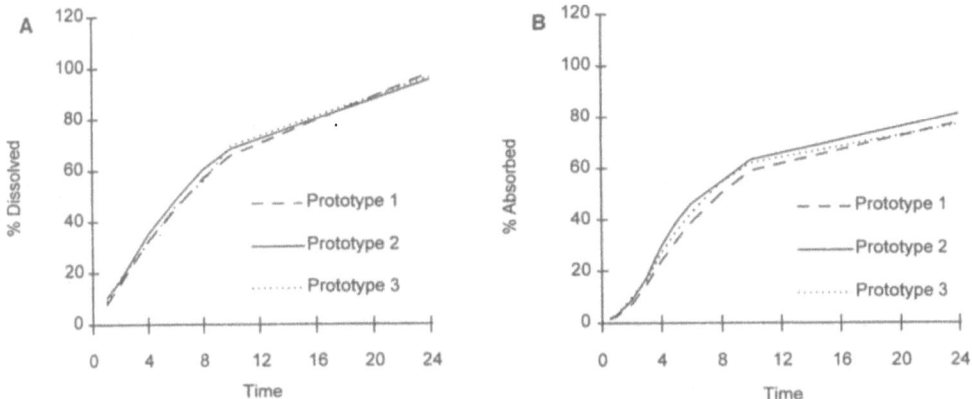

Figure 1. Formulation 1- standard dissolution method. (a) Dissolution; (b) Absorption.

absorption could also be mimicked by setting the initial conditions to pH 1, which can be easily achieved using the Biodis system, this would provide little additional information as this compound is practically insoluble at this pH and dissolution would be minimal. The length of time that the product would be left at this low pH would also have to be arbitrarily chosen and therefore incubation at pH 1 for a length of time would not be any more relevant to the dissolution test than timeshifting the dissolution curve by the same period. The choice of the two-hour timeshift was based on the observed lag in the *in-vivo* data which is presumably related in some way to gastric emptying. Although in our studies, controls on feeding and posture in particular are tightly controlled to minimise the variability in gastric emptying, complete elimination of the inter-subject and intra-subject variability is not possible and therefore any in-vitro/in-vivo relationship developed using

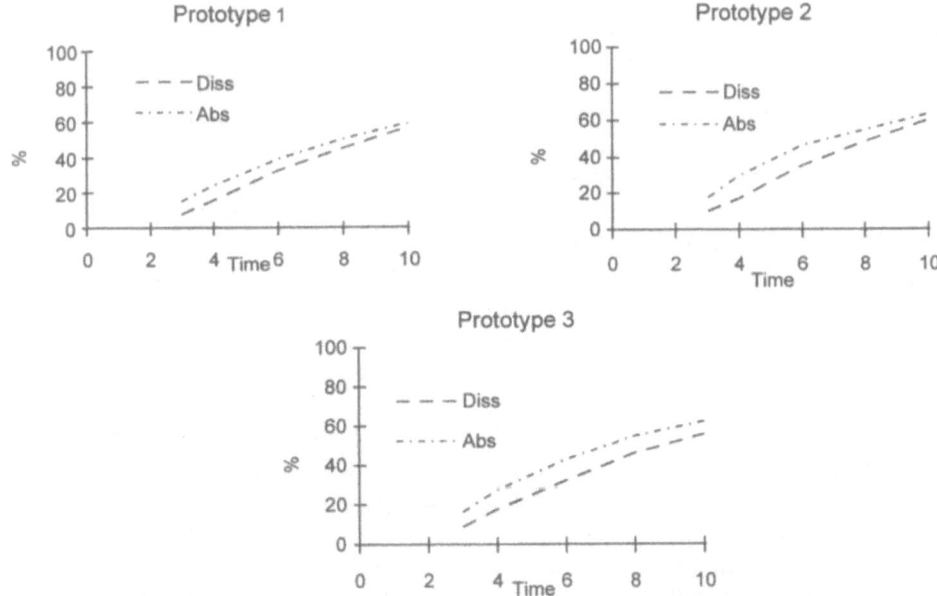

Figure 2. Formulation 1-standard dissolutionmethod timeshifted profiles.

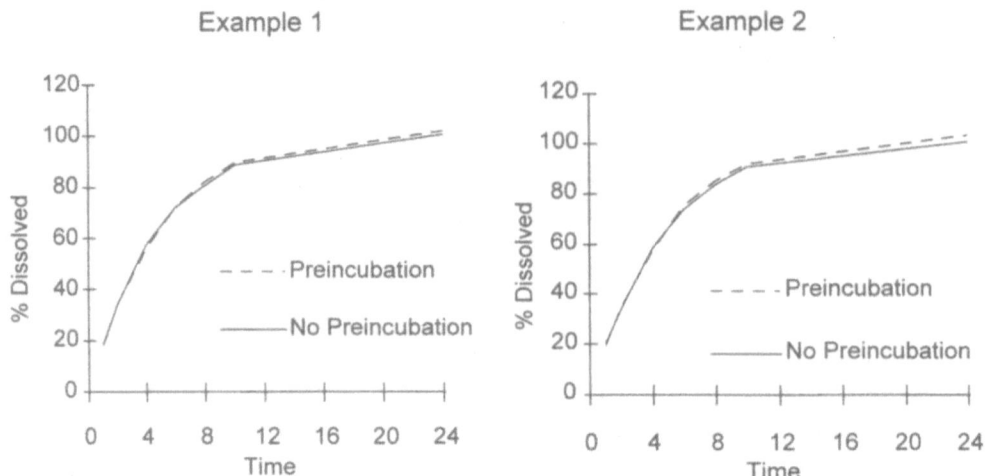

Figure 3. Formulation 1-check of pre-incubation at pH 1 for 2 hours.

this system might not hold on an individual subject level. However the purpose of IVIVR is to develop a mean relationship between absorption and dissolution and this is achieved by this method based of course on a limited number of subjects. Unless one were to perform large clinical-size studies a more accurate relationship could not be achieved no matter how simple or complex the relationship between absorption and dissolution. This of course would be very expensive, time-consuming and perhaps unethical and so is not a feasible option.

For formulation 2, the standard dissolution system gave a rank order relationship to the *in-vivo* data (Figure 4). However, this system was under-discriminatory, with small changes in dissolution resulting in much larger changes in absorption (Table 3). Approximately 5 -10% differences in dissolution resulted in differences in the primary PK parameters AUC and Cmax of 25–40%. Up to a 20% range in dissolution for an extended-release product is generally considered acceptable - particularly in the case where an IVIVR cannot be established. This data shows that such a range for this particular product in this dissolution system would result in bioinequivalent product. If this dissolution methodology were to be used as a routine quality control test for release of this product, the allowed range in dissolution would be very narrow and largely confounded by analytical variability. For this reason a non-standard dissolution method was developed which was designed to improve the discrimination between the lots.

Briefly the system was a USP III system using phosphate buffer rather than water (Table 2). A more detailed description of this method and the rationale behind its development is outlined in a paper by Colm Farrell.

For the purposes of developing IVIVR for these products, where possible dissolution samples were obtained at the same timepoints as the blood samples were obtained for the biostudies. This sampling regimen was shown to fully characterise the absorption profiles for these products.

Once a potentially discriminatory dissolution test had been identified, a prospective IVIVR study was undertaken to verify this dissolution system, using a range of qualitatively similar formulations, and including a formulation, with dissolution similar to the target formulation identified from the prototype biostudy. Typically these studies are relatively small studies.

For Formulations 1 and 2, these studies were designed as five-period randomised crossover studies in ten subjects, evaluating four test formulations, differing only in the level of the rate-controlling factor, and an immediately releasing preparation to provide the individual subject impulse response for deconvolution (Table 4). Deconvolution was

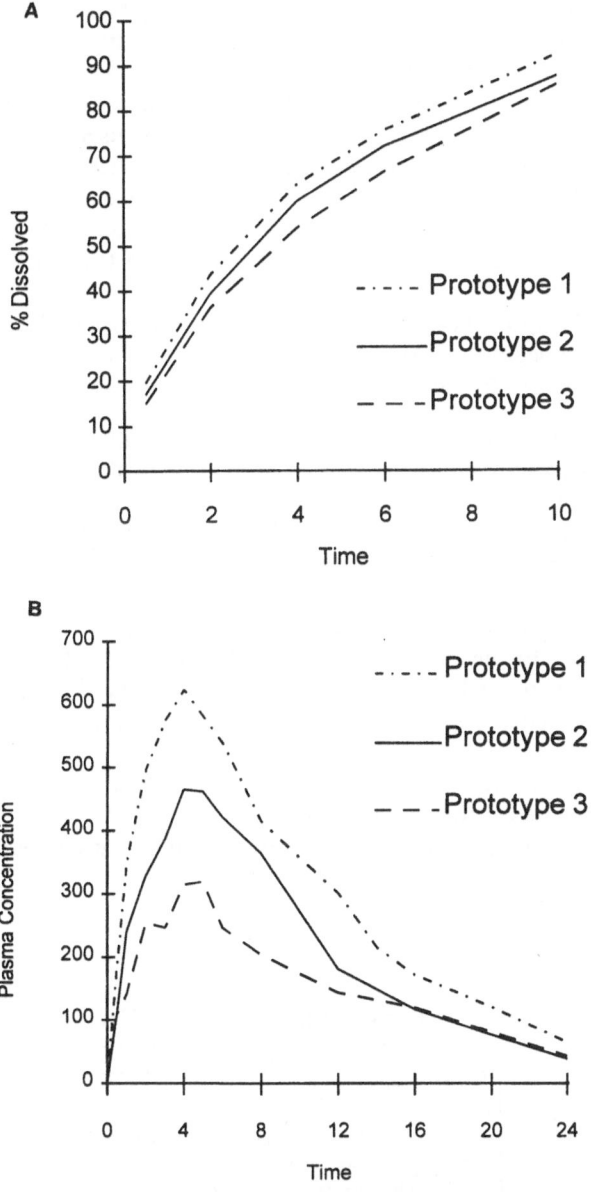

Figure 4. Formulation 2-standard dissolution method. (a) Dissolution; (b) *In vivo* plasma concentrations.

Table 3. Formulation 2 standard dissolution method effect of disoultion on AVC and Cmax

Dissolution Δ	AUC Δ % of target formulation	Cmax Δ % of target formulation
5-10 Average % difference in dissolution	75 - 128	76 - 138

performed using the PCDCON programme of Gillepsie. The four test formulations were chosen to reflect a wide range in dissolution and expected to produce *in-vivo* profiles outside the bioequivalence limits. The number of test formulations was chosen as the maximum number of formulations which could be evaluated while maintaining the balance in the study design. A five-period study is considered the maximum acceptable exposure and duration of a study due i) to the number of blood samples and volume of blood required and ii) the potential for drop-outs in the study. Alternative designs using more formulations, for example incomplete block designs might also be used for this type of study, but some information would be lost with these study designs. The number of subjects in the study was chosen based on the intra-subject variability of the drugs and/or formulations determined from the prototype biostudies. The number of subjects was chosen to ensure adequate power to determine gross differences, but not to test for bioequivalence of the formulations.

A number of approaches to developing relationships between *in-vitro* dissolution and *in-vivo* performance were taken (Table 5). This is by no means a comprehensive list of the approaches to IVIVR, but hopefully represents a cross-section of the methods.

1. Level C correlations were evaluated based on regressing the mean key pharmacokinetic parameters, AUC and Cmax on the mean amount dissolved at each of the sampling times.

A significant linear relationship ($p < 0.05$) was taken as evidence for a Level C correlation.

2. Level A correlations were evaluated based on regression of the mean *in-vivo* absorption on the mean *in-vitro* dissolution. In this case, which is considered to be the traditional Level A method, the slope was tested to be one and the intercept to be zero. In the case where a significant linear relationship was found, the slope=1 and the intercept=0, then a Level A correlation was accepted.
3. Level A as described previously, but with the intercept not necessarily equal to zero.
4. Linear Mixed Effects model as described by Adrian Dunne in his paper. Individual data was used for this method.
5. Generalised linear mixed models as described by Adrian Dunne in his paper.

Table 4. IVIVR evaluation prospective IVIVR study

In-Vivo Study Design
• 10 Subjects
• 5 Treatment Periods
• 4 test formulations with different levels of rate-controlling factor
• IR included to assist deconvolution
• Randomised, Crossover

Table 5. Approaches to developing IVIVR

Developing IVIV Relationships
1. Level C: AUC, Cmax vs Amount dissolved (mean data) in-vivo =slope (in-vitro) + intercept
2. Level A: Absorption vs Dissolution (mean data) in-vivo = in-vitro
3. Level A: Absorption vs Dissolution (mean data) in-vivo = in-vitro + C
4. Linear Mixed Effects Model: Absorption vs Dissolution (individual data) Ho: $\mu_{vivo} = \mu_{vitro}$
5. Generalised Linear Mixed Models (GLMM) Absorption vs Dissolution (individual data) - Proportional Hazards - Proportional Odds - Proportional Reverse Odds

For methods 2, 3 and 4, the % dissolved and the % absorbed were determined by expressing the cumulative mg released as a percent of the actual drug content of the formulation.

Figure 5 shows the plasma concentration versus time profiles for the four test formulations of Formulation 1 evaluated. The target *in-vivo* profile is the batch with medium dissolution. The percent deviation for the other products from the medium batch in terms of AUC and Cmax are also shown. Dissolution had no impact on the AUC of this product. However, the Cmax data indicates that a wide enough range of formulations was evaluated in this study, with the extremes of Cmax similar to the confidence interval limits allowed for bioequivalence, based on mean data.

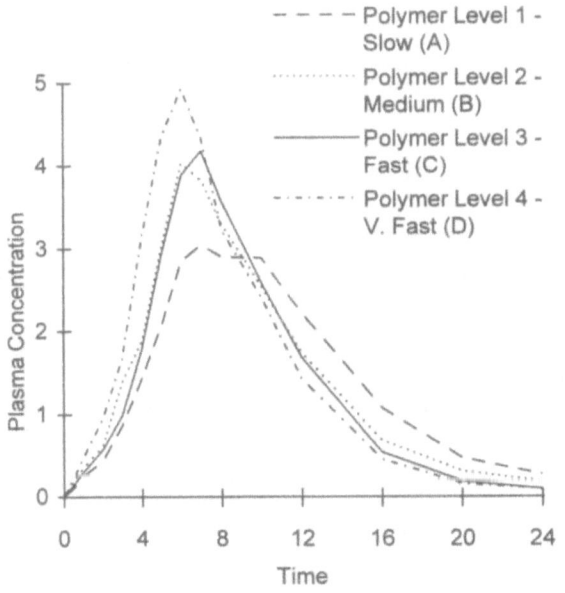

% of target formulation

Lot	AUC	Cmax
Slow	100	82
Medium	100	100
Fast	95	102
V. Fast	106	118

Figure 5. Formulation 1-prospective IVIVR study.

Figure 6 shows the % dissolved data using the standard dissolution method and the % absorbed profiles obtained by deconvolution of the plasma concentration versus time profiles. The shape of the dissolution and absorption curves are quite different as expected, with the lag in the absorption profiles which is not evident in the dissolution profiles.

Figure 7 shows the % dissolved data using the non-standard dissolution method and the % absorbed profiles. The shape of the dissolution curves using the non-standard dissolution method are now similar to the shape of the absorption curves and the spread in the data for dissolution and absorption is also similar.

Table 6 shows the level C correlations for this product, using the standard dissolution system.

Level C correlations could not be developed between AUC and dissolution as dissolution had no effect on AUC for this product. This was to be expected as the permeability of the compound is reasonably independent of site in the GI tract. However, there was a significant relationship between the peak concentrations achieved by this formulation and the dissolution at the early timepoints.

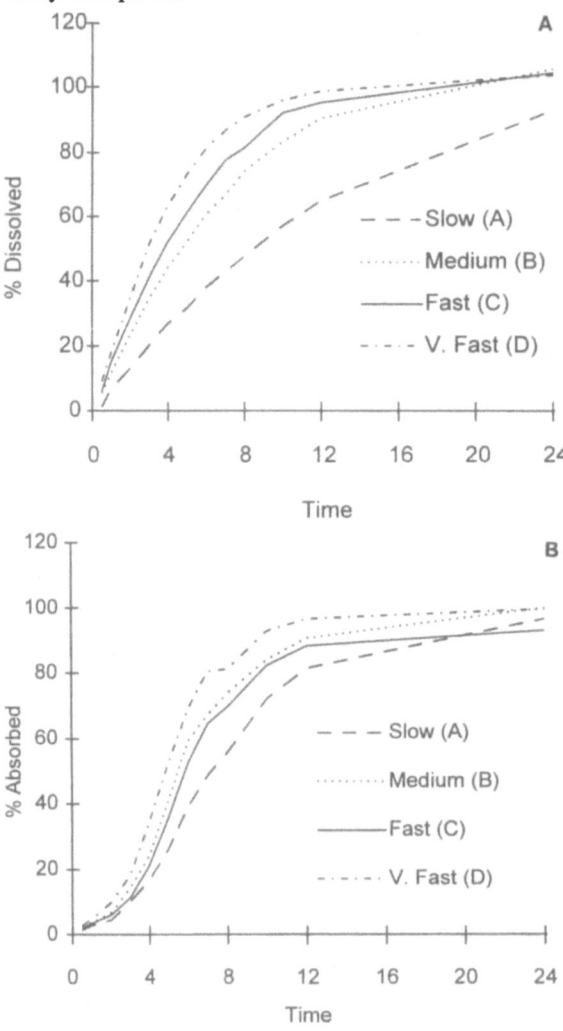

Figure 6. Formulation 1-standard dissolution method. (a) Dissolution; (b) Absorption.

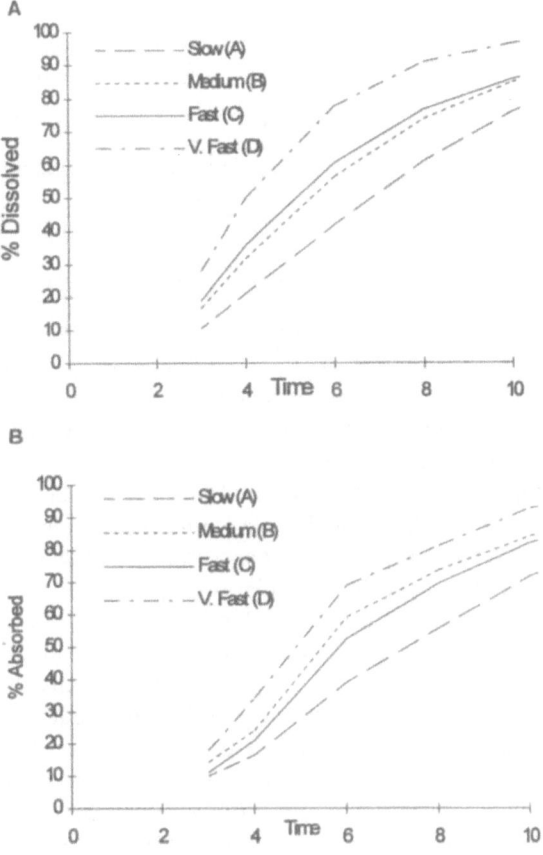

Figure 7. Formulation 1-non-standard dissolution method. (a) dissoultion; (b) absorption.

Table 6. Formulation 1 standard dissolution method level C correlations

	AUC (Geo. Mean)		Cmax (Geo. Mean)			
	r^2	p	r^2	p	Slope	Intercept
Q1	0.25	0.50	0.99	0.01*	0.22	3.13
Q2	0.11	0.67	0.88	0.06		
Q3	0.14	0.63	0.92	0.04*	0.08	2.46
Q4	0.17	0.59	0.94	0.03*	0.05	2.38
Q5	0.15	0.61	0.93	0.04*	0.05	2.23
Q6	0.14	0.63	0.92	0.04*	0.04	2.19
Q7	0.12	0.65	0.92	0.04*	0.04	1.97
Q8	0.09	0.71	0.89	0.06		
Q9	0.10	0.69	0.90	0.05		
Q10	0.04	0.80	0.83	0.09		
Q11	0.04	0.80	0.82	0.10		
Q12	0.01	0.93	0.60	0.23		

Table 7. Formulation 1 non-standard dissolution level C correlations

| | AUC (Geo. Mean) | | Cmax (Geo. Mean) | | | |
	r^2	p	r^2	p	Slope	Intercept
Q1	0.35	0.41	0.96	0.02*	0.10	2.52
Q2	0.32	0.43	0.96	0.02*	0.06	2.28
Q3	0.31	0.44	0.97	0.01*	0.05	1.51
Q4	0.31	0.44	0.98	0.01*	0.05	-0.06
Q5	0.36	0.40	0.99	0.01*	0.08	-2.88
Q6	0.54	0.27	0.94	0.03*	0.13	-7.48

Table 7 shows the level C correlations for this product, using the non-standard dissolution system. Again no relationship could be developed with AUC, but Cmax was correlated with dissolution at all the sampling times.

The relationship between dissolution and absorption using the standard disolution method is shown in Figure 8. All plots deviate from linearity and there appears to be a different relationship between dissolution and absorption for at least two of the four formulations. It is obvious from this data that the relationship between absorption and dissolution using the standard method is not a linear 1:1 relationship. Both the Level A and Linear Mixed Effects analyses confirm this, although the Level A analysiś does indicate a linear relationship for all lots. Using a Level A relationship where only the slope is tested finds a linear relationship, with slope = 1 for three of the four lots, but this is achieved mainly by the fact that all relationships begin around zero and end around 100. With these types of relationships it is expected that a Level A correlation will in some cases be achieved by chance, particularly when the residual sum of squares of the models is quite high. However, the relationship for these datasets is clearly not linear and therefore the GLMM type models as described by Adrian Dunne fitted this non-linear relationship best. Figure 9 shows the GLMM fits for the four lots evaluated in this study.

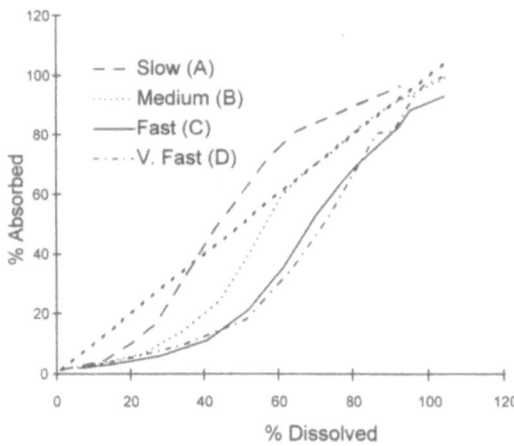

| | Lot | | | |
Method	A	B	C	D
Level A (1)	No	No	No	No
Level A (2)	No	Yes	Yes	Yes
Linear Mixed Effects	No	No	No	No
GLMM	Yes	Yes	Yes	Yes

Figure 8. Formulation 1-standard dissolution method IVIV relationships.

Although the four lots fit this model well, the parameter estimates are slightly, but significantly different for each lot. In order to obtain a unifying model, it may be necessary in this particular case to include for example some formulation variables as covariates in the model. This approach is currently being investigated.

Figure 10 shows the relationships between absorption and dissolution using the nonstandard method, which are reasonably linear, with the relationships close to the unity line, indicating the suitability of this dissolution system for this product. Both the Level A (Slope = 1) and the Linear Mixed Effects model confirm this, although two of the four batches fail on the traditional Level A method. Combining all lots indicates that the same relationship holds across the batches evaluated in this study. Therefore by changing the dissolution system, the sigmoidal relationship between absorption and dissolution, which requires complex mathematical modelling is transformed into a linear relationship, which is explained using simpler mathematical equations. As linear models fitted this dataset well, GLMM was not applied especially as the Linear Mixed Effects model is a special case of GLMM.

Figure 11 shows the plasma concentration versus time profiles for the four test formulations of Formulation 2 evaluated. The target *in-vivo* profile is the batch with medium dissolution. The percent deviation for the other products from the medium batch in terms of AUC and Cmax are also shown. Dissolution had an impact on both AUC and Cmax of this product. The data indicates that a wide enough range of formulations was evaluated in this study, with the extremes wider than the confidence interval limits allowed for bioequivalence, even based on mean data.

The Level C analysis (Table 8) on the standard dissolution found a good relationship between both AUC and Cmax for all lots and *in-vitro* dissolution at a number of the sampling timepoints. However the slope of the regression lines was very steep for these relationships, indicating that very small changes in dissolution had major implications on

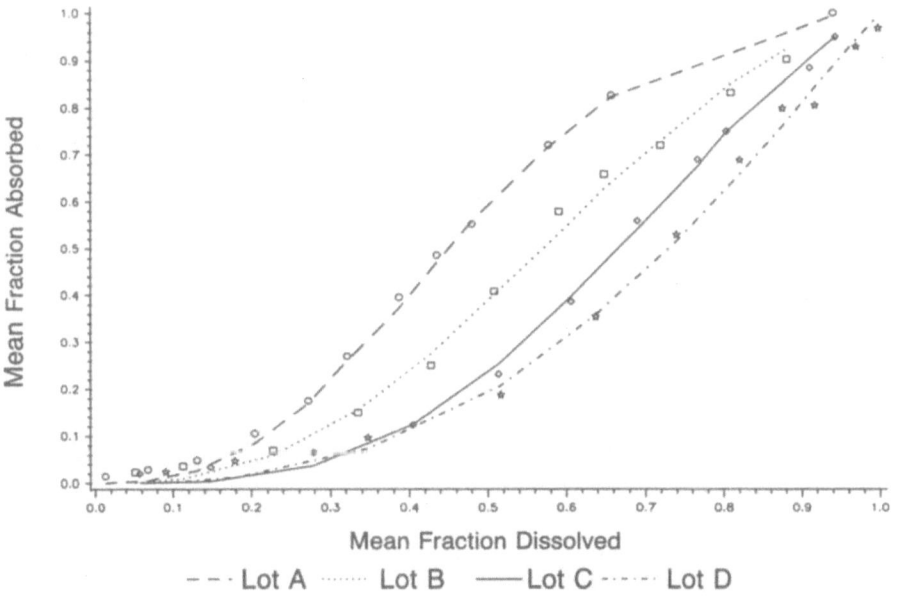

Figure 9. Proportional reverse model for formulation 1-standard dissolution method.

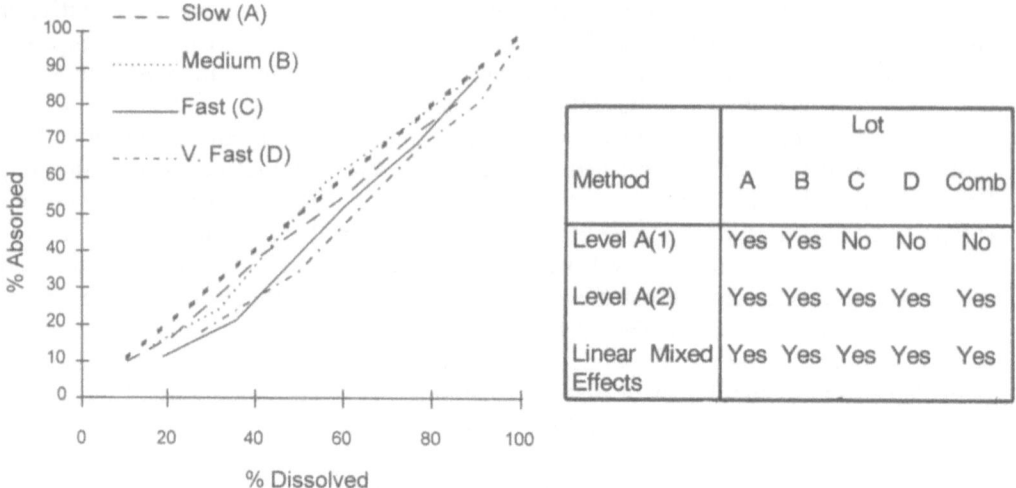

Figure 10. Formulation 1- non-standard dissolution method IVIV relationship.

in-vivo performance. Using the non-standard dissolution method, a Level C correlation with Cmax could be developed in particular at the early sampling times (Table 9). The lack of correlation at the later timepoints was at least in part due to the lack of data for Lot A at these timepoints, therefore reducing the degrees of freedom of the analysis. Significant relationships at $p<0.05$ could not be developed between AUC and dissolution at any of the timepoints using the data from all four lots. However, significant relationships with AUC at $p<0.1$ could be developed at most timepoints. The lack of significance at $p<0.05$ is probably at least partly due to the slightly anomalous data for Lot A. Regression analy-

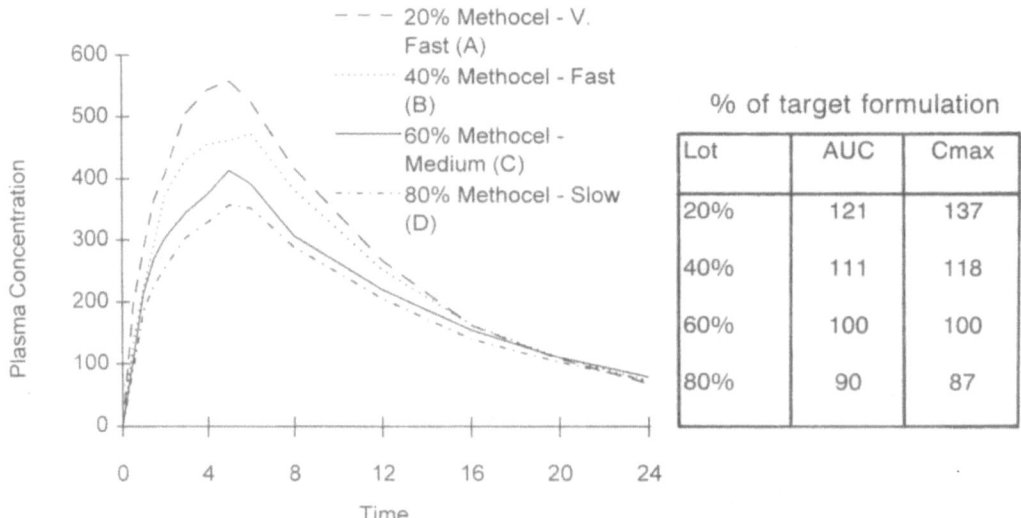

Figure 11. Formulation 2- prospective IVIVR study.

Table 8. Formulation 2- standard dissolution method levelC correlations

	AUC (Geo. Mean)				Cmax (Geo. Mean)			
	r^2	p	slope	Intercept	r^2	p	Slope	Intercept
Q1	0.79	0.11			0.86	0.07		
Q2	0.87	0.07			0.93	0.04*	13.32	-115.2
Q3	0.94	0.03*	104.8	-334.1	0.97	0.01*	12.62	-322.4
Q4	0.96	0.02*	117.9	-2697.9	0.98	0.01*	14.08	-598.3
Q5	1.00	0.002*	218.6	-13925	0.99	0.01*	25.71	-1902.8
Q6	0.08	0.71			0.15	0.62		

sis excluding Lot A results in a reduction in the degrees of freedom of the test as described above.

The % dissolved data using the standard dissolution method and the % absorbed profiles obtained by deconvolution of the plasma concentration versus time profiles are shown in Figure 12. Although the shape of the dissolution and absorption curves are similar, and a rank order relationship exists between the *in-vitro* and *in-vivo* data, small changes in the *in-vitro* dissolution result in large differences *in-vivo*. In addition, although the dissolution continues up to the 24-hour sampling time, the absorption is completed in approximately 6 hours, indicating the possibility of an absorption window for this product. This is surprising as this compound is highly permeable throughout the gi tract and may indicate some gut metabolism in the lower areas of the gut for example in the colon, which is not reflected in the *in-vitro* rat gut permeability model.

Plotting absorption against dissolution, using the standard dissolution for Formulation 2 (Figure 13) indicates that all four formulations show a good relationship between absorption and dissolution, in particular at the early timepoints. However, the usefulness of these relationships is questionable as the dissolution system is underdiscriminatory. For this reason, other approaches to developing IVIVR were not applied using the standard dissolution, as it is not practical for use either in Product Development or Quality Control.

Figure 14 shows the % dissolved and % absorbed plots for formulation 2 using the non-standard dissolution system. The spread in the *in-vitro* data is now similar to the spread in the *in-vivo* data, in particular for lots B, C, D. The dissolution for Lot A, the fastest lot is considerably faster than absorption. Apart from Lot A, dissolution continues for significantly longer

Table 9. Formulation 2- non-standard dissolution method level C correlations

	AUC (Geo. Mean)		Cmax (Geo. Mean)			
	r^2	p	r^2	p	Slope	Intercept
Q1	0.85	0.08	0.92	0.04*	9.25	249.5
Q2	0.87	0.07	0.93	0.03*	6.41	247.2
Q3	0.87	0.07	0.93	0.03*	5.31	242.1
Q4	0.87	0.07	0.93	0.04*	4.90	227.8
Q5	0.86	0.07	0.92	0.04*	4.62	196.0
Q6	0.86	0.07	0.92	0.04*	4.82	146.6
Q7	0.85	0.08	0.92	0.04*	5.42	77.25
Q8	0.93	0.17	0.98	0.09		
Q9	0.93	0.17	0.98	0.09		
Q10	0.92	0.18	0.97	0.10		
Q11	0.98	0.09	100	0.01*	13.64	-1006.5

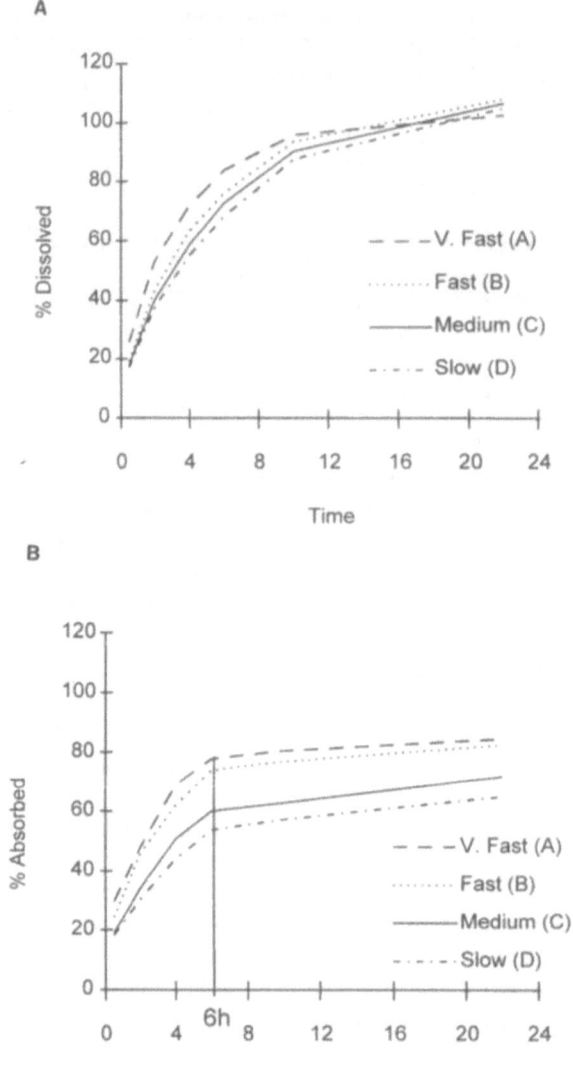

Figure 12. Formulation 2- standard dissolution method. (a) Dissolution; (b) Absorption.

than absorption. The relationship between dissolution and absorption for Formulation 2, using the non-standard dissolution is shown in Figure 15. Lots B, C and D show a good relationship between *in-vitro* dissolution, using this dissolution method and *in-vivo* absorption at the early stages. The latter part of the relationship deviates from linearity, reflecting the completion of absorption in a timeframe earlier than dissolution. The absorption for Lot A does not appear to be linearly related to the dissolution for this lot.

For clarity, Figure 16 shows these plots up to the first 6 hours, that is prior to the limitations in absorption for this product. Lots B, C and D now show a good relationship between *in-vitro* dissolution, and *in-vivo* absorption. The absorption for Lot A still does not appear to be linearly related to the dissolution for this lot.

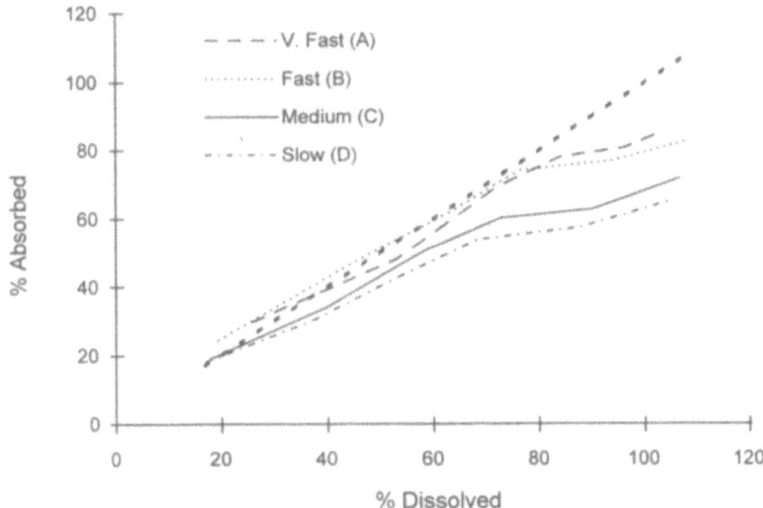

Figure 13. Formulation 2- standard dissolution method IVIV relationships.

Using only the 6h dataset for absorption and dissolution for the non-standard method, as absorption appears to be permeability rather than dissolution controlled after this time, a Level A correlation (Slope=1 ; Intercept=0) could be developed for Lots A and B. The Linear Mixed Effects model found no relationship between absorption and dissolution based on the full dataset, but a relationship for lots B and C based on the 6h data. As these relationships were primarily linear, the GLMM models were not applied.

This data shows that a Level A correlation (by either definition) exists not only for lots with apparent relationships between absorption and dissolution, but also for lots whose relationships deviate both from linearity and the unity line. The Linear Mixed Effects model on the other hand will determine a relationship between absorption and dissolution exists primarily when the relationship is closest to the unity line. Lot A in this study, which passes the Level A test, but fails the Linear Mixed Effects test has the lowest R-sq (0.936) for the Level A regression analysis and although the slope (0.789) appears to be different to 1, the residual variability in the model-fit masks this difference. Lot B passed both the Level A and the Linear Mixed Effects test and has a high R-sq (0.993) and a slope close to 1 (0.961) for the Level A test. Lot C fails the Level A test, but passes the Linear Mixed Effects test. The failure in the Level A test is surprising as the relationship line for Lot C is very close to unity. The R-sq for this test is 0.990 and the slope is 0.828, so the failure in this test is most likely caused by the bias due to ignoring the measurement error in the dissolution data, as described by Adrian Dunne in his paper. Lot D fails both the Level A and the Linear Mixed Effects test, but has the highest R-sq value for the Level A test i.e. the lowest residual sum of squares therefore increasing the significance of any deviation in the slope from 1.

Therefore this data shows that the Level A test will pass when the data is either very close to 1: 1, but also when the data deviates from this relationship. The Linear Mixed Effects model goes someway to correcting this, preferably passing datasets where the relationship is close to 1:1, but may be too stringent. It may be possible to set limits based on acceptable practical differences between absorption and dissolution for the Linear Mixed

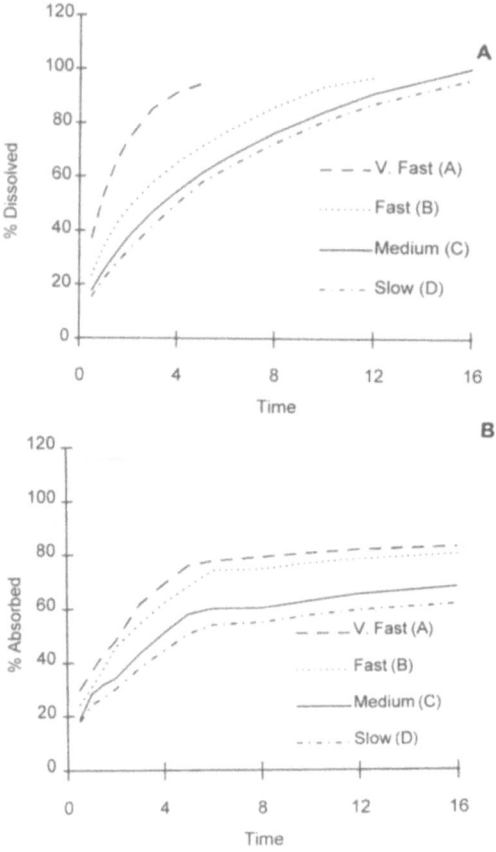

Figure 14. Formulation 2- non-standard dissolution method. (a) Dissolution; (b) Absorption.

Effects test, that is rather than test for in-vivo=in-vitro, to test instead for in-vivo plus or minus some allowed difference equals in-vitro.

In conclusion, based on only two examples illustrated in this presentation, it is obvious that there are a number of difficulties associated with developing IVIVR. Standard dissolution systems tend, for practical reasons to be simple methods, using sink conditions to ensure dissolution of the products. The dissolution in these systems often does not reflect the more complex environment of the gastrointestinal tract, with its changing pH, agitation rate, ionic strength etc. More complex mathematical models relating absorption and dissolution go some way towards overcoming the limitations in these dissolution systems in some instances.

Other approaches include developing novel dissolution systems that take into account the physicochemical properties of the drug and any variables in the formulation and its mechanism of release, that might be affected by the changing environment of the gastrointestinal tract. As indicated in the above examples this may help to develop IVIVR of a simpler linear nature and hence make interpretation and prediction of data much easier, helping in the development and quality control of products.

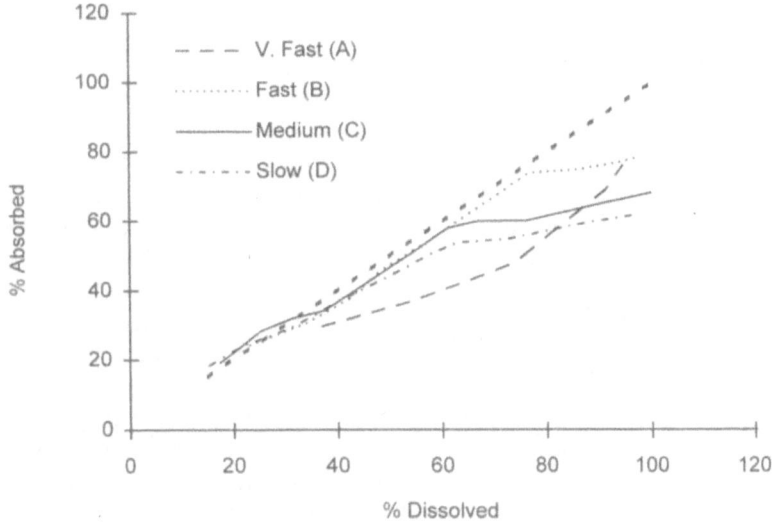

Figure 15. Formulation 2- non-standard dissolution method IVIV relationships (all data).

Of course, this will not always be possible, in particular with novel complex technologies designed to deliver drugs with inherent biopharmaceutical problems. Therefore, there is a need for complex models to be validated and standardised for use in such situations. Use of models such as GLMM also ensure that simpler fits such as the Linear Mixed Effects model will be applied as appropriate.

The importance of the Level C correlations, particularly if such relationships can be developed with the key pharmacokinetic parameters relevant to the safety and efficacy of

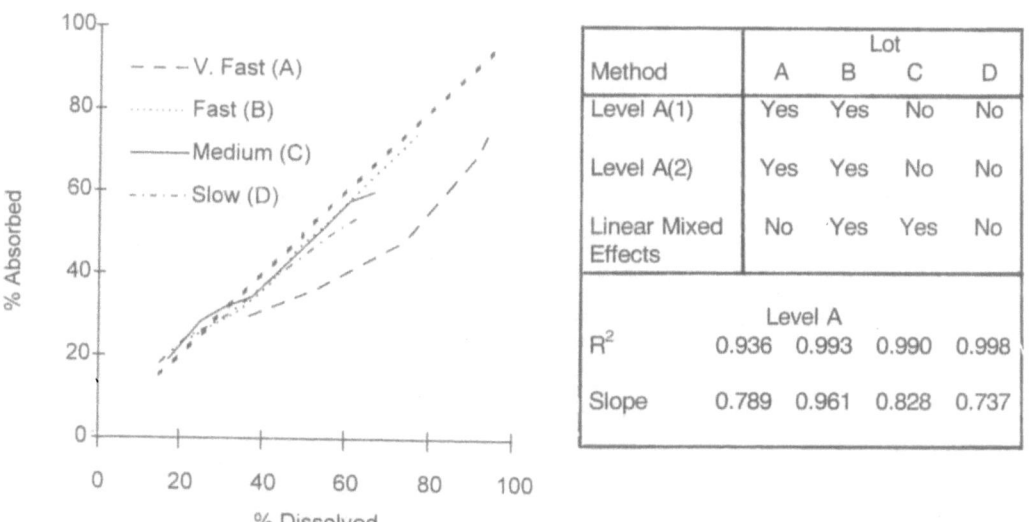

Method	Lot			
	A	B	C	D
Level A(1)	Yes	Yes	No	No
Level A(2)	Yes	Yes	No	No
Linear Mixed Effects	No	Yes	Yes	No
	Level A			
R^2	0.936	0.993	0.990	0.998
Slope	0.789	0.961	0.828	0.737

Figure 16. Formulation 2- non-standard dissolution method IVIV relationshps (6h data).

the product is also illustrated with this data. Such Level C relationships can be used to set specifications for a product even in the absence of Level A type relationships.

The data from Formulation 2 using the standard dissolution test indicates the potential for problems associated with arbitrarily accepting 15–20% ranges in dissolution, in the absence of IVIVR. For this product dissolution specifications with this range would result in bioinequivalent product, thereby potentially compromising the quality of the product. Therefore dissolution specifications developed in the absence of IVIVR should be confirmed with an in-vivo bioequivalence study evaluating the extremes in the proposed dissolution limits.

It is also important to remember that developing IVIVR is only the first step in this process. Once an IVIVR has been developed for a product, it is then necessary to validate this IVIVR by testing the predictability of the model for additional batches of the same product. Once an IVIVR has been developed and validated for a product, it can then be used for formulation and process optimisation and specifications, setting dissolution specifications, and scale-up and post-approval changes, as described in the recent draft IVIVC guidelines issued by FDA, without the need for further in-vivo studies. Therefore, despite the difficulties often associated with developing IVIVR, the usefulness of such relationships in product development, formulation and process optimisation, setting specifications, product scale-up and changes post-approval outweighs the efforts put into developing and validating these relationships.

Finally, I would like to mention some of my colleagues, without whom this presentation would not have been possible. Tom O'Hara for his help in the statistical analysis of this data, Colm Farrell for his assistance in the PK analysis of the data, Araz Raoof for her work on characterising the permeability of these compounds and Henry Madden for the in-vitro dissolution work. This work was undertaken as part of the IVIVR Cooperative Working Group Initiative.

9

THE *BIOPHARMACEUTIC DRUG CLASSIFICATION* AND DRUGS ADMINISTERED IN EXTENDED RELEASE (ER) FORMULATIONS

Owen I. Corrigan

IVIVR Cooperative Group
Department of Pharmaceutics
University of Dublin Trinity College
Dublin 2
Ireland

1. ABSTRACT

A *biopharmaceutic drug classification* scheme for correlating the *in-vitro* drug product dissolution and *in-vivo* bioavailability for IR products was proposed by Amidon et al (1995). The classification arose from drug dissolution and absorption models which identified the key parameters controlling drug absorption as the dimensionless numbers; the Absorption number (A_n), the Dissolution number (D_n) and the Dose number (D_o). This led to a biopharmaceutic classification of drugs into four groups, the establishment of a basis for determining the conditions under which *in-vitro-in-vivo* (IVIV) correlation's are expected and the use of the classification to set drug bioavailability standards for IR products. These developments raise the issue of whether the biopharmaceutic classification has relevance to ER products. In contrast to IR products, drugs selected for ER products should have good gastrointestinal (GI) permeability and an extended site of absorption. However their permeability(Papp) may change depending on the site. Solubility(C_s), effective fluid volume and hence D_o may also vary with site. Of particular relevance to both permeability and solubility is the degree of ionization of the drug. Residence time at each site, pH changes and the potential for drug degradation at different sites, the latter resulting in a restricted absorption window, will influence the time frame over which an IVIV relationship is possible. Of the drugs available in ER dosage forms ~ 63% are bases, 15% acids and the remainder either unionizable or small inorganic ions. Acidic drugs will tend to have lower solubility's high up in the gastrointestinal tract (GIT), with solubility increasing down the GIT. In contrast with increased ionization permeability should fall. Thus with acids, as the dosage form moves to a more alkaline environment down the GIT, absorption may change from dissolution control to membrane control depending on the pKa of the drug. In contrast bases will loose solubility with transit down the GIT, but become more permeable; absorption becoming more dissolution/release controlled or in extreme cases solubility controlled in the latter stages of the absorption phase. In the light of the above considerations a modified biopharmaceutic classification is proposed for ER products.

2. INTRODUCTION

The *biopharmaceutic drug classification* scheme for correlating the *in-vitro* drug product dissolution and *in-vivo* bioavailability for immediate release (IR) products was proposed by Amidon et al (1995) [1]. This led to a biopharmaceutic classification of drugs into four groups (Table 1) and the establishment of a basis for determining the conditions under which *in-vitro-in-vivo* (IVIV) correlation's are expected. This classification is being used to set drug bioavailability standards for IR products and to determine the conditions under which wavers for bioequivalence studies may be granted. The classification is incorporated into the Center for Drug Evaluation and Research (CDER) Guidance for Industry dealing with scale -up and post-approval changes (SUPAC) for IR products including site change, manufacturing process changes and equipment changes[2,3]. Furthermore application of the classification to the biopharmaceutic evaluation of IR generic products is under consideration. These developments raise the issue of whether the biopharmaceutic classification has relevance to extended release (ER) products and whether it could be applied in certain circumstances to their evaluation.

3. THE BIOPHARMACEUTIC DRUG CLASSIFICATION. THEORETICAL CONSIDERATIONS

The classification arose from a consideration of drug dissolution and absorption models[1,4-6] which identified the key parameters controlling drug absorption as the dimensionless numbers; A_n (the absorption number), D_n (the dissolution number) and D_o (the dose number). The Absorption Number (A_n) is defined as follows:

$$A_n = \frac{P_{eff} \cdot t_{res}}{R} \tag{1}$$

P_{eff} is the effective permeability coefficient, R is the tube(intestinal) radius, t_{res} the mean residence time, given by ($\pi R^2 \, l/Q$) where Q is the fluid flow rate and l is the length of the tube. In the original application to IR products the absorption site was considered to be the small intestine and t_{res} was given a value of 180 min. P_{eff} values were obtained from human intubation studies for a range of drugs and compared to human absolute bioavail-

Table 1. The biopharmaceutic drug classification for IR drug products

Class	Solubility	Permeability	IVIV correlation
I	High	High	Correlation - if dissolution slow
II	Low	High	IVIV correlation expected
III	High	Low	Little or no IVIV correlation
IV	Low	Low	Little or no IVIV correlation

ability data. The relationship obtained is reproduced in Fig. 1. As a result drugs were classified into 'high' and 'low' permeability categories, drugs with P_{eff} above 4×10^{-4} cm/sec
were classed as highly permeable and those below 1×10^{-4} cm/sec as having low permeability and consequently poor bioavailability. The dissolution number (D_n) was determined assuming a multiparticulate dosage form and considered dissolution to occur in
accordance with the diffusion controlled dissolution model of Higuchi & Hiestand[7], the
aqueous boundary layer being equal to the radius of the particle. Under these conditions
D_n is given by:

$$D_n = \frac{D.C_s}{r_0} \cdot \frac{4 \pi r_0^2}{4/3. \pi r_0^3 \rho} \cdot t_{res}$$

(2)

where D is the drug diffusivity, C_s is the drug solubility, r_0 the initial particle radius, ρ is
the density and t_{res} the mean residence time at the site of absorption. D_n is in effect the ratio of the residence time to the dissolution time, the terms to the left in equation 2 representing the dissolution time for particulates. The appropriate parameters and their
interrelationships representing the dissolution/release time for an ER product will depend
on the mechanism of release appropriate to the specific ER dosage form. For a zero order
delivery system, e.g. an osmotic pump system, the release time can be determined from
the zero order release rate and the dose.

The dose number (D_o) is given by:

$$D_o = \frac{M_o}{V_o . C_s}$$

(3)

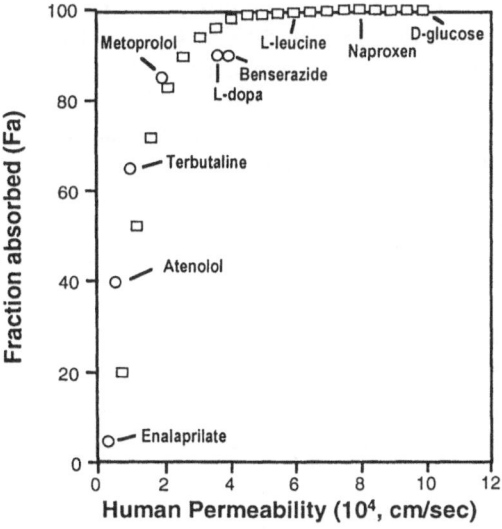

Figure 1. Relationship between human permeability (P_{eff}) and absolute bioavailability. From ref. 1 with permission.

M_o is the dose of drug, V_o is the dissolution fluid volume and C_s the drug solubility and is in effect the ratio of the dose concentration to the solubility. The lower the dose number the greater the likelihood of good bioavailability. D_o greater than 20 are considered to approximate to the solubility limited region of drug absorption. This approach led to the biopharmaceutic classification of drugs[1] shown in Table 1.

Thus the absorption of drugs from rapidly dissolving dosage forms containing Class I drugs will be controlled by gastric emptying since the half life of gastric emptying is of the order of 12 to 22min. As a consequence a dissolution specification of 85% dissolved in less than 15min was suggested as ensuring bioavailability of such drugs. Class II drugs, on the other hand, should give good IVIV correlation since dissolution should be the rate determining step controlling absorption. The application of the *biopharmaceutic classification* to IR products raises the issue of whether the biopharmaceutic classification could be applied in similar circumstances to ER type products. Since the objective of ER formulation is to extend the duration of absorption without significant loss in bioavailability it is unlikely that drugs falling into classes III and IV in Table 1 would be considered for inclusion in an ER formulation.

4. DRUGS USED IN EXTENDED RELEASE (ER) PRODUCTS

The drugs which are marketed as ER products represent a small subgroup of those drugs which are administered orally to man. In contrast to many of the drugs contained in IR products, the drugs selected for ER products should have good gastrointestinal (GI) permeability and an extended site of absorption. They generally have short half lives and, when manufactured in conventional (IR) dosage forms require multiple daily administration. The dosage interval considered ideal of ER products is 24 hr. An examination of the drugs listed in the PDR[8] indicates that there are some sixty drugs available in oral extended release dosage forms. These are summarized in Tables (2–5). Of these compounds 63% may be chemically classified as weak bases (Table 3), 15% as weak acids (Table 4) and the remainder (22%) as nonionizing compounds or as salts (Table 5).

Thus bases are by far the most common group and the vast majority are incorporated in a salt form. The most common salt forms used are hydrochlorides, sulphates and maleates (Table 3). Many of those not in the salt form have pKa outside the physiological pH range and behave as non ionizable compounds.

The acidic drugs in ER products are few in number ands tend to be mainly non-steroidal anti-inflammatory agents(NSAID) and are more likely to be in the free acid form. Where salts are used the sodium counterion is employed (Table 4). Is their a scientific rationale behind these differences. Could bases be more suitable for ER products than acids?

Table 2. Drugs in ER products

	n	%
Total	60	100
acids	9	15
bases	38	63
others	13	22

Table 3. Drugs contained in ER products. Basic compounds

Basic compounds	n=38

Free base:
Terphenidine, Nifedipine, Lansoprazole, Omeprazole, Mesalamine, Erythromycin, Isradipine, Felodipine.

Salt form:
Albuterol Sulfate, Atropine Sulfate, Verapamil HCl, Scopolamine Hydrobromide, Nicardipine HCl, Hyoscyamine Sulfate, Phendimetrazine Tartrate, Chlorpheniramine maleate, Propranolol HCl, Pseudoephedrine HCl, Metoprolol Succinate, Bromopheniramine maleate, Diltiazem HCl, Tripelennamine HCl, Phenylpropanolamine HCl, Morphine Sulphate, Phenylephrine HCl, Disopyramide Phosphate, Procainamide Hcl, Prochlorperazine Maleate, Diethylpropion HCl, Methylamphetamine HCl, Dextroamphetamine HCl, Azatadine Maleate, Quinidine Gluconate/sulphate, Carbinoxamine Maleate, Methscopolamine nitrate, Pyridostigmine Bromide, Methylphenidate Hcl, Phenindamine Tartrate.

Table 4. Drugs contained in ER products. Acidic compounds

Acidic compounds	n=9
Acetazolamide	Diclofenac Sodium
Aspirin	Naproxen Sodium
Indomethacin	Phenytoin Sodium
Ketoprofen	Divalproex Sodium
Nicotinic acid	

Table 5. Drugs contained in ER products. Other compounds

Other compounds	n=13
Acetaminophen	Potassium Chloride
Isosorbide mononitrate	Lithium Carbonate
Isosorbide dinitrate	Ferrous Sulphate
Nitroglycerin	
Guaifenesin	
Carbidopa	
Levadopa	
Theophylline	
Oxpentifylline	
Glipizide	

Those drugs which are not classified as either acids or bases are listed in Table 5 and include nonionizable compounds such as acetominophen and the nitrates, salts such as KCl and small ampoteric compounds such as levadopa.

5. PHYSIOLOGICAL VARIABLES PARTICULARLY IMPORTANT FOR ER PRODUCTS

The absorption number (A_n) contains two of the key parameters influencing the rate and extent of drug absorption, namely the effective permeability coefficient (P_{eff}) and the mean residence time (t_{res}). The key parameter in both the dissolution (D_n) and dose (D_o) numbers is the solubility which, in the case of ionizable drugs, is influenced by pH and hence the gastrointestinal site and the fluid volume. Furthermore the potential for drug degradation and hence reduced bioavailability varies with site. Hence the following four physiological factors may vary substantially during the release and absorption of a drug from an ER product: 1) Residence time at each site, 2) The effective permeability coefficient, 3) pH/buffer capacity of the intestinal fluid and 4) the potential for drug degradation at different sites.

5.1. Residence Times

If drug absorption from ER products were confined to the small intestine an absorption window of approximately 3–6[9-11] hours would be operative i.e. approximately a quarter of the ideal dosage interval. The evidence suggests that for a range of drugs absorption occurs beyond this time frame. There is increasing evidence that many drugs may be absorbed significantly in the colon. The time course of dosage form transport down the gastrointestinal tract, based on gamma scintigraphy studies is shown in Fig(2)[10]. Generally multiple pellets tend to lag behind single units.

Approximate residence times at the three primary locations in the GIT are summarized in Table 6. Colonic residence times are substantially larger than those in the small intestine. Thus even if a significant reduction in P_{eff} (see below) occurs for a drug on transfer to the colon, the much longer residence time may still result in substantial absorption. However their is considerable variation in colonic residence time, which can greatly alter plasma levels and the extent of absorption[11] (Fig. 3).

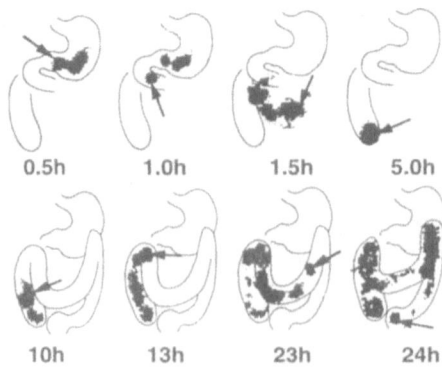

| 0.5h | 1.0h | 1.5h | 5.0h |

| 10h | 13h | 23h | 24h |

Figure 2. Transit of pellets and a capsule (İ) through the gastrointestinal tract of one subject. From ref. 10 with permission.

Table 6. Residence times in human gastrointestinal tract

Site	Residence times (h)
Stomach	0-3
Small intestine	2-5
Colon	~18

5.2. Permeability

Drugs generally are considered to be optimally absorbed from the upper part of the small intestine and consequently the P_{eff} values used in the development of the Biopharmaceutic Classification were obtained from the small intestine of human volunteers following intubation[1]. There is little published direct date on the passive permeability of drugs lower down the gastrointestinal. However the indirect evidence available suggests a significant reduction in P_{eff} with distance down the GIT. The P_{eff} is related to the absorption rate constant (k_a) as follows:

$$k_a = A. P_{eff}/V \qquad (4)$$

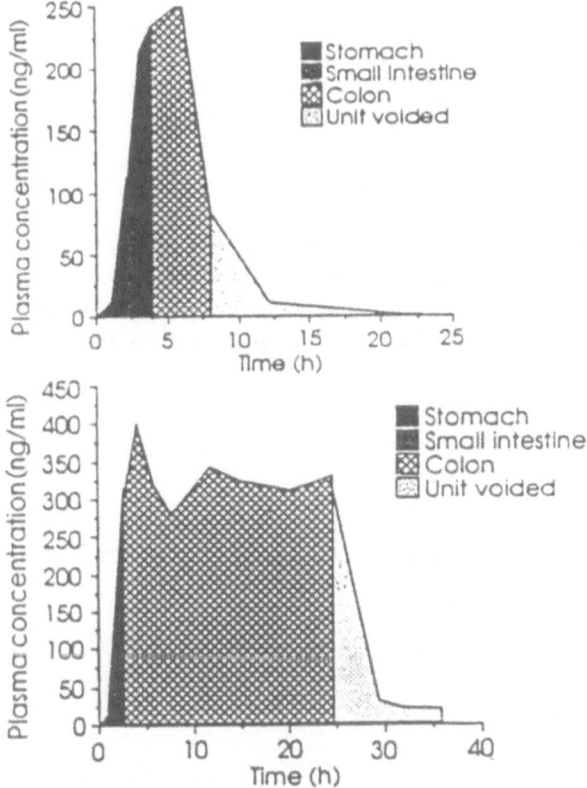

Figure 3. Effect of colon transit time on oxprenolol plasma levels, data from two subjects. From ref. 16 with permission.

where A is the effective surface area and V is the fluid volume. From a physicochemical point of view the P_{eff} may be further subdivided as follows:

$$P_{eff} = \frac{P_{aq}}{1+ P_{aq}/(P_o.X_s+P_p)}$$

Where P_{aq} is the aqueous boundary layer permeability, P_o the lipoidal pathway permeability, P_p the aqueous pore permeability, X_s the fraction of nondissociated drug[12].

Lipophilic drug will have a high P_o and hence a high P_{eff}. The P_{eff} of drugs which ionize will vary with pH since this dictates the fraction unionized and hence available for absorption through the P_o route. The ionized drug may be absorbed through the negatively charged aqueous pore pathway, however this pathway is highly size dependent and the evidence suggests that the pore size decreases with distance down the intestine. Thus calculated pore radii in the jejunum of 6.7–8.8 Å and Ileum of 2.9–3.8 Å have been suggested[13].

As part of a program to develop ER products of the hydrophilic drug isosorbide-5-mononitrate (IS5MN) Kramer[14] evaluated the rate and extent of drug absorption after site specific delivery to human volunteers i.e. via a nasogastric tube to the jejunum, terminal ileum and ascending colon. A reduction in extent and a slowing of the rate of absorption, relative to the two small intestine sites, was observed following the colonic administration. These findings were consistent with those of Fisher et al, using ^{111}In-labeled IS5MN in controlled release pellets, who found 19% of the dose absorbed from the colon the product being 84% bioavailable[15]. Despite the poorer colonic absorption Kramer concluded that the development of a once daily product was feasible.

5.3. pH/Buffer Capacity Changes

Substantial hydrogen ion changes, reflected in pH changes, occur with distance down the GIT. A typical pH versus time profile for the GIT is shown in Fig 4(16). There is the initial rise in pH on exiting the stomach, with pH continuing to rise down the small intestine until the caecum is reached at which point a drop of the order of one pH unit ap-

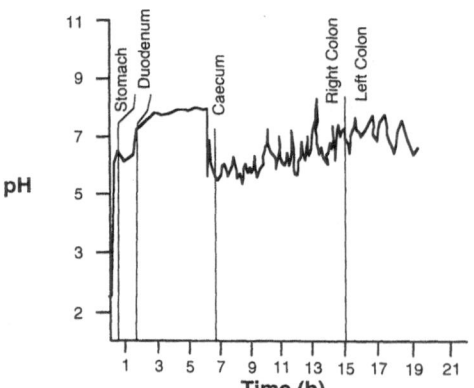

Figure 4. Changes in pH Throughout the gastrointestinal tract from stomach to colon. From ref. 16 with permission.

Table 7. Gastrointestinal pH changes under fasted and fed conditions

Site	Fasted pH		Fed pH
Stomach	1.4-2.1		4.3-5.4
Small intestine			
duodenum	4.9-6.4		4.2-6.1
jejunum	4.4-6.6		5.2-6.2
ileum	6.5-7.4		6.8-7.5
Large intestine			
caecum		6.4	
upper colon		6.0	
lower colon		7.5	

From ref 17.

pears to occur. The pH then rises gradually down the colon. These pH changes are summarised in Table 7, and can have dramatic effects on ionizable drugs altering the proportion of ionized to unionized drug and hence solubility and permeability.

The solubility versus pH profile for a typical weak acidic drug, ibuprofen, is shown in Fig 5, the large increase in solubility beyond the pK_a reflects the increased ionization of the drug. Such changes in C_s of drugs having pK_a in the physiological range will result in dramatic changes in D_o with site pH. Examples of these changes for a number of drugs are shown in Table (8). Also included in Table 8 are the volumes of aqueous dissolution media necessary to dissolve the ER dose. Solubility change in the physiological pH range, as reflected by the volume required to dissolve a dose, range from 38 litres to 40ml, in the case of naproxen, as the pH increases from 1 to 7.4. With ibuprofen the range is 12 litres to 100ml. The major challenge of ER formulation development is to control the rate of dissolution, generally at a zero order rate, against this background of widely fluctuating solubility. As the pH increases so does the solubility of acidic compounds, however the P_{eff} is very likely to decreasing due to ionization i.e. conversion to the hydrophilic, negatively charged, less permeable species. It is not just the pH but rather the buffer capacity of the fluid which will in many cases effect drug release from the delivery system. These factors affect the dissolution rate of both the drug and also any ionizing excipients, and these release rates may be interdependent. Buffer capacities of duodenal samples tend to be low but relatively constant over the pH range 4–8 in contrast to common media used in dissolution tests[18].

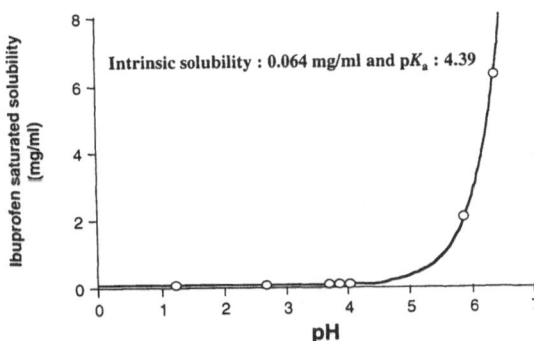

Figure 5. The pH-solubility profile of ibuprofen at 37°C.

Table 8. Dose numbers (D_0) and solvent volumes required to dissolve a dose (V_{sol})

	D_0* pH range 1 -7.4	V_{sol} (litres)	Ratio
Ibuprofen	48 -0.38	12-0.1	126
Naproxen	150-0.14	38-0.04	1060
Nifedipine	43	10.6	
IS-5-MN	0.00314	1ml	

Dose numbers (D_0) in the physiological pH range using 250ml as reference volume*.

5.4. Drug Degradation in the Colon

A further factor influencing drug bioavailability from ER products and consequently the likelihood of an IVIVR is the potential for deactivation at the gastrointestinal site prior to absorption. Many drugs are metabolized in the colon, in many cases by the enteral bacteria. Thus the potential for extended and device controlled drug release and absorption is limited. Typical drugs which are metabolised in the colon are listed in Table 9 and include compounds contained in ER products.

The above factors may therefore greatly modify the rate and extent of absorption of a drug from an ER product. Permeability(P_{eff}) may change depending on the site. Solubility (C_s), effective fluid volume and hence D_o may also vary with site. Of particular relevance to both permeability and solubility is the degree of ionisation of the drug. Residence time at each site, pH changes and the potential for drug degradation at different sites, the latter being reflected in a restricted absorption window, will influence the time frame over which an IVIV relationship is possible. Consequently acidic drugs will tend to have lower solubility's high up in GIT, with solubility increasing down the GIT. In contrast with increased ionisation intestinal permeability should fall. Thus with acids, as the dosage form moves to a more alkaline environment down the GIT absorption may move from dissolution control to membrane control. In contrast bases will loose solubility with transit down the GIT, but become more permeable; absorption becoming more dissolution/release controlled or in extreme cases solubility controlled in the latter stages of the absorption phase.

6. EXTENDED RELEASE PRODUCTS

It should be stated at the outset that the application of the above approach, employing dimensionless numbers, to IR products has focused on changes in F i.e. changes in the extent of bioavailability. However, bioavailability encompasses both the concepts of rate of absorption as well as extent of absorption. In the case of ER product formulation the

Table 9. Typical drugs which are metabolised in the colon

Morphine	Lactulose
Indomethacin	Sulphasalazine
l-dopa	Steroids
digoxin	Phenacetin
Atropine	Drugs with nito or azo groups
Chloramphenicol	Polypeptide drugs

Based on Fara (1985)[19].

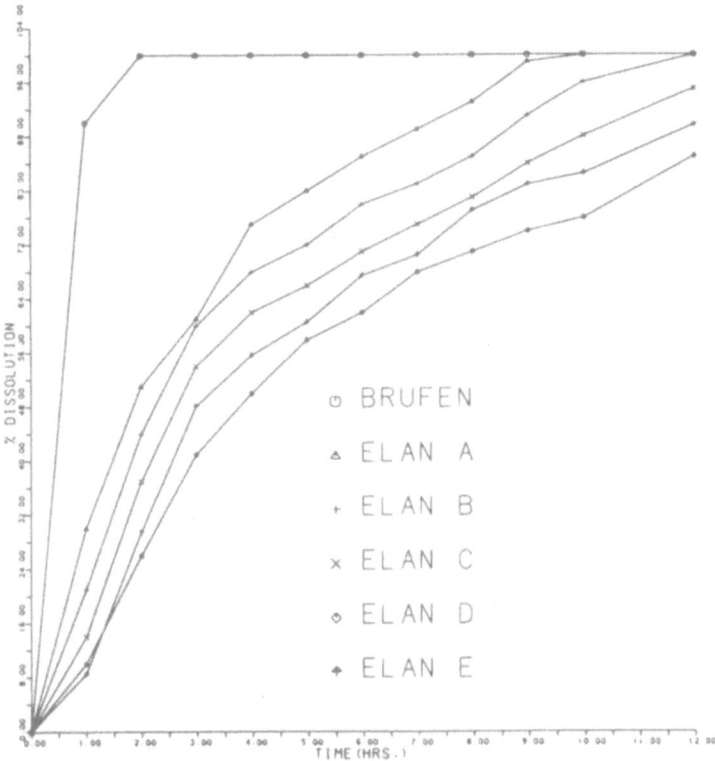

Figure 6. Effect of formulation on the *in-vitro* release of ibuprofen. From ref. 20 with permission.

aim is to control/slow down the rate of absorption, but maintain the bioavailability i.e. extent of absorption. Thus a classification system for ER products is unlikely to include a category for drugs with low P_{eff}. Furthermore, in the development of the biopharmaceutical classification for IR products the absorption site was considered to be limited to the small intestine and a single value of P_{eff}. In contrast with ER products a more extended site of absorption may be involved, P_{eff} may change depending on the site while solubility, effective fluid volume i.e. dose number (D_n) may also vary with site. Of particular relevance to both permeability and solubility will be the degree of ionization of the drug.

Our studies of a range of drugs in ER products appear to be consistent with the above observations. We examined the dissolution and absorption of a range of coated multiparticulate dosage forms of ibuprofen, having a range of dissolution profiles (Fig 6)[20]. Release approximated to a first order process, k_d the dissolution rate constants of the dosage forms decreasing systematically (Fig. 6). The rate of absorption as reflected in the time to peak plasma level and the extent of absorption, as reflected by the AUC, declined with a slowing in drug release/dissolution rate (Fig. 7).

The data was fitted to an "absorption window" model, assuming first order dissolution and first order absorption (Fig. 8). Within the confines of the window the absorption process was dissolution limited. Typical model fits of the Wagner-Nelson[21] absorption profiles for one of the products are shown in Fig 9. At early time points absorption is consistent with dissolution control while at later times absorption control predominates. Over-

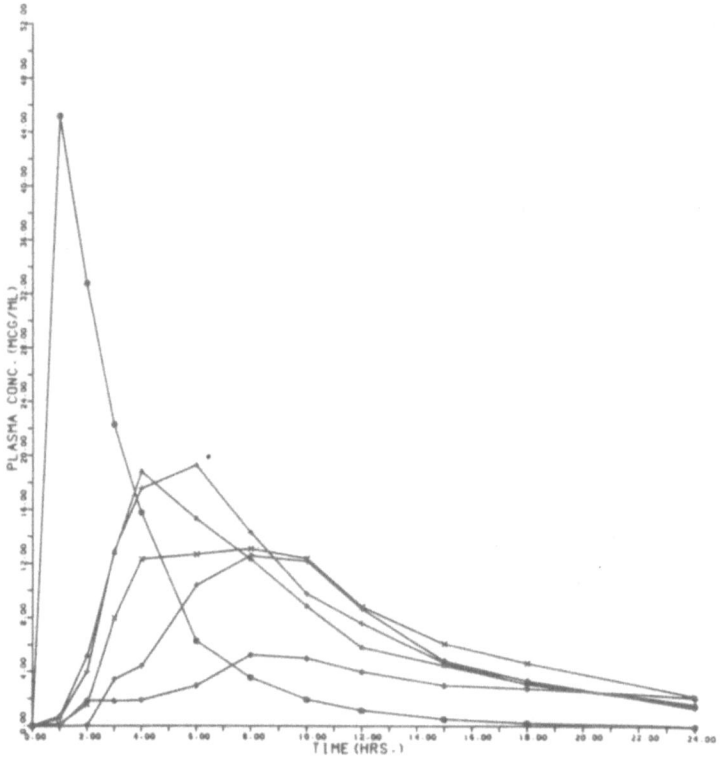

Figure 7. Effect of formulation on the *in-vivo* plasma level versus time profiles of ibuprofen. From ref. 20 with permission.

all the data was consistent with a model of dissolution limited absorption in an absorption window of 4–5 hrs.

Bioavailability and gamma scintigraphy studies[22] on another NSAID, E 123, show absorption also occurring during the small intestine transit phase (Fig 10). Barr et al. measured differences in the extent of Amoxicillin absorption from various regions of the GIT[24]. They found that amoxicillin absorption was rate and site dependent. The drug was well absorbed in the duodenum and jejunum, absorption was decreased and rate dependent in the ileum. the drug was unabsorbed when infused in all colonic regions. The authors concluded that ER formulations whose passage through the GIT is not impeded would meet with limited success because total absorption would be limited by the transit time of the drug through the small intestine which is estimated to be 3–5 hours.

In contrast to the above, the neutral drug theophylline appears to be more extensively absorbed along the GIT. Using paracetamol as a marker for stomach emptying and the absorption of sulphapyridine produced from sulfasalazine in the colon, as a marker of small intestine transit Peh (1996)[25] was able to identify the regions of absorption profiles associated with absorption from; the three regions stomach, small intestine, colon. The results of both fasting and food studies indicated significant absorption in the colon region, of the order of 20–35% of absorption occurring in the colon depending on food status.

Gamma scintigraphy studies of the low solubility hydrophobic drug nifedipine also indicates significant colonic absorption[26]. For example tablets coated with Eudragit S, were found to disintegrate after 700–800 min in-vivo and only gave significant plasma

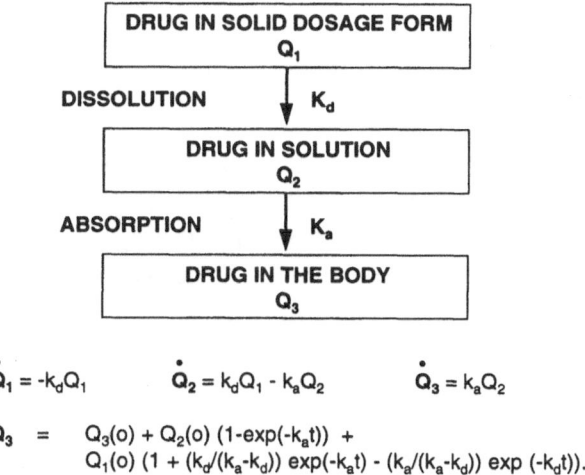

$$\dot{Q}_1 = -k_d Q_1 \qquad \dot{Q}_2 = k_d Q_1 - k_a Q_2 \qquad \dot{Q}_3 = k_a Q_2$$

$$Q_3 = Q_3(o) + Q_2(o) (1-\exp(-k_a t)) +$$
$$Q_1(o) (1 + (k_d/(k_a-k_d)) \exp(-k_a t) - (k_a/(k_a-k_d)) \exp(-k_d t)).$$

Figure 8. Absorption window model. From ref. 20 with permission.

levels after this period, peak plasma levels occurring in the 18–32 hr period post administration (Fig. 11).

Studies of the beta blockers, metoprolol and oxprenolol, both basic drugs also suggest significant capacity for colonic absorption as evidenced by the absorption data obtained using the Oros osmotic drug delivery system with input rates of metoprolol of 19mg/h[27] and oxprenolol 16mg/hr[28]. These zero order delivery systems effectively isolate the dissolution process from the lumen environment. The in-vivo absorption profile for metoprolol was linear from 1–16h, the time for 90% absorption being 18hr. The authors concluded, given the normal small intestine transit times that uniform absorption of metoprolol from the device occurred in all segments of the gut including the colon. Data obtained from the oxprenolol system was also consistent with extensive colonic absorption.

If one plots, for a range of test products containing a range of drugs; Ibuprofen[20], Diphylline[26], Quinidine[29], Indomethacin[26], methyldopa[26], the relative bioavailability vs the time for 80% dissolution ($T_{80\%}$) , one can see, from the slope of the plots the sensitivity of bioavailability (extent) to dissolution/release (Fig 12). The results are consistent with poorer absorption lower down the GIT, acid compounds tending to have the more steep profiles consistent with a narrower absorption window. Application of a simple absorption window model approach, with dissolution controlled first order absorption occurring only in the window, the relationship between relative bioavailability(F_{rel} %) and $t_{80\%}$ is given by;

$$F(\%) = 100.(1-(1/(ka-kd)). Ka.\exp(-kd.T)-kd.\exp(-ka.T))) \qquad (5)$$

and

$$kd - 1.604/t_{80\%}$$

where ka and kd are the absorption and dissolution rate constants respectively and T is the residence time in the absorption window. If it is assumed that permeability is high (i.e. P_{eff}

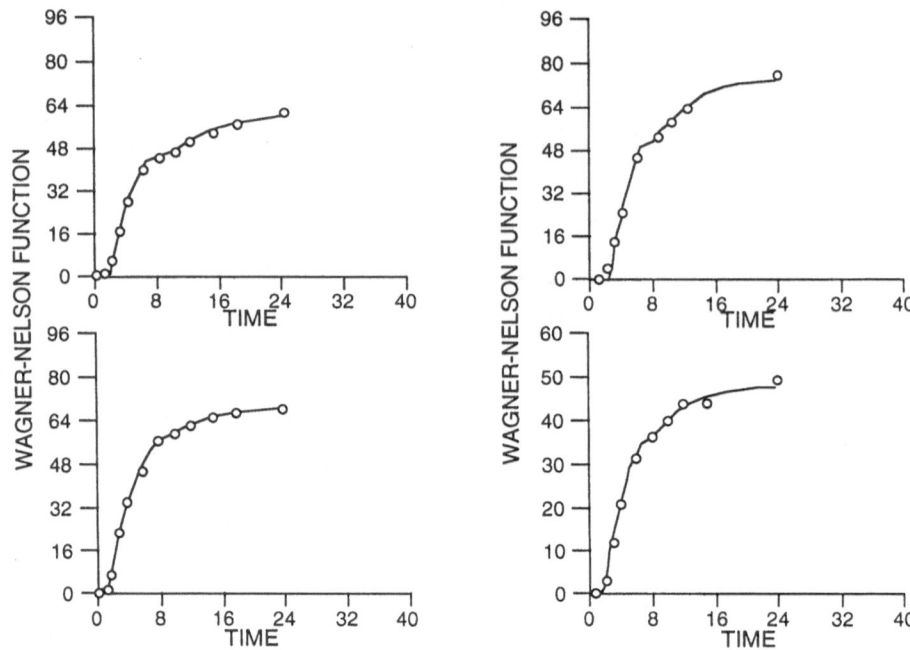

Figure 9. Absorption window model fitted to individual Wagner-Nelson transformed *in-vivo* absorption date. (Product B). From ref. 20 with permiasion.

$= 4 \times 10^{-4}$ cm.sec^{-1}) and the intestine has a radius of 1.75cm^{30} a reasonable estimate of k_a is 1.6 hr^{-1} (Eq 4). The data in Fig (12) were fitted to Eq 5 and the resulting estimates of T for each drug contained in the multiparticulate dosage forms are given in Table (10). The values range from 2.6 to12.3, the ionizing acidic compounds tending to have the lower values in contrast to those drugs which are unionized in the physiological pH range.

7. A BIOPHARMACEUTIC CLASSIFICATION FOR ER DRUGS

In the light of the above considerations the following modified biopharmaceutic classification containing the following three classes, high aqueous solubility, low aqueous solubility and variable solubility is proposed for ER products. Since drugs in ER products will have good permeability characteristics two classes i.e. III and IV, the low permeability groups in the original classification, have been deleted.

Class I   Solubility high and independent of site.
            a) permeability- high and independent of site.
            b) permeability dependent on site - narrow absorption window.
Class II   Solubility low and independent of site.
            a) permeability high and independent of site.
            b) permeability dependent on site - narrow absorption window.
Class V   Variable solubility and permeability
            a) High C_s in upper GIT, low C_s in lower GIT - converse effect on Papp (BASES).

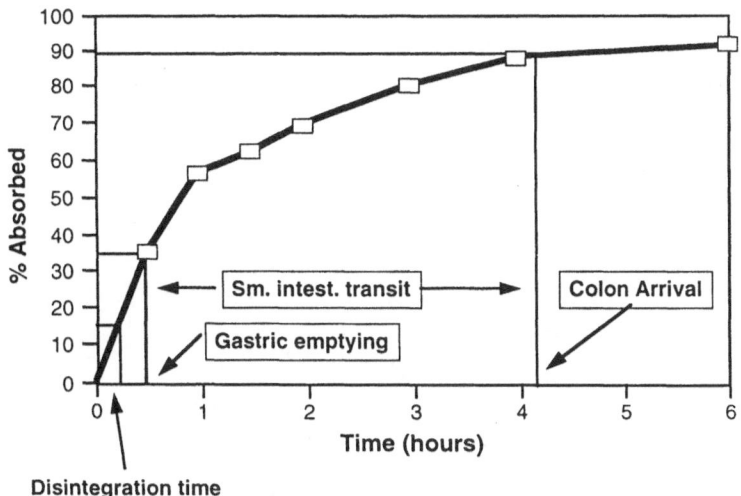

Figure 10. Relationship between absorption profile and site of actual absorption for the NSAID E6117. The physiological sites were identified by gamma scintigraphy.

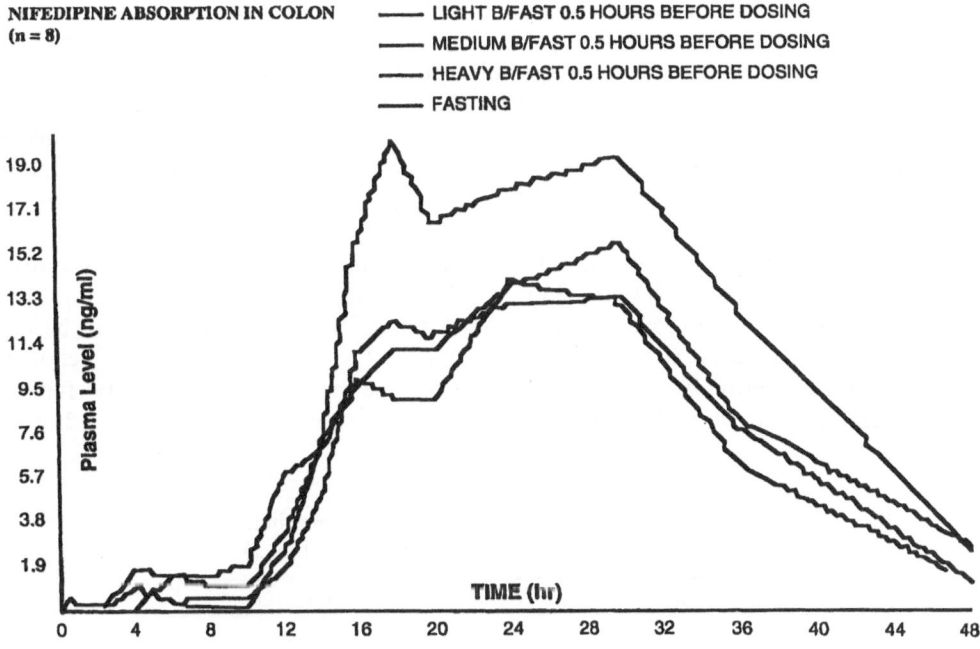

Figure 11. Colonic absorption of nifedipine.

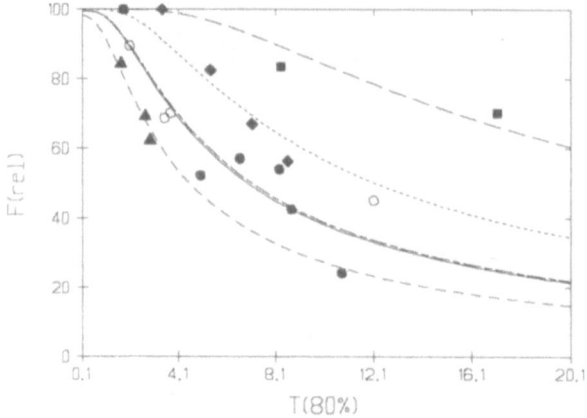

Figure 12. Relationship between relative bioavailability of ER products of a range of drugs and the time for 80% dissolution. The fitted lines were obtained assuming first order drug release occurring in an absorption window. Key: ■ Diphylline, O   Methyldopa, ◆ Quinidine Sulphate, ● Ibuprofen and ▲ Indomethacin.

b) Low C_s in upper GIT, high C_s in lower GIT - converse effect on Papp (ACIDS).

Class I. Class Ia will contain the more soluble compounds in Table (5) i.e. drugs such as acetominophen and KCl. Good IVIV relationships i.e. Level A type correlations are to be expected and a single dissolution medium should suffice. However compounds in class 1b, having a narrow absorption window are unlikely to give a good linear correlation throughout the full time course of absorption, absorption becoming permeability controlled in the latter part of the absorption phase. Thus Level C type correlations may only be possible. Drugs which are degraded in the colon will behave similarly in terms of IVIV correlation. With Class I drugs provision of sink conditions in dissolution testing should not prove problematic.

Class II. In contrast to Class I difficulties in the provision of sink conditions during dissolution may arise with Class II drugs. However in the case of class IIa drugs, provided appropriate sink conditions prevail, level A type IVIV relationships are to be expected without the need for a range of dissolution media. Care must be taken in the choice of additives used to provide sink conditions, since these may alter the solubility/ functionality of excipients present in the dosage form and thereby undermine the likelihood of an IVIV

Table 10. Esimated residence times in the "absorption window" (from equation 5)

Drug	Residence time
Ibuprofen	3.68 hr
Diphylline	12.27 hr
Indomethacin	2.65 hr
Quinidine Sulphate	5.92 hr
Methyldopa	3.74 hr

correlation[31,32]. In contrast Class IIb drugs are less likely to give Level A type correlations and in extreme cases absorption could become solubility controlled.

Class V. Class V contains those drugs which ionize in the gastrointestinal pH range and are subdivided into Va: acidic compounds and Vb basic compounds for the reasons outlined above. It is expected that products containing these compounds will require testing in a range of dissolution media of different pH. Level A type correlations are more likely to be obtained in the case of the basic compounds particularly if release from the product has been designed to be pH independent. Some bases and /or their salts become extremely insoluble at pH values above their pKa and absorption could change from dissolution to solubility control resulting in a loss of a Level A type IVIV relationship.

In the case of acidic drugs, if significant conversion to an ionized and much less permeable species occurs with passage down the gastrointestinal tract, a Level A type correlation is less likely and a narrow absorption window effect will result. Amoxicillin which shows negligible colonic absorption(24) could also fit into this category.

The major limitation of any two parameter (C_s, P_{eff}) classification based on a modification of that proposed for IR products (Table 1) is the absence of parameters which incorporate properties of the dosage form/delivery system. With ER products the absorption defining characteristics of the drug are modified and become to a large extent dictated by the delivery system i.e. the delivery system, rather than the drug properties dictate the release rate versus time profile. While these delivery system based attributes may be incorporated into the dissolution number(Dn) the latter is insufficient to encompass the release rate profile generated by the majority of delivery systems in-vivo.

8. CONCLUSIONS

Much more data is needed on the regional variations in drug permeability along the GIT and the relationship between these variations and physicochemical properties of the drugs. In developing a classification for IR products the primary concern was with the extent of absorption. However in developing a classification for ER products the time course of drug input, the variation in input rate, is of equal importance.

9. REFERENCES

1. Amidon, G.L., Lennernas, H. Shah, V.P. and Crison, J.R., Pharm. Res. 12, 413–420 (1995).
2. Rudman, A and Williams, R.L. Guidance for Industry Immediate Release Solid Dosage Forms. CDER Nov. 1995.
3. Federal Register. 60, 21638- 21643.
4. Sinko, P.J. Leesman,G.D., and Amidon, G.L. Pharm. Res. 8, 979–988 (1991).
5. Oh, D.M., Curl,R.L. and Amidon,G.L. Pharm. Res. 10, 264–270 (1993).. Ho, N.F.H., Merkle, H.P. and Higuchi, W.I. Drug Devel. & Ind. Pharm. (1111–1184) (1983).
7. Higuchi, W.I. & Hiestand, E.N., J.Pharm. Sci., 52, 67–71(1963).
8. Physicians Desk Reference, 50th Ed , Medical Economics Company, N.J., 1996.
9. Davis, S.S., Hardy, J.G. and Fara, J.W. Gut, 27, 886–892(1986)
10. Hardy, J.G., Wilson, C.G. and Wood,E., J.Pharm. Pharmac., 37, 874–877 (1985).
11. Steed K.P., Wilson C.G., and Washington, N. Ch. 6 in Physiological Pharmaceutics, Biological Barriers to Drug Absorption, Ed. Wilson, C.G. and Washington N., 1989, Ellis Horwood Ltd., Series in Pharmaceutical Technology, Chichester,

12. Ho, N.F.H., Park, J.Y., Morozowich,W. and Higuchi, W.I., Chapter 8 in Design of Biopharmaceutical Properties through Prodrugs and Analogues Ed. B. Roche A.Ph.A. 1977.
13. Soergel, K.H., Gastroenterology, 105, 1247–1249 (1993).
14. Kramer, W.G., J.Clin. Pharmacol. 34, 1218–1221(1994).
15. Fischer,W., Boertz,A. Davis, S.S., Khosla, R., Caawello, W., Sandrock, K., Cordes, G. Pharm. Res., 4,480–485 (1987)
16. Wilson C.G. in drug Delivery to the Gastrointestinal Tract. Chapter 13, Eds., J.G. Hardy, S.S. Davis and C.G. Wilson, Ellis Horwood Ltd., Series in Pharmaceutical Technology, Chichester, p.161, 1989.
17. Uch, A.S. & Dressman, J.B. In Formulation of poorly - Available Drugs For Oral Administration, Minutes, Ed. P. Couvrer, D. Duchene and I. Kalles, Editions de Santé, p.152, Paris 1996.
18. Ramtoola, Z. and Corrigan O.I., Drug Devel. and Ind. Pharm.,15, 2359- 2374(1989).
19. Fara, J.W., Colonic drug absorption and metabolism. Ch. 10, 103–112. in Novel Drug Delivery and its Therapeutic Applications. Ed. Prescott, JE & Nimmo,WS 1989 Wiley &Sons Ltd.
20. Dunne, A.,Corrigan, O.I., Bottini P.B. and Geoghegan, E.J. Proc. 2nd European Congress of Biopharmaceutics and Pharmacokinetics. Volume 1., 421–427,(1984).
21. Wagner, J.G. and Nelson, E., J. Pharm. Sci 1963.
22. Devane J., unpublished.
23. Barr, W. H., Zola, E.M., Candler, E.L.Hwang, S.M., Tendolkar, A.V., Shamburek,R. Parker, B. Hilty M.D., Clin.Pharmacol. Therap. 56, 279–85 (1994).
24. Peh, K.K., Ph. D. Thesis, Universiti Sains Malaysia, 1996.
25. Devane, J. unpublished.
26. Godbillon, J., Gerardin, A. Richard, J. Leroy, D. and Moppert, J., Br. J. Clin. Pharmac. 19, 213–2185, (1985).
27. Bradbrook, D. John, V.A. Morrison, P.J.Rogers, H.J. & Spector, R.G., Br. J. Clin Pharmac. 19, 163s-169s (1985).
28. Corrigan, O.I. , Dunne, A., McLoughlin, H. Geoghegan, E., Killian,C.A. and Panoz, D.E. Controlled Release Delivery Systems, Chapter 10, 163–167, Ed Roseman, T.J. and Mansdorf, S.Z., Marcel Dekker, Inc. (1983).
29. Fagerholm, U. and Lennernas, H., Eu. J. Pharm. Sci. 3, 247–253(1995).
30. Corrigan, O.I. Drug Devel. Ind. Pharm 18, 695–708, (1992).
31. Levy, G. J.Mond.Pharm. 3 237–254(1967).

10

INVESTIGATING *IN VITRO* DRUG RELEASE MECHANISMS INSIDE DOSAGE FORMS

Monitoring Liquid Ingress in HPMC Hydrophilic Matrices Using Confocal Microscopy

C. D. Melia,[1*] P. Marshall,[1] P. Stark,[2] S. Cunningham,[2] A. Kinahan,[2] and J. Devane[2]

IVIVR Cooperative Working Group
[1]Department of Pharmaceutical Sciences
Nottingham University, Nottingham NG7 2RD
United Kingdom
[2]Elan Corporation PLC
Athlone, Ireland

1. INTRODUCTION

The internal mechanisms that drive drug release underpin both the *in-vitro* and the *in-vivo* performance of oral dosage forms and therefore underlie any derived *in-vivo/in-vitro* relationships. Until relatively recently, it has been difficult to directly observe the processes inside dosage forms, and drug release mechanisms have been largely inferred from analyses of *in-vitro* release kinetics. It is still extremely difficult to obtain information about release mechanisms *in-vivo*, but methologies using imaging microscopies are currently being developed at Nottingham that enable us to directly visualise the changes taking place inside dosage forms *in-vitro* as they hydrate and perform. Techniques such as cryogenic SEM, NMR microscopy and laser scanning confocal microscopy have given valuable insights into the changing internal patterns of hydration, drug dissolution, and water and solute diffusion in, for example, hydrophilic matrix [1-4] and pellet [5,6] extended release systems. In the present work, confocal laser scanning microscopy has been used to provide evidence for a difference in release mechanism between two complex HPMC hydrophilic matrix formulations which showed unexpectedly different release profiles *in-vivo*. In a wider context this work exemplifies how advanced imaging technologies may be

* Correspondence to Dr. C. D. Melia

In Vitro–in Vivo Correlations, edited by Young *et al.*
Plenum Press, New York, 1997

used to provide some understanding of the complex phenomena that may underlie the process of drug release inside controlled release devices.

2. RATIONALE FOR UNDERTAKING THE STUDY

Hydrophilic matrices are a widely-used technology for preparing extended-release oral dosage forms. They are conventionally-prepared tablets or capsules containing a water-swelling, high viscosity polymer such as hydroxypropyl methylcellulose (HPMC) incorporated uniformly into the matrix[7]. When immersed in an aqueous environment, the polymer at the tablet surface hydrates rapidly and forms a viscous mucilage or "gel" which becomes a retarding barrier to further hydration and drug release (Figure 1). In the centre of the matrix is a dry core which provides a further supply of drug and polymer as the outer gel erodes and hydration slowly proceeds.

Drug release in these dosage forms is believed to occur principally through a combination of diffusion or erosion, the ratio being dependent on drug solubility. For soluble drugs release is primarily diffusion-controlled and the drug release rate is related to the polymer content of the matrix. At low levels of polymer (15–25%) small increases in polymer content cause a marked slowing of the drug release rate, but at higher levels this effect becomes progressively less pronounced (Figure 2) . Above 60% polymer, a point is usually reached where there is very little further change [8] as the diffusion barrier properties and gel strength of the surface gel layer are approaching the maximum attainable.

It was therefore surprising when an *in-vivo* study revealed a substantially greater retardation of drug release from an 80% HPMC matrix than an equivalent 60% HPMC tablet; the drop in Cmax being similar to that observed between 30 and 60% polymer formulations. Differentiation between the 60 and 80% formulations could only be reproduced *in-vitro* under dissolution conditions which subjected the matrix to conditions of considerably enhanced erosion, suggesting a shift from diffusion to a more erosion-modulated drug release mechanism in the case of the 60% matrix.

3. MATERIALS

Hydrophilic matrix tablets were prepared by dry compression under carefully controlled conditions according to the formula shown in Table 1. These multicomponent matrices would be typical of a commercially developed formulation intended to delay release of a highly water soluble drug.

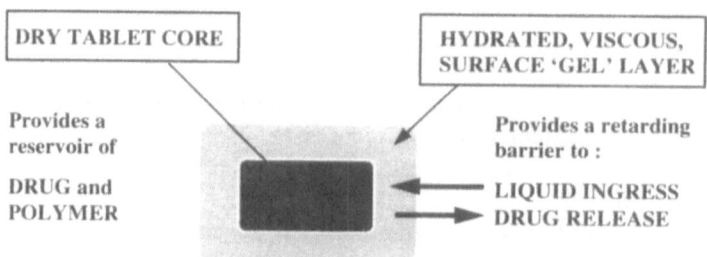

Figure 1. The principle of the hydrophilic matrix extended-release dosage form.

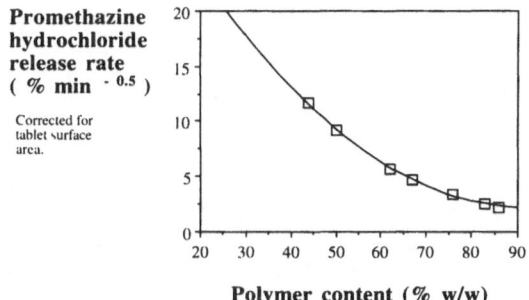

Figure 2. The relationship of drug release rate to polymer content in hydrophilic matrix tablets containing a high viscosity HPMC (Methocel K15M) and a highly soluble drug. (From the data of Ford et al [8]).

4. METHODS

The pattern and progress of hydration was investigated by obtaining a time series of images which followed the ingress of a fluorophore solution inside the matrix. The matrices were hydrated in degassed aqueous 0.05%w/v congo red solution in a sealed sample cell maintained at 37°C, using a geometry which allowed the tablets to be observed from above, whilst undergoing hydration at their radial surfaces. The tablets were ground flat to ensure that their upper surface was flush fit with the sample cell, but imaging was undertaken at depth well away from this area. Planar images of the distribution of the fluorophore were obtained at a depth of 300 μm within the developing gel layer and core, using a Bio-Rad MRC 600 confocal laser scanning microscope and an excitation wavelength of 488 nm.

5. RESULTS

5.1. The Pattern of Liquid Ingress

Figure 3 shows typical results taken from a time sequence of images. The solution marker fluoresces strongly in the presence of hydrated HPMC and cellulose, and the bright areas of the image therefore represent the hydration of these materials. It is important to remember that the z-discrimination of the confocal microscope provides pictures

Table 1. The composition of the HPMC matrix tablets used in the study

COMPOSITION OF THE MATRIX	POLYMER CONTENT	
	60%	80%
Drug	18.75	18.75 %
HPMC - High viscosity	60.00	80.75
Microcrystalline cellulose	30.65	0.00
Lubricant	0.20	0.20
Glidant	0.30	0.30

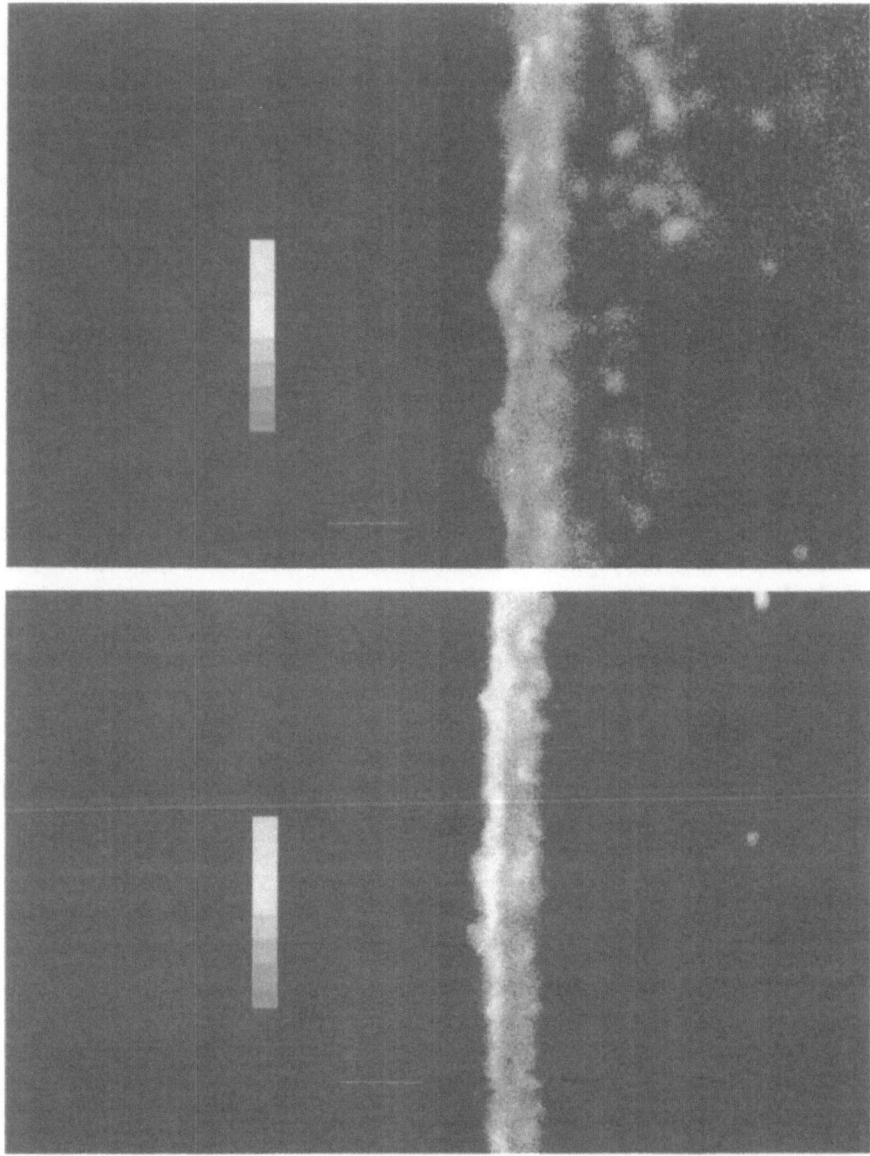

Figure 3. Typical images obtained at the edge of 60% (upper image) and 80% (lower image) matrices after 1 minute.The bright regions indicate areas of polymer or cellulose hydration. The vertical band is the growing gel layer and the area to the right is the tablet core. The images are coded for fluorescence intensity from white (highest) to black (lowest) as indicated by the wedge. Scale bar = 250 μm.

which are images of polymer hydration *inside* the gel layer without interference from the tablet surface or other out of focus regions. The images are greyscale coded for fluorescent intensity and the areas of highest intensity therefore show the greatest degree of liquid penetration into the hydrating polymer matrix. Both images clearly show the formation of the gel layer at the surface of the tablet, visible as a bright band at the surface of the tablet growing in thickness as hydration proceeds (Figure 5). However, in the case

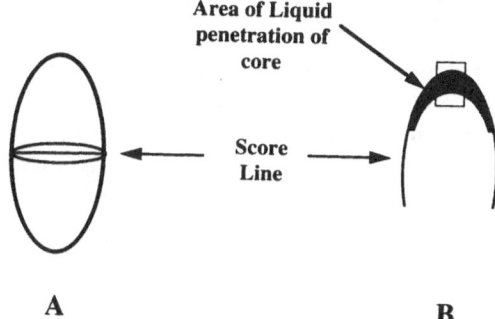

Figure 4. (A) The tablet as viewed from above and (B) the area in which liquid ingress into the core beyond the gel layer was observed in the 60% matrix (coloured grey). The rectangle represents the area from which the images in figure 3 were taken.

of the 60% HPMC matrix, a distinct network of liquid ingress reaching deeper into the tablet core was also observed. This phenomenon was reproducible and was apparent in the very earliest images (t=30 seconds). It remained visible throughout the timescale of the experiment. No such phenomenon was observed in the 80% matrix.

5.2. Site Dependence of the Internal Hydration Pattern

The images in Figure 3 were taken in the area around the tablet tip. However, because the tablet was oval in shape and had a deep score line at its centre, imaging was also undertaken at other sites around the tablet periphery. The results showed that the liquid penetration beyond the gel layer observed in the 60% tablet, occurred principally in the area of the tablet tip and never under the tablet score line (Figure 4). In the 80% matrix, no ingress into the core was seen at any point on the tablet periphery. The score line is where tablet material is most heavily compressed and a decreased tablet porosity is likely at this site. This suggests that ingress into the core is compression-dependent and that in areas that are highly compressed, the available pathway for liquid ingress into the inner core had been destroyed.

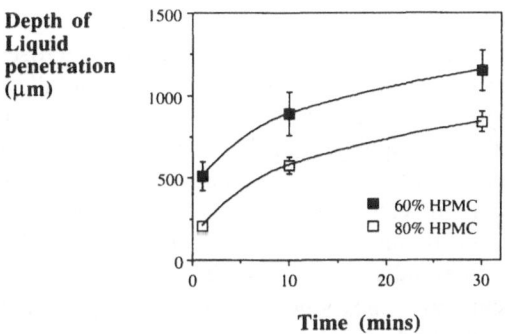

Figure 5. The time course of liquid penetration measured as the combined thickness of the gel layer and any deeper hydration network. Mean ± 1 SD (n=30).

5.3. Time Course of the Depth of Penetration

Figure 5 shows the progress of hydration with time, measured as the combined thickness of the gel layer and any deeper hydration network. The profiles show that differentiation between the two formulations occurs at the very beginning of hydration but thereafter the time course of liquid penetration follows a parallel path.

6. DISCUSSION

Deeper water penetration into hydrophilic matrices leads to increased erosion of the surface gel layer [9] and will increase the rate of drug loading into the gel layer, thereby increasing drug release. The pattern of hydration observed in the 60% matrix suggests an additional internal mechanism that promotes liquid penetration in the earliest stages of hydration, and this could arise either as a result of the differences in composition or processing. The differences in formulation between the 60% and 80% matrix certainly provide such potential. The 60% tablet contains a decreased polymer content accompanied by the addition of significant amounts of microcrystalline cellulose, a material with well-known wicking properties and considerable potential for enhancing water penetration. Microcrystalline cellulose it would not contribute to the diffusion barrier properties of the gel layer as it is non-gelling, and has the potential to channel water through the gel layer and into the core. A first explanation for these results is therefore enhanced water penetration through the gel layer and into the core, via an internal cellulose wicking network. The second possibility is that the decreased HPMC content slows the rate of closure of internal tablet pores and thereby allows capillary ingress through the pores to proceed for a longer time at the beginning of matrix hydration. The third possibility is that the 60% tablet has a more extensive porous internal network which allows further penetration in the initial stages of hydration. Whilst this cannot be ruled out, it is thought unlikely as microcrystalline cellulose exhibits extensive plastic deformation during dry compression, bridging and filling the tablet voids [11]

7. CONCLUSIONS

The non-invasive imaging capabilities of confocal microscopy have allowed us to observe the progress of hydration within two HPMC hydrophilic matrix formulations and have highlighted a significant difference in the internal pattern of polymer hydration. Polymer hydration is the principal process by which these dosage forms sustain release, and the observed differences therefore have considerable ramifications for drug release and could explain a shift in release mechanism as suggested by *in-vitro* release studies. Several mechanisms could be involved but the inclusion of a wicking diluent, microcrystalline cellulose, could potentially cause this effect.

More globally, this work shows how advanced imaging techniques may be used to provide direct evidence of the internal processes inside real-life commercial dosage forms as they perform. At the most fundamental level, it is these internal mechanisms that will ultimately determine the shape of their drug release profiles, both *in-vitro* and *in-vivo*.

REFERENCES

1. Hodsdon AC, Mitchell JR, Davies MC, Melia CD " Structure and behaviour in hydrophilic matrix sustained release dosage forms : 3. The influence of pH on the sustained release performance and internal gel structure of sodium alginate matrices"J Controlled Release 33:143–152 (1995)
2. Bowtell RW, Sharp JC, Mansfield P, Rajabi-Siahboomi AR, Davies MC, Melia CD "NMR Microscopy of hydrating hydrophilic matrix pharmaceutical tablets" J.Magnetic Resonance Imaging. 12:361–364 (1994)
3. Rajabi-Siahboomi AR, Bowtell RW, Mansfield P, Davies MC, Melia CD "Structure and behaviour in hydrophilic matrix sustained release dosage forms : 4. Studies of self-diffusion water mobility and diffusion coefficients in the gel layer of HPMC tablets using NMR imaging." Pharm. Research 13 (3):376–380 (1996)
4. Cutts LS, Bowtell R, Paterson-Stephens I, Davies MC, Melia CD "Solute and water transport across the gel layer of hydrating HPMC matrices. Proceed.Intern.Symp.Control.Rel.Bioact.Mater. 22: 1104 (1995)
5. Melia CD, Khan KA, Wilding IR "Changes in drug distribution and internal structure of oral pellet systems during in-vitro dissolution: cryogenic scanning electron microscopy and X-ray microanalysis" Pharm.Technol.16 (3): 56–64 (1992)
6. Cutts LS, Hibberd S, Adler J, Davies MC, Melia CD "Characterising drug release processes within controlled release dosage forms using the confocal laser scanning microscope" J. Controlled Release 42:115–124 1(1996)
7. Melia CD "Hydrophilic Matrix Sustained Release Systems based on Polysaccharide Carriers" Critical Reviews in Therapeutic Drug Carrier Systems 8(4) : 395–421 (1991)
8. Ford JL, Rubinstein MH, Hogan JE "Formulation of sustained release promethazine HCl tablets using hydroxypropylmethylcellulose" Intern.J.Pharmaceutics. 24: 327–338 (1985)
9. Hodsdon AC. PhD Thesis. University of Nottingham (1994)
10. Cartensen J "Solid pharmaceutics : Mechanical properties and rate phenomena" Academic Press, London (1980)

11

DETERMINATION OF CRITICAL MANUFACTURING AND FORMULATION VARIABLES FOR A HYDROPHILIC MATRIX TABLET FORMULATION USING AN *IN VITRO* DISCRIMINATORY DISSOLUTION METHOD

P. Stark,[1] A. Kinahan,[1] S. Cunningham,[1] J. Butler,[1] T. O'Hara,[1] A. Dunne,[2] J. Connolly,[2] and J. Devane[1]

IVIVR Co-operative Working Group
[1]Elan Corporation plc, Athlone, Co.
Westmeath, Ireland
[2]University College Dublin
Dublin, Ireland

1. INTRODUCTION

Formulation and manufacturing variables have been reported to influence the *in-vitro* release from hydrophilic matrix formulations. Studies on formulation and manufacturing variables have examined drug concentration[1–4] and particle size[1,3,4] polymer viscosity grade[1–5] and particle size[2] tablet compression force[1,6] and formulation additives[2,4,5,7].

A controlled release hydrophilic matrix tablet formulation of a Class I drug[8] containing Hydroxypropyl methylcellulose (HPMC) as the rate-controlling agent has been developed in this laboratory. A novel *in-vitro* discriminatory dissolution method has been developed for this product by examining formulations of differing HPMC concentrations[9]. This *in-vitro* dissolution method has been used to identify parameters which influence the release of this drug from the HPMC matrix system.

This study is concerned with the determination of the critical manufacturing and formulation variables for a hydrophilic matrix tablet formulation using an *in-vitro in-vivo* relationship developed using a novel *in-vitro* discriminatory dissolution method. The experimental work was carried out in two stages. The first study was performed according to a Design of Experiments plan to indicate the critical variables for this formulation. The critical factors identified in the first study were then included in a second study in conjunction with another potentially critical formulation variable to produce products of different *in-vitro* profiles which were then evaluated *in-vivo*. Both experiments are discussed individually.

2. MATERIALS AND METHODS - EXPERIMENT 1

2.1. Formulation Details

The hydrophilic matrix formulation evaluated in this study contains 60% Hydroxypropyl methylcellulose (HPMC) as the rate controlling agent. Formulation details of the hydrophilic matrix tablet are found in Table 1.

2.2. Experimental Design

A randomised Factorial Design of Experiments plan was used to determine critical manufacturing and formulation variables for the hydrophilic matrix formulation. The manufacturing and formulation parameters investigated were particle size of active, HPMC particle size, diluent grade and target tablet hardness. The particle size of the active, HPMC particle size and diluent grade factors were investigated at two levels using a 2^3 Factorial design which was replicated twice to give a total of 16 runs i.e. two runs for each combination of the factors evaluated. Each run was split into two and tabletted at low and high levels of tablet hardness to give a total of 32 runs in the experiment. Table 2 gives details of the experimental runs.

2.3. Method of Manufacture

2.3.1. Manufacture of Tablet Blend. The active and excipients (except the lubricant) were blended for 10 minutes in a V-blender. Addition of the lubricant to the blend was followed by blending for 5 minutes and potency of the blend was determined. Two blends of 2kg were produced for each of the eight combinations of drug particle size, HPMC particle size and diluent grade resulting in 16 tablet blends in total.

2.3.2. Manufacture of Tablets. Each tablet blend was split into two and one half tabletted at a target tablet hardness of 150N and the other half tabletted at a target tablet hardness of 250N on a Horn-Noack Rotary Tablet Press resulting in 32 tablet batches in total. This yielded two genuine replicates of each of the 16 combinations of the levels of the four factors. Tablet hardness was determined at set-up and at intervals through-out tableting using a Checkmaster hardness tester. Full details of the tablet batches manufactured and the associated factor levels may be found in Table 3.

Table 1. Formulation details of a hydrophilic matrix tablet

Component	Function	% w/w
Class I drug	Active	—
Hydroxypropyl methylcellulose (HPMC)	CR polymer	60
Diluent	Filler	q.s
Glidant	Glidant	<1%
Lubricant	Lubricant	<1%

Table 2. Factors and levels at which they were investigated (experiment 1)

Factor	Low level	High level
Class I drug particle size	$D_{90} = 160\mu m$	$D_{90} = 216\mu m$
HPMC particle size	$D_{90} = 185\mu m$	$D_{90} = 304\mu m$
Diluent grade	$D_{90} = 170\mu m$	$D_{90} = 230\mu m$
Target Tablet Hardness	150 Newtons	250 Newtons

2.4. *In Vitro* Dissolution Testing

One batch of each of the sixteen combinations of hydrophilic matrix tablets manufactured were evaluated *in-vitro* using the discriminatory dissolution method according to a Balanced Incomplete Block Design of Experiments plan. In-vitro dissolution testing was carried out using a BioDis Apparatus (USP III) with a 20 mesh polypropylene screen, at 30 dips/minute in 250ml potassium di-hydrogen phosphate buffer (KH_2PO_4, 0.32M) main-

Table 3. Tablet batches manufactured and the associated factor levels (experiment 1)

Tablet lot number	Blend lot number	Drug particle size	HPMC particle size	Diluent grade	Target hardness (N)
PD14082	PD14070	High	Low	High	250
PD14083	PD14065	High	High	Low	250
PD14084	PD14077	High	High	High	150
PD14085	PD14078	Low	High	Low	150
PD14086	PD14068	High	Low	High	250
PD14087	PD14067	Low	Low	Low	150
PD14088	PD14070	High	Low	High	150
PD14089	PD14065	High	High	Low	150
PD14090	PD14066	Low	High	Low	150
PD14091	PD14069	High	High	Low	150
PD14092	PD14076	Low	High	High	150
PD14093	PD14069	High	High	Low	250
PD14094	PD14079	Low	Low	High	250
PD14095	PD14068	High	Low	High	150
PD14096	PD14067	Low	Low	Low	250
PD14097	PD14075	Low	High	Low	250
PD14098	PD14074	High	Low	Low	150
PD14099	PD14079	Low	Low	High	150
PD14100	PD14072	High	High	High	150
PD14101	PD14073	Low	Low	High	150
PD14102	PD14072	High	High	High	250
PD14103	PD14076	Low	High	High	250
PD14104	PD14080	High	Low	Low	150
PD14105	PD14075	Low	High	Low	150
PD14106	PD14078	Low	High	Low	250
PD14107	PD14080	High	Low	Low	250
PD14108	PD14071	Low	Low	Low	250
PD14109	PD14066	Low	High	Low	250
PD14110	PD14073	Low	Low	High	250
PD14111	PD14071	Low	Low	Low	150
PD14112	PD14077	High	High	High	250
PD14113	PD14074	High	Low	Low	250

tained at 37°C. Analysis of the dissolution samples was performed using a Waters 480 HPLC system with UV detection at 220nm.

3. RESULTS AND DISCUSSION - EXPERIMENT 1

Table 4 summarizes the *in-vitro* dissolution obtained for each of the batches examined ordered by $AUC_{1-8 \text{ hours}}$. The mean effects of each of the factors examined in Experiment 1 are discussed.

3.1. HPMC Particle Size

The *in-vitro* method indicated HPMC particle size as the primary factor associated with the *in-vitro* release of the active drug from the matrix tablets at a fixed HPMC concentration. Increasing the HPMC particle size from a low level to a high level resulted in a faster dissolution profile, as illustrated in Figure 1.

3.2. Tablet Hardness

Tablet hardness also was observed to be important with faster dissolution profiles associated with the lower tablet hardness as shown in Figure 2.

3.3. Active Particle Size and Diluent Grade

The main effects of particle size of the Class I drug and the diluent grade used were not shown, to be important using this method, as displayed in Figures 3 and 4, respectively.

Table 4. Batches tested using USP III dissolution test ordered by $AUC_{1-8 \text{ hours}}$

Lot No.	$AUC_{1-8 \text{ hours}}$	HPMC particle size	Tablet target hardness	Diluent grade	Drug particle size
PD14108	441.16	Low	High	PH101	Low
PD14094	442.29	Low	High	PH102	Low
PD14113	446.30	Low	High	PH101	High
PD14101	449.93	Low	Low	PH102	Low
PD14111	450.38	Low	Low	PH101	Low
PD14082	456.91	Low	High	PH102	High
PD14087	462.32	Low	Low	PH101	Low
PD14095	467.93	Low	Low	PH102	High
PD14103	478.31	High	High	PH102	Low
PD14097	478.49	High	High	PH101	Low
PD14104	482.18	Low	Low	PH101	High
PD14102	485.59	High	High	PH102	High
PD14093	506.81	High	High	PH101	High
PD14084	509.84	High	Low	PH102	High
PD14089	515.30	High	Low	PH101	High
PD14085	528.95	High	Low	PH101	Low
PD14090	531.74	High	Low	PH102	Low

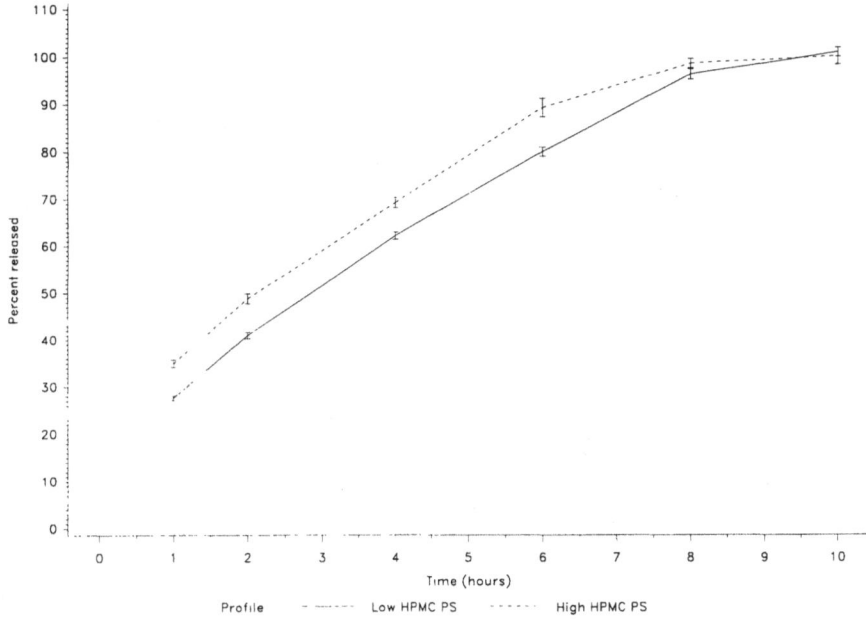

Error bars extend to 2 Standard Errors either side of mean

Figure 1. Mean *in-vitro* dissolution profiles of tablets containing low/high levels of HPMC particle size.

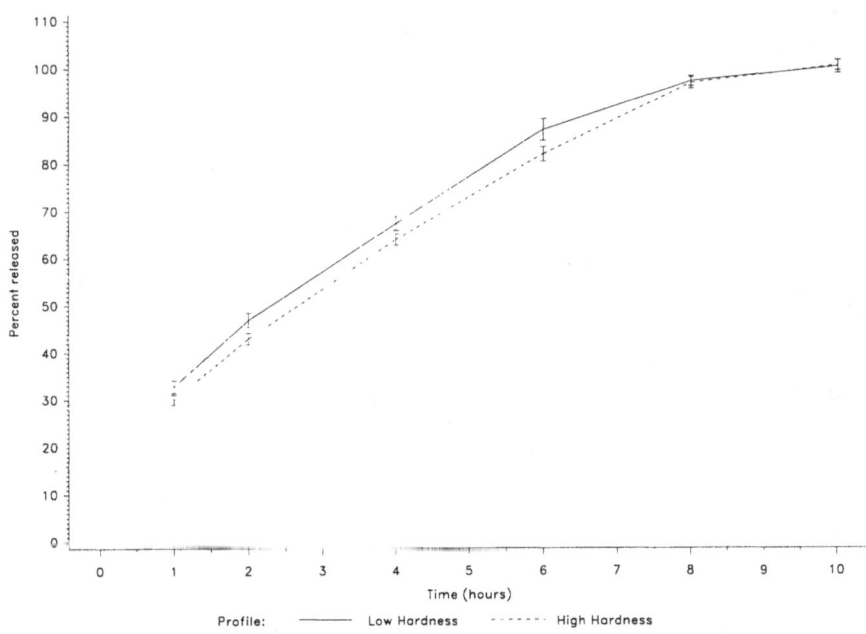

Error bars extend to 2 Standard Errors either side of mean

Figure 2. Mean *in-vitro* dissolution profiles of tablets manufactured at low/high levels of tablet hardness.

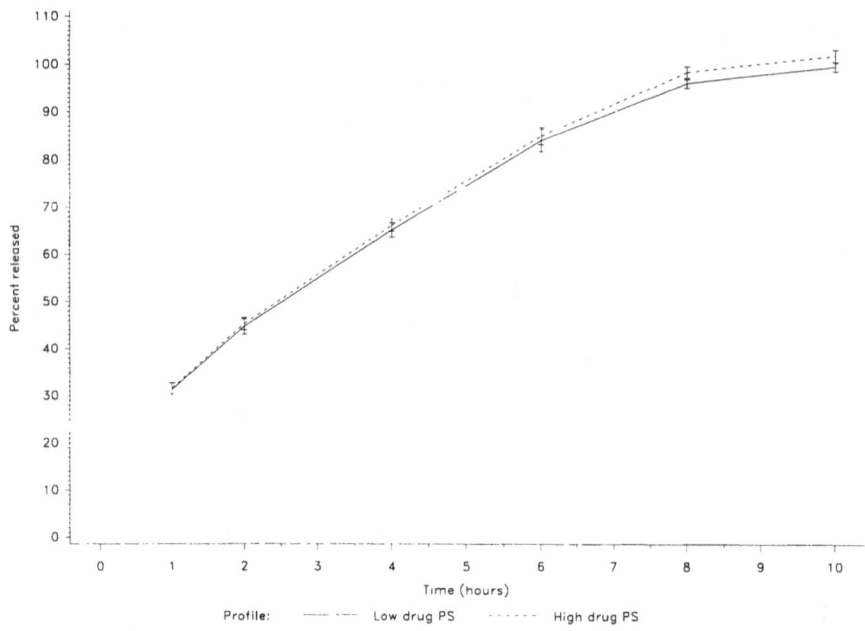

Error bars extend to 2 Standard Errors either side of mean

Figure 3. Mean *in-vitro* dissolution profiles of tablets containing low/high levels of drug particle size.

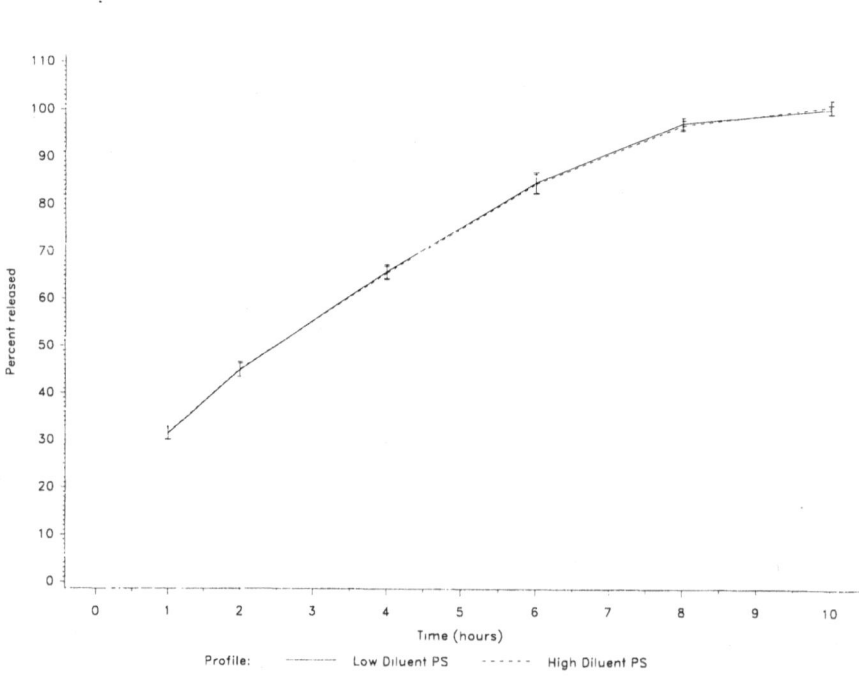

Error bars extend to 2 Standard Errors either side of mean

Figure 4. Mean *in-vitro* dissolution profiles of tablets containing low/high levels of diluent particle size.

3.4. HPMC Particle Size and Target Hardness

The HPMC particle size was demonstrated to have an effect at both levels of hardness i.e increasing HPMC particle size resulted in an accelerated dissolution profile for the low and high levels of hardness. Figure 5 indicates the faster dissolution profile for the combination high HPMC particle size/low tablet hardness. Conversely, tablets of low HPMC particle size and high tablet hardness had the slower dissolution profile. An interaction between HPMC particle size and target tablet hardness is the difference between the effect of HPMC particle size at high target hardness and the effect of HPMC particle size at low target hardness. This interaction was not considered to be important.

4. CONCLUSIONS - EXPERIMENT 1

HPMC particle size and tablet hardness were identified as potential critical manufacturing variables using the *in-vitro* discriminatory dissolution method, while the main effects of particle size of the active and of the diluent grade were not found to be important. The interaction between HPMC particle size and target tablet hardness was not considered to be important. At this stage, a confirmatory *in-vivo* study would normally be carried out using products which produced the slowest and fastest *in-vitro* dissolution profiles respectively. However, it was decided to first evaluate an additional factor which could potentially impact on the *in-vivo* performance of the product in a second experiment.

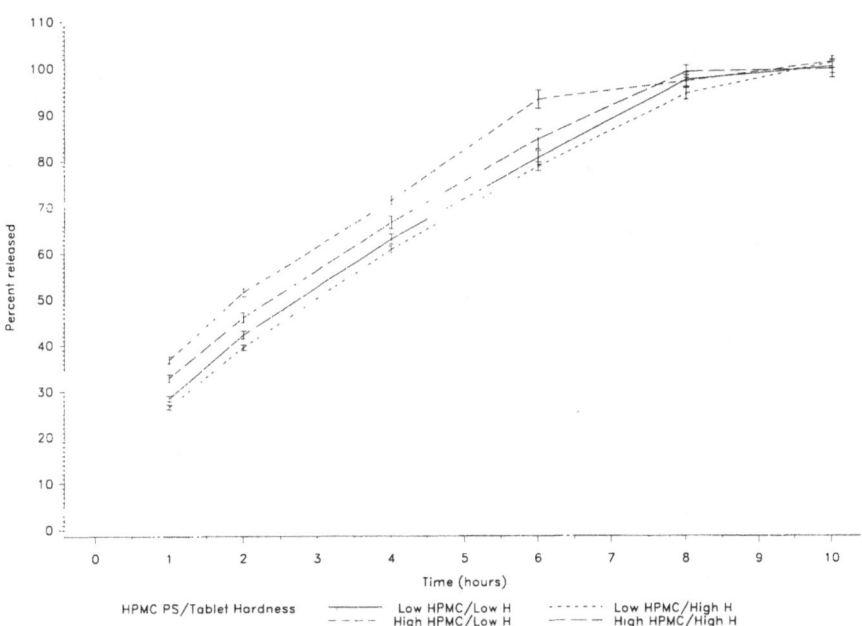

Figure 5. Mean *in-vitro* dissolution profiles of tablets containing low/high HPMC particle size manufactured at low/high tablet hardness.

5. INTRODUCTION - EXPERIMENT 2

Prior to carrying out a confirmatory *in-vivo* evaluation of the critical manufacturing and formulation variables identified in Experiment 1, the chemical substitution of the HPMC was identified as another potentially important variable. Dahl et al. demonstrated that the *in-vitro* release of Naproxen from a HPMC matrix tablet formulation was directly proportional to the hydroxypropyl substitution[10]. Typically, the manufacturers supply HPMC with the methoxy concentration specification range of between 19% - 24% and hydroxypropoxy concentration range of 7% - 12%. The chemical structure of HPMC is shown in Figure 6. The extremes of the first experiment and this additional factor were evaluated together in this second experiment.

6. MATERIALS AND METHODS - EXPERIMENT 2

6.1. Experimental Design

Four batches of the hydrophilic matrix formulations were manufactured using HPMC material which had different ratios of methoxy substitution to hydroxypropyl substitution, denoted here as the MeO/HOPrO ratio. HPMC Lot Nos. CO8909 and C07978 were considered to have high and low MeO/HOPrO ratios respectively as calculated in Table 5.

Four batches were manufactured which evaluated high and low methoxy/hydroxypropyl ratios. Based on the first experiment, high HPMC particle size was confounded with low target tablet hardness and low HPMC particle size with high target tablet hardness. Tablet manufacture and *in-vitro* dissolution testing was completed as described for Experiment 1. Formulation details for each of the batches manufactured are shown in Table 6.

7. RESULTS AND DISCUSSION - EXPERIMENT 2

7.1. *In Vitro* Evaluation

It can be seen from Figure 7 that changing the ratios of methoxy/hydroxypropyl substitution (MeO/HOPrO ratio) from a low to the high level results in faster dissolution pro-

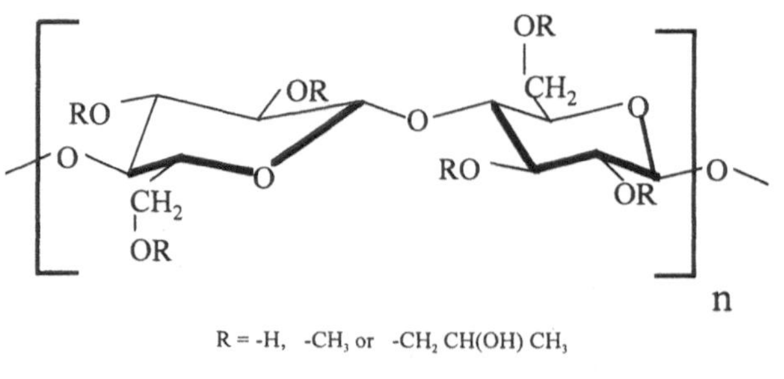

R = -H, -CH$_3$ or -CH$_2$CH(OH)CH$_3$

Figure 6. The chemical structure of hydroxypropyl methylcellulose.

Table 5. The chemical substitution of hydroxypropyl methylcellulose

Hydroxypropyl methylcellulose Lot No.	C08089	C07978	Manufacturers' specification
Methoxy Substitution (MeO)	22.9%	22.6%	19.0% - 24.0 %
Hydroxypropyl Substitution (HOPrO)	9.6%	11.5%	7.0% - 12.0 %
MEO/HOPrO ratio	2.39(high)	1.97(low)	—

files for both low HPMC particle size /high tablet hardness and high HPMC particle size /low tablet hardness batches. This effect was observed to be greater than the effects of the HPMC particle size or the tablet hardness. The *in-vitro* dissolution profiles are ranked from fastest to slowest in Table 7. From Figure 7, it can also be seen that the spread of the resultant dissolution profiles for the four batches were wider than the spread of dissolution results from the 40% - 80% HPMC batches previously evaluated in this laboratory[9]. These *in-vitro* results therefore indicated that batches manufactured using HPMC material with high or low extremes of methoxy/hydroxypropyl ratio would potentially be bio-inequivalent, assuming that the *in-vitro* test was indeed predictive of *in-vivo* performance.

7.2. *In Vivo* Evaluation

In-vivo evaluation examined the products which produced the fastest and slowest *in-vitro* profiles. A 60% HPMC product, evaluated in a previous study[9], was also included as a reference. Its *in-vitro* dissolution profile was similar to the product with a low substitution ratio/ high HPMC particle size/low tablet hardness, ranked third in Table 7. The fourth product (ranked second in Table 7) had a high methoxy/hydroxypropyl substitution, a low HPMC particle size and a high tablet hardness and was included on the basis that the shape of its dissolution profile differed from that observed for the other lots examined. An immediate release product was included for deconvolution purposes.

The *in-vivo* study demonstrated that all products are bioequivalent to the reference product in terms of extent and rate of absorption with no significant difference in the Cmax or AUC reported between the treatments. Figure 8 illustrates the mean plasma profiles after administration. The *in-vitro* test is over-sensitive and thus produced false positives, while the formulation is more robust to the manufacturing and formulation variables evaluated than originally considered.

8. CONCLUSIONS - EXPERIMENT 2

The critical manufacturing variables of HPMC particle size, tablet hardness and chemical substitution of the HPMC do not impact on the *in-vivo* performance of this hydrophilic matrix formulation. These data show that novel *in-vitro* dissolution method is

Table 6. Tablet batches manufactured and the associated factor levels (experiment 2)

MeO/HOPrO ratio	HPMC particle size	Target hardness (N)
High	Low	250
High	High	150
Low	Low	250
Low	High	150

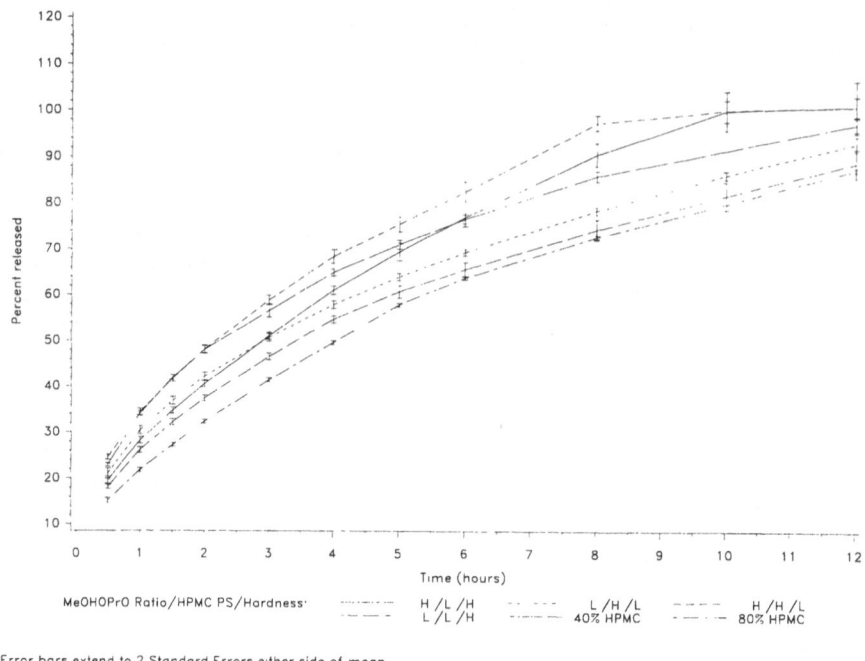

Figure 7. Mean *in vitro* dissolution profiles of tablets containing low and high levels of MeO/HOPrO, HPMC particle size.

overselective to the formulation and manufacturing variables evaluated in these experiments. Further optimisation of this *in-vitro* dissolution system is on-going to optimise the selectivity of the *in-vitro* dissolution method.

9. CONCLUSIONS

The use of Design of Experiments plans has proved to be an effective strategy in determining the potential effect of critical formulation and manufacturing variables. A discriminatory *in-vitro* test method is required to screen products with different formulation and manufacturing variables in order to determine the most appropriate candidates for *in-vivo* biostudy. In-vivo evaluation of products which have different *in-vitro* profiles is required to be completed in order to confirm the predictive nature of the *in-vitro* test and to avoid erroneous assumptions which might lead to a overly restrictive manufacturing process.

Table 7. *In-vitro* rank order of tablet batches (experiment 2)

MeO/HOPrO ratio	HPMC particle size	Target hardness (N)	Rank order
High	Low	High	2
High	High	Low	1
Low	Low	High	4
Low	High	Low	3

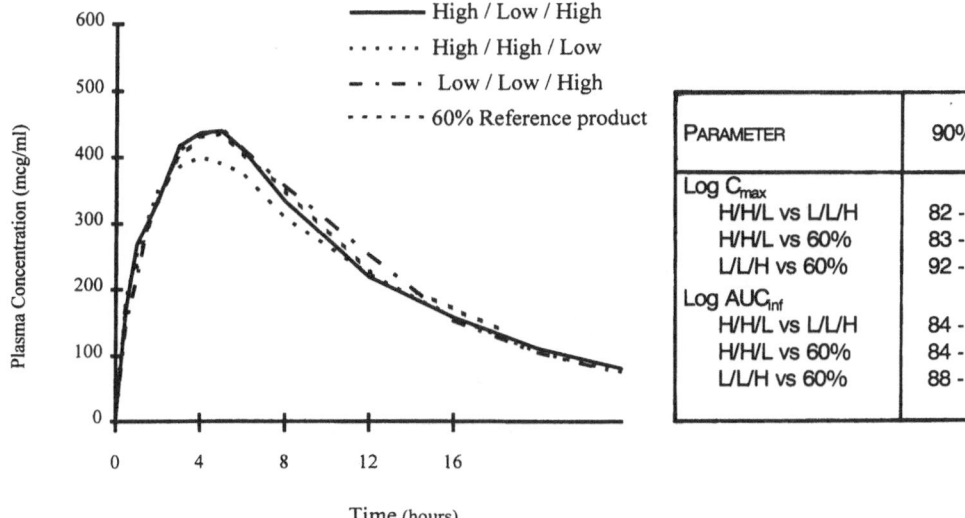

Figure 8. Mean plasma profiles of tablets which had differing *in vitro* profiles.

10. REFERENCES

1. Ford, M.H. Rubinstein and J.E. Hogan, Formulation of sustained release promethazine hydrochloride tablets using hydroxpropyl methylcellulose matrices, Int. J. Pharm., 24 (1985) 327–338.
2. Alderman, A review of cellulose ethers in hydrophilic matrices for oral controlled-release dosage forms, Int. J. Pharm. Technol. Prod. Manuf., 5 (1984) 1–9.
3. Ford, M.H. Rubinstein and J. Hogan, Propranolol hydrochloride and aminophylline release from matrix tablets containing hydroxpropyl methylcellulose, Int. J. Pharm., 24 (1985) 339–350.
4. Ford, M.H. Rubinstein and J.E. Hogan, Dissolution of a poorly water soluble drug, indomethacin, from hydroxypropyl methylcellulose controlled release tablets, J. Pharm. Pharmacol., 37-supplement (1986) 33P.
5. Ford, M.H. Rubinstein, A. Changela and J.E. Hogan, Influence of pH on the dissolution of promethazine hydrochloride from hydroxypropyl methylcellulose controlled release tablets, J. Pharm, Pharmacol., 37-supplement (1986) 115P.
6. Korsmeyer, R. Gurny, E. Doelker, P. Buri and N.A. Peppas, Mechanisms of potassium chloride release from compressed, hydrophilic, polymeric matrices: Effect of entrapped air, J. Pharm. Sci., 72 (1983) 1189–1191.
7. Ford, M.H. Rubinstein, F. McCaul, J.E. Hogan and P.J. Edgar, Importance of drug type, tablet shape and added diluents on drug release kinetics from hydroxypropyl methylcellulose matrix tablets, Int. J. Pharm., 40 (1987) 223–234.
8. Amidon, H.L. Lennernas, V.P. Shah and J.R. Crison, A Theoretical Basis for a Biopharmaceutic Drug Classification: The Correlation of in Vitro Drug Product Dissolution and in Vivo Bioavailability, Pharm. Res., 12 (1995) 413–420.
9. C . Farrell, J. Butler, P. Stark H. Madden and J. Devane, The Development of a Novel In-Vitro Discriminatory Dissolution Method for a Class I Drug in a Matrix Tablet Formulation, in press.
10. Dahl, T. Calderwood, A. Bormeth, K. Trimble and E. Piepmeier, Influence of Physico-chemical Properties of Hydroxypropyl Methylcellulose on Naproxen Release from Sustained Release Matrix Tablets, J. Controlled Release, 14 (1990) 1–10.

12

ARTIFICIAL NEURAL NETWORK BASED *IN VITRO–IN VIVO* CORRELATIONS

Ajaz S. Hussain

Division of Product Quality Research
Office of Testing and Research
Center for Drug Evaluation and Research
Food and Drug Administration
5600 Fishers Lane, HFD-900
Rockville, Maryland 20857

1. INTRODUCTION

In vitro drug dissolution test facilitates design, development, and quality assurance of solid (oral) products. During product development dissolution tests are used to characterize impact of formulation variables on drug release and to optimize formulations to achieve a target drug release profile (assuming the *in vitro* test is predictive of *in vivo* drug release). Every manufactured batch must meet pre-set specifications prior to being released for distribution. Certain "minor" post-approval changes in the manufacturing process and/or formulation may be justified by demonstrating equivalent *in vitro* dissolution profiles[1]. For certain "major" post-approval changes the FDA may grant requests for waiver of *in vivo* bioequivalence studies when a sponsor demonstrates acceptable predictive ability of the selected *in vitro* test (i.e., availability of an validated *in vitro* to *in vivo* correlation -IVIVC)[2].

Generally, the following steps are involved in establishing an IVIVC: (1) design of formulations with differing *in vitro* drug release profiles, (2) *in vivo* evaluation of these formulation (pilot bioavailability study with an oral or intravenous solution as a reference), (3) estimation of *in vivo* drug release profile by a suitable deconvolution procedure, and (4) developing a suitable statistical or pharmacokinetic function to relate *in vitro* dissolution to *in vivo* drug release or blood level profiles. If this process fails to provide an acceptable IVIVC, different *in vitro* test conditions and/or different empirical or mechanistic functions are investigated. Instead of the two step deconvolution procedure (steps 3 and 4, above) it may be desirable to directly map *in vitro* release profile to the plasma level profile using a convolution-based (discussed in this volume by Gillespie) or other suitable procedures. Irrespective of the procedure, from the regulatory perspective, it is desirable to predict the entire *in vivo* time course from the *in vitro* data[2].

This chapter explores the potential role artificial neural networks (ANN) may play in predicting *in vivo* drug concentration time profile from *in vitro* dissolution profile. ANN offer capabilities that are different from those offered by traditional regression/statistical approaches, therefore, its applications in pharmaceutical product development process merits investigation. This chapter provides brief background information for a class of ANNs, the mapping networks, and explores potential advantages and limitations of ANN based IVIVC models. The data used to demonstrate IVIVC application was abstracted form three recent presentations.[3-5]

1.1. Background Information

Neural Computing technology is vast, rapidly developing, and multi-disciplinary in scope. The focus of this chapter is on the backpropagation network that belongs to class of ANN called the Mapping Networks which are widely used for modeling or function approximation[6]. Other examples of mapping networks include; the counterpropagation network, radial basis function network, and the general regression network[6,7].

ANNs consists of a network of interconnected processing elements (PE) or nodes (analogous to neurons). A typical PE and a network of PE is shown in figures 1 and 2, respectively. Arrangement of PE and their interconnections is referred to as the network architecture. PE are arranged in layers according to their function. Layers of PE that brings information in the network and those that distribute information to the outside are the input and output layers, respectively. A layer of PE located between the input and output layers is called the hidden or intermediate layer.

Most frequently used PE connection pattern is such that all input PE are connected only to PE in the hidden layer neurons which in turn are connected to PE in the output layer (as shown in figure 2). Each connection in the network has an associated weight parameter which is analogous to a synapse in biological neural network and its numerical value emulates the synaptic strength. Two common ANN connection patterns include the feedforward and the recurrent architectures.

The feedforward architecture (see figure 2) allows data flow to occur in one direction and the network output is solely based on the current set of inputs. This architecture is suitable for situations when all the necessary information (inputs) is available to estimate the network response (output).

Recurrent architecture may be applicable when sequences (e.g., time-series) of inputs are important and/or when information of prior states are necessary to estimate a current response. Information may be recirculated and processed along with in-coming inputs by means of feedback connections. Limited or partial recurrent networks provide feedback

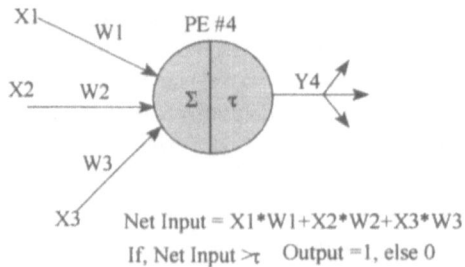

Net Input = X1*W1+X2*W2+X3*W3

If, Net Input $>_\tau$ Output =1, else 0

Figure 1. A typical processing element.

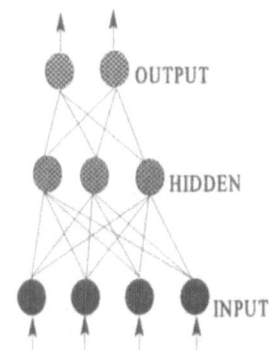

Figure 2. Multi-layer feed forward architecture.

connections from the hidden or output units to a set of additional input PE (referred to as context units).

The computation process initiates as a signal (input) is received at the input layer. Input layer PE transfer this signal to all PE in the hidden layer (in most ANNs this occurs without any modification of the input signal). Each of PE in the hidden layer then computes a weighted sum of all input signals it receives. This sum constitutes the argument to a transform or activation function and the resulting value of this function is the output. Several types of transform/activation function are in use, for example, threshold logic (i.e., if weighed sum is greater or equal to a threshold value, α, the output equals 1 else it equals 0) and sigmoidal functions.

The output of hidden layer PE gets distributed to all output layer PE (assuming there is only one hidden layer) and similar computations occur to generate signals that leave the network. For an ANN to provide appropriate (or desired) response it first needs to be trained to recognize underlying patterns that define functional relationships between input and output data vectors. Prior to any training, weights have small random values and the resulting output reflects this random state. Training process involves adjusting connection weights and methods used to accomplish this are referred to as learning rules or algorithms. Several different types of learning algorithms have been developed and these can be classified as "supervised" or "unsupervised". Supervised learning uses a training data set consisting of inputs and the desired outputs vectors. Unsupervised training is carried out with a training set of only the input vectors (i.e., the desired output is not defined) and the ANN learns to categorize/cluster these input patterns. The process of adjusting the weights (supervised learning rule) is generally a variation of one of the following: (1) Hebbian learning: a connection weight in the input path of a PE is incremented if both the input and the desired output value is high, (2) Delta rule: is based on reducing the difference between the network output and the desired output, or (3) Competitive learning: PE compete among themselves and the one which yields the strongest response to a given input is activated.

A properly trained ANN can *generalize* to previously unseen (but within the domain of its training) input vectors. A poorly trained ANN may perform well on input/output examples within its training set but is unable to generalize - this is similar to what is termed as "overfitting" in statistics and is also referred to as "memorization."

The term "backpropagation" describes a supervised training algorithm for multilayer networks. It computes a gradient, with respect to network parameters -weights and thresholds, of a measure (such as; the mean squared error) of the difference between the current output and the desired output, then modify the parameters so as to minimize this gradient. Training involves; (1) propagation of activations forward through the network to provide an output for the current input vector, (2) determine the error and propagate it backward, and (3) modify weights to reduce the error.

A number of modification to the basic backpropagation algorithm have been proposed that improve weight optimization process by avoiding local minima and converge quickly. However, several problems remain, such as; (1) initial selection of number of hidden layers and PE in these layers is guided by 'rules of thumb' that may not have strong scientific rationale, (2) convergence criteria is difficult to define, (3) high connectivity may lead to problems of "memorization" and/or increase the computational power requirement. Practical solutions are available to solve these problems, but these appear to be based on individual preference and experience. In spite of these problems mapping networks have been successfully applied to a variety of complex problems. The backpropagation networks are capable of approximating any continuous function to any desired degree of accuracy (if sufficient number of hidden PE are allowed) and offers several advantages over traditional regression analysis. They are able to map more general functional forms without having to define the function and unlike linear regression these do not require assumptions of linear superposition and orthogonal functions. Several applications of ANN to pharmaceutical problems have been presented[8-16] and this list continues to grow.

2. ANN-BASED IVIVC DEVELOPMENT

The proposed hypothesis is that a mapping network, such as the backpropagation network, may be used to relate meaningful *in vitro* dissolution profiles directly to *in vivo* blood level profiles.

Since ANN are data driven, both content and format of the training set are key for any successful application. The mapping ability of ANNs is not an issues. Questions we are confronted with include: In a typical IVIVC development study the number of different formulations evaluated is small, will this information be sufficient to develop and validate an ANN-based IVIVC? Do we also need a oral solution or intravenous reference? Will this approach be limited to drugs that obey linear kinetics? How should the information be presented (feedforward or recurrent network design)?

To address some of these issues three examples of ANN-based IVIVC approaches are summarized. Detailed description of one ANN-based IVIVC development project is described.

2.1. Data and ANN Development

Metoprolol Extended Release Tablets (hydrophilic matrix system).[3] *In vitro* drug release (under different test conditions) and *in vivo* serum data for three different formulations (differing in their release rates, referred to as: fast, medium, and slow) of metoprolol tartrate (100 mg) were generated. The bioavailability study was a four way cross-over study in nine healthy volunteers (who were screened to be extensive metabolizers) with 50 mg oral solution dose as the reference. Initial IVIVC development efforts using the traditional approach utilized the USP paddle apparatus at 50 rpm and failed to establish a meaningful correlation. Several other dissolution test conditions were evaluated and one method was selected (USP basket, 150 rpm, pH 6.8 buffer) with the help of ANN (ability to predict the blood levels of the medium release formulation) to be the most promising. Using the selected *in vitro* method Level A correlation was established using the traditional two stage deconvolution approach and validated using internal crossvalidation method[2,3].

ANN-based IVIVC were investigated using the most simple feedforward network (similar to figure 2) trained by an modified version of the backpropagation algorithm (the Enhanced Delta Bar Delta algorithm with hyperbolic tangent (Tanh) transform function for both hidden and output nodes). The ANN was emulated on a digital computer with the NeuralWare Professional II/plus software package.

The number of input nodes were set equal to the number of *in vitro* dissolution samples (n=12, including time zero) collected at pre-established fixed sampling schedule. Each input node represent a particular sampling time and the fractional drug release (for example: 0, 0.05, 0.1, 0.15,....) data format was used (summation of the fractional drug release provides the total drug released at the last sample time which was 12 hours). Solution data (serum levels normalized to 100 mg dose) were also included (format: 1, 0, 0, 0, 0, ...). Six dissolution runs were available for each tablet formulation and all six individual tablet data were included in the training set.

In the initial analysis specific patient demographic information were also included as inputs, in the network described here only the body weight was retained. The network output were the serum metoprolol concentrations at fixed times (similar to the input layer). A typical input/output data sequence was as : D1, D2, D3, D4, ..., Dn, BW, S1, S2, S3, S4, ..., Sm; where D refers to dissolved fraction (at times 1 to n), BW is the body weight and S is the serum metoprolol concentration (at times 1 to m). It should be noted that dissolution and pharmacokinetic times were not synchronized, i.e., D1(e.g., 0.5 hr) does not correspond to S1 (e.g., 1 hr), this forces the network to automatically perform time-scaling to develop functional relationships. For each subject six dissolution data for the three formulations was included in the data set.

One hidden layer was included and the number of hidden PE was set arbitrarily to 13 (equal to the number of output PE). This number of hidden PE were found to be sufficient to "memorize" the training set (i.e., was able to emulate the mean data to withing +/- 2% of its value) when the entire set was presented 200 times (epochs). This suggested that the network was sufficiently large for the problem at hand. Application failure in this situation would then be due to insufficient data, inappropriate input/output format, and/or memorization.

Three training sets were created by leaving data of one formulation out (test set) and networks were trained (Note: another option would be to create a test set by randomly selecting 10–20% of the examples from the training set of two formulations, the third formulation would then serve as an external-validation set). The training time was selected by the "Save-Best" option of the software package. Invoking this option trains and tests the network periodically while monitoring the test set prediction error. Training is stopped when the test set prediction error start to increase after reaching a minimum value, at this point the network starts to memorize the data (note: prediction error for the training set will continue to decline as training continues). This option identified the following optimum number of training epochs: 6, 16, and 10 for fast, medium, and slow release formulations, respectively. At these conditions the root mean square error (RMS) were 7.3, 4.3, and 17.0 for the fast, medium, and slow formulations, respectively. One needs to identify one value of training iterations that satisfies all formulations. The prediction ability of the three networks were then evaluated at 6, 10 and 16 epochs and 16 epochs was selected for all networks. At this setting no improvement (or deterioration) in the prediction of slow formulation was noticed (RMS = 17.5), RMS for fast and medium formulations were 9.9 and 4.4, respectively. It should be noted that the ob-

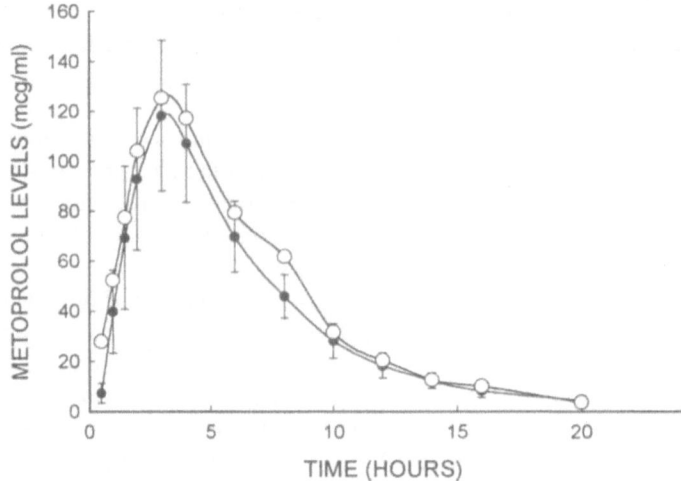

Figure 3. Cross-validation results for the fast release metoprolol formulations (open circles - predicted values and closed circles observed with standard deviations).

served training period were relatively short, other algorithms such as the basic backpropagation generally require longer training (thousands of epochs).

The predicted profiles (leave-one-out) for the three formulations are shown in figures 4, 5, and 6. When compared to the traditional two stage deconvolution (TA) approach, the root mean square error and the mean absolute prediction error (MAPE) were similar for the medium release formulation (RMS: TA = 5.7, ANN = 4.4; MAPE: TA = 4.2, ANN = 3.3) but not for the fast or the slow release formulations. The traditional approach was superior for prediction of the slow formulation (RMS: TA = 4.9, ANN = 17.5; MAPE: TA = 3.8, ANN = 11.4) and the ANN provided better prediction for the fast formulation

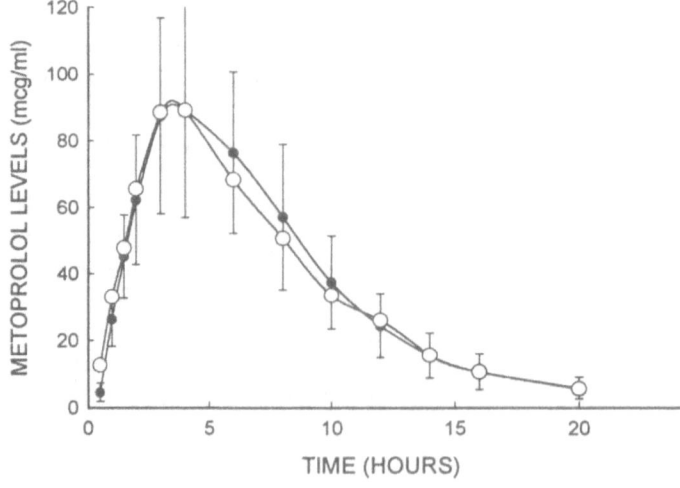

Figure 4. Cross-validation results for the medium release metoprolol formulations (open circles - predicted values and closed circles observed with standard deviations).

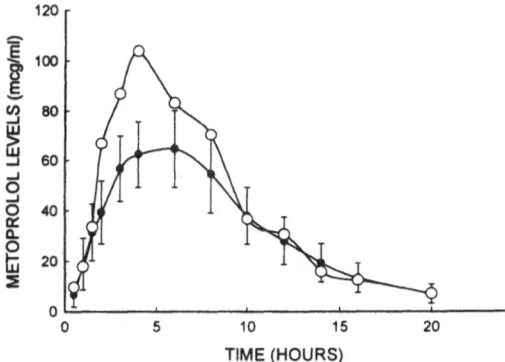

Figure 5. Cross-validation results for the slow release metoprolol formulations (open circles - predicted values and closed circles observed with standard deviations). Diltiazem Extended Release Capsules (coated beads)[4].

(RMS: TA = 17.1, ANN = 9.9; MAPE: TA = 10.2, ANN = 8.0). During this analysis, another data set became available (medium release formulation - manufactured at different scale) and when used for "external- validation" both the ANN and the traditional approach provided similar prediction.

Ditiazem Extended Release Capsules (coated beads).[4] Another data set consisting of three diltiazem extended release formulations (fast, medium, and slow release) were evaluated using ANN. The network architecture and development strategy were similar to that described above. In this case, however, the simple two stage deconvolution approach using the standard linear model was not acceptable. Other convolution approaches using the *in vitro* release data and solution data serving as the unit impulse response were investigated. In one model, the apparent rate of input was nonlinearly modeled as a function of the amount dissolved at any given time, consistent with a saturable first-pass elimination process. In the second nonlinear model the apparent rate of input was truncated (assuming an absorption window), and the third model was a combination of the saturable first-pass and the truncated absorption processes. These nonlinear models described the data better than the standard linear methods (discussed in this volume by Gillespie). ANN cross-validation results of this analysis are presented in figures 6, 7, and 8. The MAPE were 10.7, 4.2, 6.3 and the RMS were 15.4, 6.1, and 9.3, for fast, medium, and slow formulations.

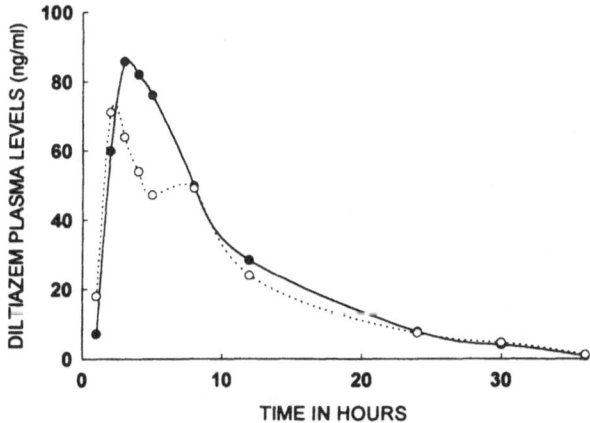

Figure 6. Cross-validation results for the fast release diltiazem formulations (open circles - predicted values and closed circles observed values).

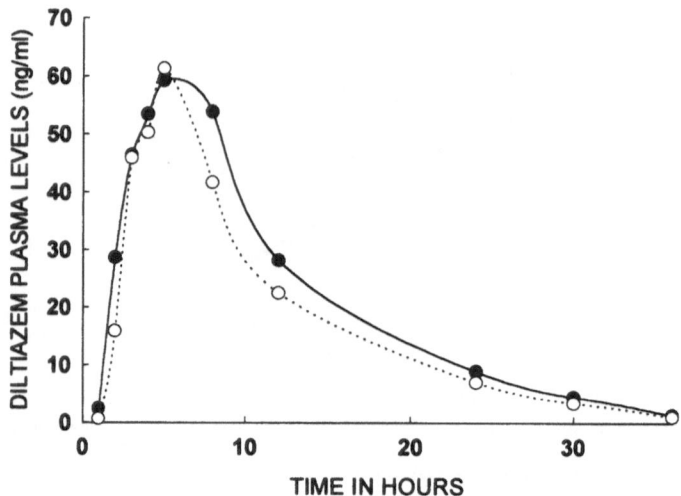

Figure 7. Cross-validation results for the medium release diltiazem formulations (open circles - predicted values and closed circles observed values).

The third study that report an ANN approach for IVIVC evaluated several different input/output formats and ANN models (feedforward backpropagation, recurrent networks, and the general regression network). In this study data from three different extended release formulations were used (two for training and one for network evaluation). The input/output formats investigated included: (1) functional approximation format as discussed above for metoprolol and diltiazem (direct mapping of dissolution (D) profiles to serum levels (S) - Input: D1, D2,....., Dn; Output: S1, S2, ...,Sm), (2) mapping individual dissolution points to individual serum levels (Input: Pharmacokinetic time (Tpk), D1, D2, ..., Dm; Output: S(Tpk)), (3) For data where the pharmacokinetic time and the disso-

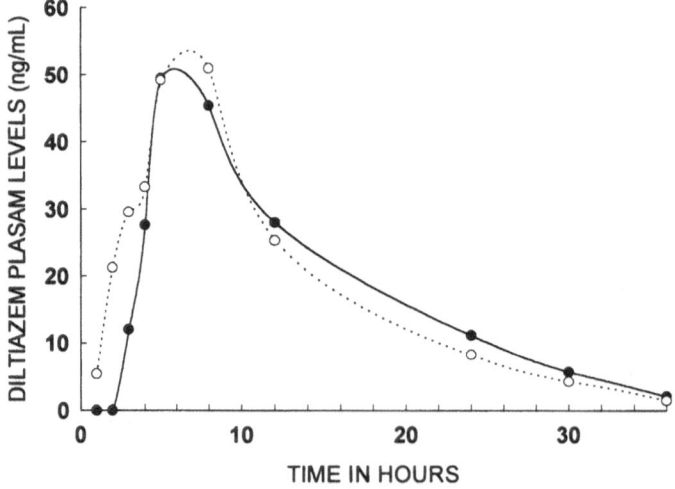

Figure 8. Cross-validation results for the slow release diltiazem formulations (open circles - predicted values and closed circles observed values). Evaluation of various input/output formats and network architecture [5].

lution time were the same- Input: Tpk/diss, D(Tpk/diss); Output: S(Tpk)), and (4) D = 0 for Tdiss > Tpk and D = D for Tdiss < Tpk, Input: Tpk, D1, D2, ...,Dn, Output: S(Tpk). In this study cumulative % drug released values were used and no reference formulation was included. The pharmacokinetic profiles seems to suggest the "flip-flop" situation (i.e., input rate < elimination rate).

Feedforward network and the general regression network (a technique based on kernel regression) with the functional approximation input/output format (#1) provided good predictive capabilities. Several other configuration also proved to be suitable for IVIVC development.

3. CONCLUSIONS

ANN applications are generally successful when sufficient data are available for network development and evaluation. Development and validation of IVIVC is generally carried out with limited data. In the studies described three formulations were available for IVIVC development. The developed ANN models provided excellent predictions for the medium formulation (interpolation). The two studies that evaluated the predictive ability of the slow and fast release formulation clearly demonstrated the inability of the ANN to extrapolate in these situation. In case of metoprolol (see figure 5) the slow release formulation could not be predicted by a network trained on solution, fast and medium release formulation data. In case of diltiazem, poor predictions were observed for fast release formulation (figure 8). If only the metoprolol data were available one could argue that reasonable prediction of the fast release metoprolol formulation was due to the inclusion of the solution data (this does not appear to be the case with diltiazem where solution data was also included in the training set). It is also interesting to note that only the slow release diltiazem formulation had significant lag-time (*in vitro* and *in vivo*) and the network fails to predict the early time-point.

ANN are quite flexible with respect to input/output format and network architecture. The simplest function approximation type of input/output format and the feedforward architecture were found to be suitable for three different IVIVC problems. It is quite remarkable that this generic network was applicable to drugs/products with different pharmacokinetic characteristics (approximate linear kinetics in this study - metoprolol, nonlinear kinetics - diltiazem, and "flip-flop" situation). Inclusion of a unit impulse reference did not appear to be critical for ANN development, however this issue needs to be further evaluated.

Several other advantages of ANN-based IVIVC were also noted; (1) repeated measures nature of IVIVC data is addressed as all available dissolution and serum level data can easily be incorporated in the training data sets, (2) no assumptions were necessary with respect to the IVIVC function or pharmacokinetic models, therefore model missspecification problems are avoided, (3) subject-specific information may be easily incorporated, (4) formulation specific-information could also be incorporated to strengthen the link to formulation development (critical variable assessment and optimization database), and (5) direct mapping of *in vitro* data to *in vivo* response using a generic network saves time.

4. ACKNOWLEDGMENTS

The author wishes to thank Drs. Jim Dowell, Patrick Marroum, Natalie Eddington, and David Young for their valuable contributions.

The opinions expressed here are those of the author and may not reflect the views and policies of the FDA.

5. REFERENCES

1. "Draft Guidance for Industry: Modified Release Solid Oral Dosage Forms; Scale-Up and Post Approval Changes: Chemistry, Manufacturing and Controls, In Vitro Dissolution Testing and In Vivo Bioequivalence Documentation", U. S. Department of Health and Human Services, Food and Drug Administration, Center for Drug Evaluation and Research, July 1996.
2. "Draft Guidance for Industry: Extended Release Solid Oral Dosage Forms; Development, Evaluation and Application of *In Vitro/In Vivo* Correlations". U. S. Department of Health and Human Services, Food and Drug Administration, Center for Drug Evaluation and Research, July 1996.
3. Hussain, A. S., et. al. Artificial neural network (ANN) approach for establishing a predictive relationship between *in vitro* drug release and *in vitro* serum drug levels. Pharm. Res. 10: S-489 (1996)
4. Piscitelli, D. A., et. al. Investigation of an Level A correlation for diltiazem. Pharm. Res. 10: S-475 (1996).
5. Dowell, J. A.., et. al. Neural network model development for *in vitro - in vivo* relationships. Pharm. Res. 10: S-450 (1996).
6. Robert Hecht-Nielsen. Neurocomputing, Addison-Wesley Publishing Company, Reading, Massachusetts, 1989.
7. Neural Computing: A Technology Handbook for Professional II/Plus and NeuralWorks Explorer. Neural-Ware, Inc., Pittsburgh, PA. (1993).
8. Shadmehr, R. and D'Argenio, D. Z. A neural network for nonlinear Bayesian estimation in drug therapy. Neural Comput. 2: 216–225 (1990).
9. Hussain, A. S., Yu, X., Johnson, R. D. Application of neural computing to pharmaceutical product development, Pharm. Res. 8, 1248–1252 (1991).
10. Veng-Pedersen, P., Modi, N. B. Neural networks in pharmacodynamic modeling. Is current modeling practice of complex kinetic system at a dead end? J. Pharmacokin. Biopharm., 20, 397–412 (1992).
11. Hussain, A. S., et. al., Feasibility of developing a neural network for prediction of human pharmacokinetic parameters from animal data. Pharm. Res. 10, 466–469 (1993).
12. Erb, R. J. Introduction to backpropagation neural network computations. Pharm. Res. 10, 165–170 (1993).
13. Hussain, A. S., Shivanand, P., Johnson, R. D. Application of neural computing in pharmaceutical product development: Computer-Aided formulation design. Drug. Dev. Ind. Pharm.20, 1739–1752 (1994).
14. Murtoniemi, E. et. al., The advantages by the use of neural networks in modeling the fluidized bed granulation process. Int. J. Pharm. 108, 155–164 (1994).
15. Brier, M. E., Zurada, J. M., Aronoff, G. R. Neural network predicted gentamicin peak and trough concentrations. Pharm. Res. 12, 406–412, (1995).
16. Gobburu, J. V. S., Chen, E. P. Artificial neural networks as a novel approach to integrated pharmacokinetic-pharmacodynamic analysis. J. Pharm. Sci. 85, 505- 510 (1996).
17. Kesavan, J. G. And Peck, G. E. Pharmaceutical granulation and tablet formulation using neural networks. Pharm. Develop. And Tech. 1, 391–404 (1996).
18. Hussain A. S.: Application of artificial neural network to population pharmacokinetic and pharmacodynamic data analysis. In: Pharmacokinetic/Pharmacodynamic Analysis: Accelerating Drug Discovery and Development. IBC Biomedical Library Series, Southborough, MA., J. Schlegel (Ed.), July 1996

13

SETTING DISSOLUTION SPECIFICATIONS FOR MODIFIED-RELEASE DOSAGE FORMS

Deborah A. Piscitelli and David Young

IVIVC Cooperative Working Group
Pharmacokinetics-Biopharmaceutics Laboratory
School of Pharmacy
University of Maryland at Baltimore
Baltimore, Maryland 21201

1. BACKGROUND

Dissolution specifications are limits for the percent of drug released at specific times. Setting these boundaries assures that all formulations which meet these limits perform similarly. For years dissolution specifications have served as an *in vitro* quality assurance (e.g., in stability testing). As a quality control measure, dissolution specifications are defined by the Sponsor and the Food and Drug Administration (FDA). Acceptance criteria is often based on three time points: 1) an early time point to identify if dose dumping occurs, 2) a time point to characterize the release profile and demonstrate the extension of release, and 3) a time point to prove that most of the intended entire dose is delivered. The USP dissolution acceptance criteria for diltiazem hydrochloride extended release capsules is based on time points at 3, 9, and 12 hours. The following chart describes the USP dissolution specifications for diltiazem hydrochloride[1];

Time (hours)	Amount dissolved
3	between 10% and 25%
9	between 45% and 85%
12	not less than 70%

In the past, dissolution specifications were based primarily on the *in vitro* performance of the product without any consideration to *in vivo* significance. More recently, the FDA has expanded the role of dissolution specifications in the guidelines for scale up and post approval changes.[2,3] In these guidances, a scaled-up formulation or a minor post approval change to the formulation may not require a bioavailability study, if it can be shown to fall within the dissolution specifications. Thus, the ideal situation would be to set the dissolution specifications such that scaled up and post approval changed formula-

In Vitro–in Vivo Correlations, edited by Young *et al.*
Plenum Press, New York, 1997

tions as well as the stability lots fall within the dissolution boundaries.[4] Dissolution specifications can also serve a similar role in pre-approval formulation development where formulations can be optimized using the dissolution specifications as a guide rather than
depending on multiple bioavailability studies.

At the present time, this broaden goal requires that the dissolution system no longer
serves as a quality control dissolution test but instead is a surrogate for *in vivo* bioavailability. By doing this, the *in vitro* dissolution test becomes a predictor of *in vivo* performance of the formulation, hence, the specifications can be set to minimize the possibility of
releasing formulations that would be different in their *in vivo* performance. In order to accomplish this, it is crucial to incorporate *in vivo* data when setting dissolution limits. Dissolution specifications can be set so that all formulations that have dissolution profiles
within the specifications are bioequivalent, or minimally, all formulations should be bioequivalent to the pivotal (bio) batch. In this chapter, various methods for setting dissolution specifications which serve as predictors of *in vivo* performance are reviewed. First,
however, it is important to review the significant characteristics when choosing a method
for setting dissolution specifications.

2. CHARACTERISTICS FOR A DISSOLUTION SPECIFICATION METHOD

2.1. Discriminating Dissolution System

The dissolution system is not as important as the resulting dissolution profiles. If the
system is nondiscriminating, this may create a problem when developing the specifications.
Formulations may appear similar *in vitro*, yet result in bioinequivalence *in vivo*. Shah and
Williams demonstrate this point with quinidine gluconate.[5] Two quinidine gluconate formulations produced similar *in vitro* profiles when the dissolution test was performed in an acidic
medium. However, the formulations produced different plasma concentration-time curves and
were in fact bioinequivalent. When the dissolution medium was changed to pH 5.4 acetate
buffer, the dissolution profiles became different; thus more accurately representing the *in vivo*
process. It is best to have a discriminating system so that if a formulation is bioinequivalent *in
vivo*, this may be identified *in vitro*. Dissolution testing may need to be performed under a variety of conditions: changing medium (e.g., pH, osmolarity, surfactant), mixing speed, or apparatus (e.g., USP Apparatus I, II, and III). If the system is over-discriminating, the resultant
dissolution specifications will be wider when encompassing all bioequivalent formulations.

2.2. Incorporation of *in Vivo* Data

In order for the dissolution system to serve as a surrogate, the dissolution data must
be correlated to the *in vivo* data. There are numerous chapters in this book that describe *in
vitro-in vivo* correlations (IVIVC) methods and this chapter will not expand on this topic
further. However, it is important to emphasize that only with a defined IVIVC can one be
confident that formulations within the dissolution specifications will most likely result in
bioequivalent products.

2.3. Including Variability

Ideally, variability should be included in setting dissolution specifications. This includes the variability in the dissolution data; however, this is usually small. More importantly, this includes variability in the *in vivo* data, since this is usually larger than the *in vitro* variability. Since each drug has a certain amount of variability when administered to patients, it would be beneficial to quantify this variability and include it when setting dissolution specifications. For example, a Level A correlation may be determined for 10 subjects. The mean and variance can then be calculated for the parameters of the correlation (i.e., slope and intercept). This intersubject variability can then be incorporated into the method used for setting the dissolution specifications. In this way the specifications would be wider for those drugs with a larger amount of variability and narrower for those drugs with a smaller amount of variability.

2.4. Plasma Concentration-Time Curves

A method for setting dissolution specifications should demonstrate in some fashion what these dissolution profiles would look like as plasma concentration-time curves. If the specifications are being used to predict *in vivo* performance, it is necessary to see the *in vivo* profiles. This may be done by administering the upper and lower formulations in a bioavailability study or perhaps more efficiently by convoluting the upper and lower limits to achieve the associated plasma concentration-time curves. Additionally, if some form of intersubject/intertablet variability can be incorporated into the plasma concentration-time profiles, bioequivalence using the typical bioequivalence matrices can be evaluated.

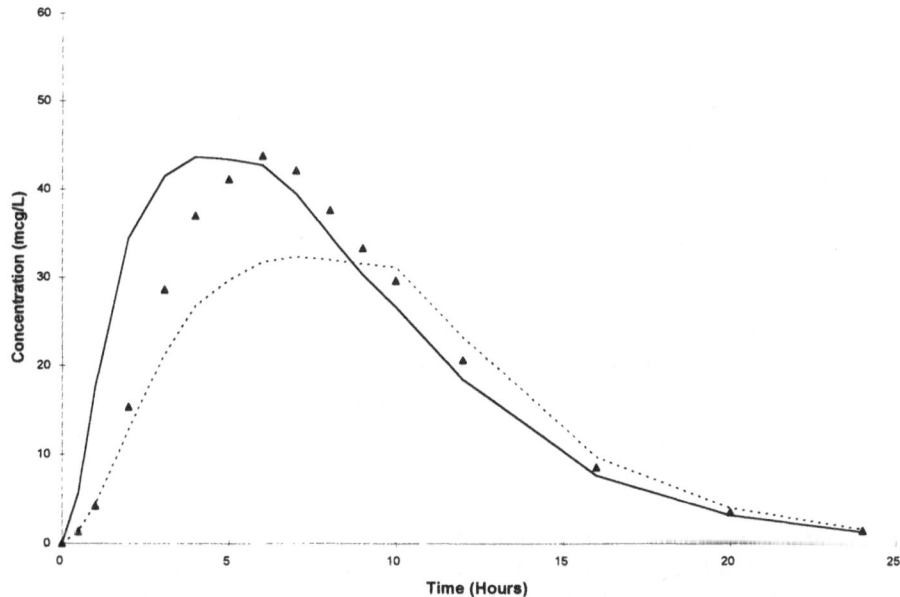

Figure 1. Plasma concentration-time profiles convoluted from dissolution specifications. The solid and broken lines represent the upper and lower dissolution specifications, respectively. The triangles represent the pivotal batch.

1.	Manufacture a range of formulations which produce a range of dissolution profiles.
2.	Administer these formulations in a bioavailability study and test for bioequivalence.
3.	The dissolution range that results in bioequivalent formulations would be the specifications.

Figure 2. Steps for setting dissolution specifications using the set range method.

An example of typical plasma concentration-time profiles convoluted from dissolution specifications can be seen in Figure 1. The upper and lower limits are represented by the solid and broken line, respectively. The triangles represent the pivotal batch.

3. METHODS

It is suggested in the FDA guidance that a minimum of three time points be required to set dissolution specifications.[4] Ideally, these time points should cover the early, middle and late segments of the dissolution profile. The final time point should occur when at least 80% of the label claim has been released. If the maximum amount released is less than 80% of the label claim, then the last time point should be the time where the plateau in the profile has been achieved. This section of this chapter outlines several methods which may be used to develop dissolution specifications.

3.1. Set Range Method

The FDA guidance outlines this method when an IVIVC does not exist.[4] This method does not incorporate *in vivo* data or intersubject variability. The steps are outlined in Figure 2. In this method, the upper and lower limits are not shown as plasma concentration profiles, thus there is no assurance that these limits result in bioequivalent plasma profiles unless a separate bioavailability study is performed.

3.2. Empirical Method

The empirical method is described in Figure 3. The empirical method includes *in vivo* data, intersubject variability, and results in plasma concentration-time profiles. This method uses several formulations; the more formulations included, the more confident one can be that the specifications are appropriate. This method is costly both in time and money. Quite often this method turns out to be a retrospective analysis of several inde-

| 1. | Perform dissolution testing on the clinical lots and mean the profile. |
| 2. | The maximum suggested range at any dissolution time point is ± 10% of the label claim deviation from the mean dissolution profile. (Deviations from the + 10% criteria may be accepted by the FDA provided that the range at any time point does not exceed 25%.) |

Figure 3. Steps for setting dissolution specifications using the empirical method.

---

1. Establish an IVIVC.
2. Calculate the 90% confidence intervals around the IVIVC parameters.
3. Using the IVIVC and the midpoint from the dissolution specifications, calculate the mean amount absorbed/released.
4. Calculate the upper and lower dissolution boundary using the mean in vivo data and the upper and lower 90% confidence intervals for the parameters.

Figure 4. Steps for setting dissolution specifications using the 90% confidence interval method.

pendent bioequivalence studies. Thus, this analysis may lead to an additional bioequivalence study to ensure that the two outside dissolution profiles (upper and lower) are bioequivalent.

3.3. 90% Confidence Interval

The 90% confidence method includes *in vivo* data and intersubject variability. It does not predict plasma concentration-time profiles, thus it is unable to test if the specifications result in bioequivalent formulations. The steps are outlined in Figure 4.

3.4. 95% Confidence Method

Dissolution specifications are justified using 95% confidence intervals around the the mean plasma concentration-time profile.[7] This profile was obtained following the administration of the pivotal batch in a bioavailability study. The 95% confidence method includes *in vivo* data, intersubject variability, and results in plasma concentration-time profiles. The steps are outlined in Figure 5.

Although this method predicts plasma concentration-time profiles for the upper and lower dissolution limits, it does not require the limits to be tested for bioequivalence.

3.5. Artificial Neural Network Method

Artificial neural network (ANN) had been described in two other chapters within this book. These ANN methods include *in vitro* and *in vivo* data. They can provide plasma concentration-time profiles for given *in vitro* profiles. The steps are outlined in Figure 6.

---

1. Establish an IVIVC.
2. Calculate the in vivo release using the working dissolution limits and the IVIVC.
3. Convolute the in vivo release to generate the upper and lower plasma concentration profiles.
4. Administer the pivotal batch in vivo.
5. Construct 95% confidence intervals around the observed data.
6. If the working dissolution specifications are within the 95% confidence intervals, then any formulation with a dissolution profile that falls within the dissolution specifications would be bioequivalent to the pivotal batch.

Figure 5. Steps for setting dissolution specifications using the 95% confidence interval method.

1. Determine appropriate data set to train the ANN.
2. Determine an optimal ANN configuration (i.e., type of ANN, training paradigm, number of nodes, etc.)
3. Train the ANN (Take steps to ensure memorization of the ANN is not occurring).
4. Validate the ANN with a data set tat was not used for training.
5. Predict in vivo plasma profile with the ANN using the proposed dissolution specifications.

Figure 6. Steps for setting dissolution specifications using the ANN method

3.6. Convolution-Based Method

In the convolution-based method the *in vivo* release/absorption is modeled as a function of the *in vitro* release.[8] Thus, this method requires an oral solution, intravenous, or immediate release reference dose. The convolution-based method uses the *in vitro* dissolution specifications, an established IVIVC, intersubject variability, and results in plasma concentration-time profiles. The steps for this method can be found in Figure 7.

The limits may be tested for bioequivalence. Intersubject variability may be added to the reference dose parameters and/or the IVIVC parameters.

4. ADDITIONAL CONSIDERATIONS

As seen, it is valuable if the method for setting the dissolution specifications can predict the plasma curves for the upper and lower dissolution limits. These plasma curves, which represent the mean plasma curves for formulations with the upper and lower dissolution limits, can then be compared to each other or the pivotal batch. Expanding the prediction to include intersubject variability would provide a more realistic representation of the *in vivo* performance: Bioequivalence could then be tested for the formulations representing the upper, lower, and/or pivotal batch dissolution profiles. In Figure 8, a population simulation for 12 subjects has been performed for the upper dissolution limit. The solid and broken line represent the mean plasma profiles for the upper and lower dissolution specifications, respectively. As shown in this figure, intersubject variability can be an important factor when evaluating the dissolution specifications.

1. Establish an IVIVC.
2. Convolute the dissolution limits and the pivotal batch.
3. Test for bioequivalence between the upper and lower limits.
4. If the limits are bioinequivalent, test for bioequivalence between each limit and the pivotal batch.
5. If the limits are bioequivalent, accept the dissolution specifications and consider expanding the limits.

Figure 7. Steps for setting dissolution specifications using the convolution based method.

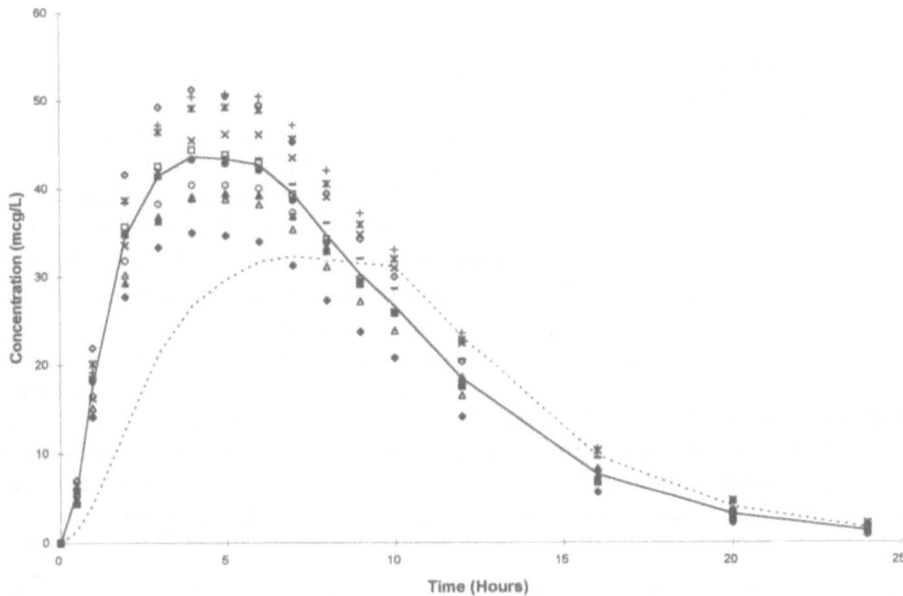

Figure 8. Population simulation for the formulation with the upper specification dissolution profile. The solid and broken lines represent the mean plasma concentration-time profiles for the mean upper and lower dissolution profiles, respectively. The individual symbols represent the simulated plasma concentrations for 12 subjects.

Table 1. Advantages and disadvantages for the dissolution specifications setting methods

Methods	Advantages	Disadvantages
Empirical	1) Includes *in vivo* data 2) Includes variability 3) Includes plasma profiles	1) Costly in time and money 2) Potential to be a retrospective analysis 3) May require an additional bioequivalence study
Set Range	1) Inexpensive	1) Does not include *in vivo* data 2) Does not include variability 3) Does not predict the plasma profiles for the specifications
90% Confidence Interval	1) Includes *in vivo* data 2) Includes variability	1) Does not predict the plasma profiles for the specifications 2) Does not prove bioequivalence between the upper and lower limits
95% Confidence Interval	1) Includes *in vivo* data 2) Includes variability	1) Does not predict the plasma profiles for the specifications 2) Does not prove bioequivalence between the upper and lower limits 3) Requires both deconvolution and convolution programs
Artificial Neural Network	1) Includes *in vivo* data 2) Predicts plasma profiles 3) May test for bioequivalence between the upper and lower limits	1) Unable to have a priori structure for inter- and intra-subject variability 2) Specification of the types of inputs, ANN configuration variables, and format of the input-output relationship can lead to an exorbitant number of possible ANN 3) The range of input values used to train a network becomes an important consideration in the successful prediction from an ANN. Desirable to predict from inputs well within range used to train; dissolution specifications are often at either end of the ranges. 4) Requires training in ANN
Convolution-Based	1) Includes *in vivo* data 2) Can include variability 3) Predicts plasma profiles 4) May test for bioequivalence between the upper and lower limits	1) Requires both deconvolution and convolution programs

5. SUMMARY

Dissolution specifications are used for quality assurance and may also serve as a surrogate for *in vivo* bioavailability. These limits can guide formulation development and eliminate the need for bioavailability studies for scale up and post approval changes. Several methods for setting dissolution specifications have been reviewed in this chapter. A summary of the advantages and disadvantages for each method can be found in Table 1. When choosing a method for setting dissolution specifications, it is important to 1) have a discriminating dissolution system, 2) incorporate *in vivo* data, 3) include intersubject variability, and 4) predict plasma concentration-time profiles. Predicting plasma concentration curves allows one to see how the change in formulation or dissolution limits perform *in vivo*. Dissolution specifications should be set so that all formulations that have dissolution profiles within the limits of the specifications are bioequivalent. This can be assured if the boundaries are tested for bioequivalence. Minimally, the formulations that have dissolution profiles within the limits of the specifications should be bioequivalent to the pivotal batch. A population prediction of the plasma concentration-time profiles for the upper and lower limit would incorporate the true intersubject variability for the formulation.

REFERENCES

1. The United States Pharmacopeia 23 and The National Formulary 18, The United States Pharmacopeial Convention, Inc. Supp. 1, 1995:2449–2450.
2. Guidance for Industry. Immediate Release Solid Oral Dosage Forms. Scale-Up and Post-Approval Changes: Chemistry, Manufacturing and Controls, In Vitro Dissolution Testing, and In Vivo Bioequivalence Documentation. Center for Drug Evaluation and Research, November 1995.
3. Guidance for Industry. Modified Release Solid Oral Dosage Forms. Scale-Up and Post-Approval Changes: Chemistry, Manufacturing and Controls, In Vitro Dissolution Testing, and In Vivo Bioequivalence Documentation. Center for Drug Evaluation and Research, May 1996.
4. Guidance for Industry. Extended Release Solid Oral Dosage Forms Development, Evaluation and Application of In Vitro/In Vivo Correlations. Center for Drug Evaluation and Research, July 1,1996.
5. Shah VP and Williams RL. *In vivo* and *In vitro* correlations: scientific and regulatory perspectives, in Generics and Bioequivalence. CRC Press, Inc., Boca Raton, Florida, 1994. 108–109.
6. In Vitro/In Vivo Correlations for Extended Release Oral Dosage Forms. Pharmacopeial Forum. July-August 1988:4160.
7. Leeson, L.J., *In vitro-In vivo* correlations. Drug Information Journal. 1995;29:903–915.
8. Gillespie WR. Convolution-based approaches for *In vivo-In vitro* correlation modeling. In Vitro-In Vivo Relationship Workshop. Baltimore, MD. September 1996.

14

REVIEW OF METHODOLOGIES FOR THE COMPARISON OF DISSOLUTION PROFILE DATA

Tom O'Hara,[1] Adrian Dunne,[2] Audrey Kinahan,[1] Sean Cunningham,[1] Paul Stark,[1] and John Devane[1]

IVIVR Co-operative Working Group
[1] Élan Corporation plc
Monksland, Athlone
Ireland
[2] Department of Statistics
University College Dublin
Dublin 4, Ireland

1. INTRODUCTION

Methods of comparing *in vitro* dissolution profile data are of interest in the development of Extended Release dosage formulations. One is usually interested in determining if the dissolution profile for the Test formulation is significantly different from that of the Reference formulation. Some methods used to compare dissolution profiles include:

i. Graphical comparisons i.e. a plot of the mean profiles of the Test and Reference Formulations with error bars at each time point,

ii. Statistical methods which compare the dissolution profiles at each time point e.g. perform a t-test comparing the Test and Reference means separately at each time point,

iii. Statistical methods which compare the entire dissolution profile for Test and Reference formulations, and

iv. Mathematical methods which compare the entire dissolution profile for Test and Reference formulations, without taking the variability or covariance structure in the data into account.

Some methods in categories (iii), and (iv) are investigated by means of simulation experiments in this chapter.

In Vitro–in Vivo Correlations, edited by Young *et al.*
Plenum Press, New York, 1997

2. NOTATION AND COINCIDENCE HYPOTHESIS

Dissolution data are collected at a number of time points denoted by t_k where $k = 1$, $2,....,p$. Let $Y_{ij}(t_k)$ represent the measured fraction of drug dissolved from dosage unit i at time t_k from Formulation j (j=1 for the Reference formulation and j=2 for the Test formulation). Consider the dissolution data from the ith dosage unit of the jth formulation to be a p dimensional random vector[1]

$$\mathbf{Y}_{ij} = (Y_{ij}(t_1), Y_{ij}(t_2),... Y_{ij}(t_p))'$$

The Reference and Test dissolution data have mean vectors μ_1 and μ_2 respectively and covariance matrices Σ_1 and Σ_2 respectively. The hypothesis to be tested (henceforth referred to as the coincidence hypothesis) is H_0: $\mu_1 = \mu_2$ i.e. the mean dissolution profile across time for the Reference formulation is coincident with the mean dissolution profile for the Test formulation.

3. THE METHODS

Three methods for testing the coincidence hypothesis are considered. These are:

    i. Multivariate Analysis of Variance (MANOVA)[1],
    ii. Linear Mixed Effects model[2,3], and
    iii. A method adopted by the SUPAC IR guidelines[4,5] known as the f_2 equation.

With the MANOVA method it is assumed that $\Sigma_1 = \Sigma_2$ and that the data are normally distributed. No assumptions are made about the structure of the covariance matrices Σ_1 and Σ_2. The coincidence hypothesis for the MANOVA method is tested using the Hotelling's T^2 test[1]. With the Linear Mixed Effects model method, it is assumed that all observations made on the same dosage unit are equally correlated with one another, hence the covariance matrices Σ_1 and Σ_2 have a compound symmetric structure. The Linear Mixed Effects model method is more powerful than the MANOVA method when the dissolution data have a compound symmetric covariance structure. The test statistics for both the MANOVA and Linear Mixed Effects model methods are proportional to the Mahalanobis distance[1] which is a measure of the distance between the Reference and Test formulation mean dissolution profile vectors ($\overline{Y}_1$ and $\overline{Y}_2$ respectively) taking the variability and covariance structure into account. It is defined mathematically as:

$$M = (\overline{Y}_1 - \overline{Y}_2)' \underline{\Sigma}^{-1} (\overline{Y}_1 - \overline{Y}_2)$$

where $\overline{Y}_1$ and $\overline{Y}_2$ are the mean dissolution profile vectors for the Reference and Test formulations respectively and $\underline{\Sigma}$ is the pooled Covariance matrix. The f_2 equation is defined as:

$$f_2 = 50 \times Log_{10} \left\{ \left[1 + \frac{1}{n} \sum_{k=1}^{n} w_k (\overline{Y}_{1k} - \overline{Y}_{2k})^2 \right]^{-0.5} \times 100 \right\}$$

where $\overline{Y}_{1k}$ and $\overline{Y}_{2k}$ are the mean fraction of drug dissolved at time k for the Reference and Test formulations respectively, and w_k is an optional weight factor (the w_k are all set equal to 1 in the simulation experiments). The mean profiles are assumed to differ by no more than 15% at any time point if f_2 lies between 50 and 100. Unlike the MANOVA and Linear Mixed Effects model methods, the f_2 equation neither takes account of the variability nor the correlation between time points in the dissolution data. This is considered to be a major disadvantage of this method.

4. SIMULATION EXPERIMENTS

The methods described above were compared by means of simulation experiments. Multivariate normal data were generated at observation times 0.5, 1, 2, 3, 4, 5, 6, 7, 8, 10, 12, 24 hours with the vector μ_1 set at (0.087, 0.171, 0.332, 0.494, 0.609, 0.706, 0.783, 0.835, 0.875, 0.925, 0.952, 0.999)' and each element of μ_2 calculated using the formula[6]

$$\mu_{2i} = 1 - (1 - \mu_{1i})^r$$

where r (more correctly $ln(r)$) is a measure of 'distance' between the two vectors. Note that when $r=1$ ($ln(r)=0$) the two vectors are coincident. Reference and Test formulation dissolution data were generated with a random 'between tablet' term which was common to all observations on the same tablet and a 'within tablet' error term which was unique to each observation. Two simulation experiments were carried out. In the first simulation experiment, the standard deviation of the 'between tablet' term was 0.02 and the standard deviation of the 'within tablet' error term was 0.015. In the second simulation experiment, the standard deviation of the 'between tablet' term was 0.01 and the standard deviation of the 'within tablet' error term was 0.0075. There were 12 dosage units for the Reference and Test formulations. For each of 57 values of r (equally spaced on a logarithmic scale) in the range 0.497 to 2.014, one thousand sets of data were generated for each simulation experiment. The values used for the various parameters in the first simulation experiment were in fact estimates derived from a set of real data and were used so that the simulated data sets would be as realistic as possible. The reason for carrying out the second simulation experiment was to determine how the three methods would perform when the variability in the data was lower than what one would expect to see in practice. Each set of data was analysed by each of the three methods:

    a. Hotelling's T^2 test of the hypothesis $\mu_1=\mu_2$.
    b. Linear Mixed Effects model analysis testing the hypothesis $\mu_1=\mu_2$.
    c. The f_2 equation to determine if the computed value of f_2 lies between 50 and 100.

A nominal 5% type I error rate was used for methods (a) and (b).

For each value of r and for each method of analysis the proportion of the 1000 data sets for which the hypothesis being tested was not rejected was noted in each of the simulation experiments. These proportions are plotted against ln(r) in the operating characteristic (OC) curves shown in figures 1 and 2.

The data were generated using SAS and the analyses conducted using the GLM and MIXED procedures in SAS.

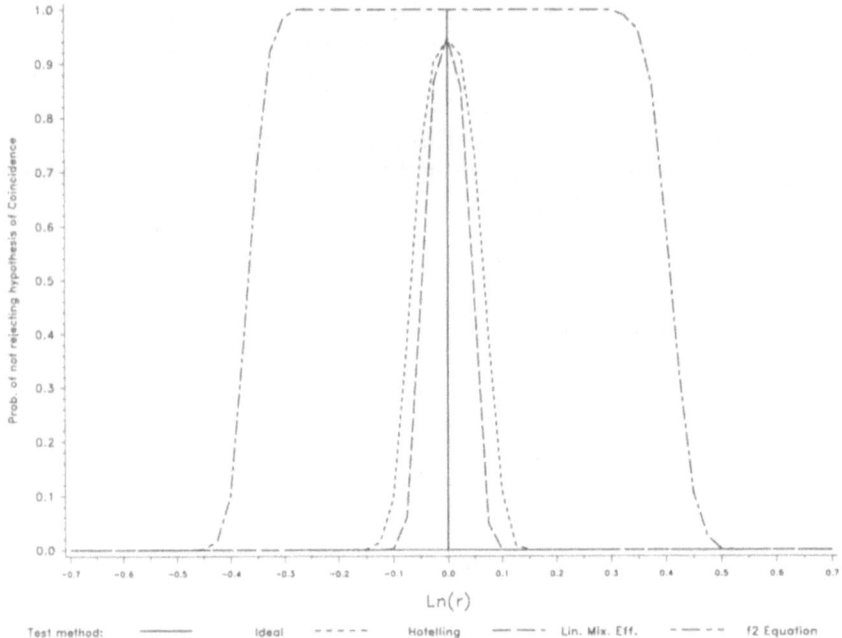

Figure 1. The operating characteristic curves for the different methods compared in the first simulation experiment. Variability in simulated data estimated from real data.

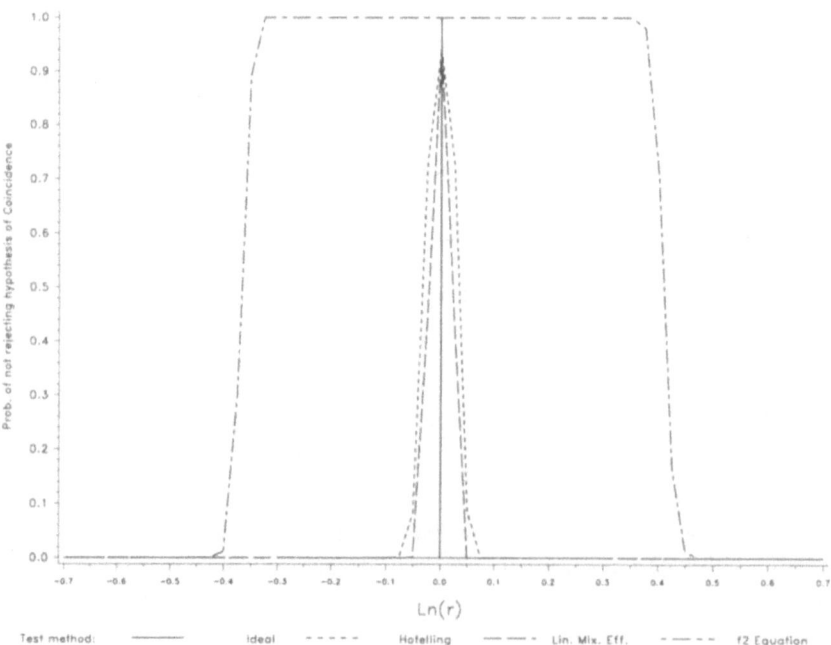

Figure 2. The operating characteristic curves for the different methods compared in the second simulation experiment. Variability in simulated data lower than the variability seen in real data.

5. DISCUSSION

The operating characteristic (OC) curves for the three methods are displayed in figures 1 and 2 for both simulation experiments. The OC curve for a method is the probability of not rejecting the hypothesis of coincidence plotted against $ln(r)$. The ideal method will have a 100% chance (i.e. probability equal to 1) of not rejecting the hypothesis of coincidence when it is true i.e. when $ln(r)=0$. When the hypothesis of coincidence is not true i.e. when $ln(r)\neq0$, the ideal method will have 0% chance (i.e. probability equal to 0) of not rejecting the hypothesis of coincidence. Unfortunately, the ideal method does not exist because of variability, therefore one must consider the Power and Type I error of a candidate method. The Power is defined as the probability of rejecting the hypothesis of coincidence when it is not true i.e. when $ln(r)\neq0$. The Type I error of a method is the probability of rejecting the hypothesis of coincidence when it is true i.e. when $ln(r)=0$. The Power of a good method should be high when the hypothesis of coincidence is not true and the Type I error should not exceed 5% when the hypothesis of coincidence is true. In other words, the OC curve of a good candidate method should lie as close as possible to that of the ideal method.

Figure 1 shows that Hotelling's T^2 and the Linear Mixed Effects model methods both have high Power when $ln(r)\neq0$ and the Type 1 error is about 5% in both cases. The Linear Mixed Effects model method is more powerful than Hotelling's T^2 when the correlation structure of the data is compound symmetric. The f_2 equation concludes that the mean dissolution profiles for the Reference and Test formulations are coincident even when $ln(r)\neq0$, $(-0.4 \leq ln(r) \leq 0.5)$. These values of $ln(r)$ correspond to maximum absolute differences between Reference and Test mean profiles of 14%-16% which roughly confirms that an f_2 value between 50 and 100 allows for differences of up to a maximum of 15% between Reference and Test mean profiles. When the variability in the data is even less than what would be expected (see figure 2), Hotelling's T^2 and the Linear Mixed Effects model methods are even more powerful when the hypothesis of coincidence in not true i.e. their OC curves lie closer to that of the ideal method than in figure 1 where the variability in the data was higher indicating that both of these methods take the variability into account when determining how far apart the mean Reference and Test dissolution profiles are. This is not the case with the f_2 equation because its OC curve is more or less the same in figures 1 and 2 indicating that this method is not taking the variability into account when evaluating the distance between the mean Reference and Test dissolution profiles i.e. the Power of the method is not affected by reducing the variability in the data. Therefore, the simulation experiments demonstrate the advantages of Hotelling's T^2 and the Linear Mixed Effects model methods over the f_2 equation.

6. REFERENCES

1. Morrison , D.F. (1978) *Multivariate Statistical Methods*, McGraw-Hill, Singapore.
2. Crowder, M.J. and Hand, D.J. (1990) *Analysis of Repeated Measures*. Chapman and Hall, London.
3. Searle, S.R., Casella, G. and McCulloch, C.E. (1992) *Variance Components*. Wiley, New York.
4. SUPAC Guidance for Industry, Immediate Release Solid Oral Dosage Forms. Center for Drug Evaluation and Research (CDER), November, 1995. Section VII, Page 23.
5. Moore, J.W. and Flanner, H.H. Mathematical Comparison of Dissolution Profiles. *Pharmaceutical Technology*, June 1996, 64–74.
6. Dunne, A. Approaches to IVIVR Modelling ans Statistical Analysis. Talk presented at IVIVR Workshop, Baltimore, Maryland, 4–6 September 1996.

15

ASSESSING WHETHER CONTROLLED RELEASE PRODUCTS WITH DIFFERING *IN VITRO* DISSOLUTION RATES HAVE THE SAME *IN VIVO–IN VITRO* RELATIONSHIP

Guoliang Cao and Charles Locke

Abbott Laboratories
100 Abbott Park Rd.
Abbott Park, Illinois 60064-3500

ABSTRACT

In order to demonstrate a complete *in vivo-in vitro* correlation (IVIVC) for a controlled release (CR) formulation, it is necessary that the relationship between *in vivo* percent absorbed and *in vitro* percent dissolved be the same whether the dissolution rate is as targeted or whether the dissolution rate is faster or slower than intended for the marketed product. This is essential if the IVIVC will be used in the future for a decision on the acceptability of a change in the formulation. Suppose that a four period cross-over study is performed, with each subject receiving the to-be-marketed CR product, a product with a faster *in vitro* dissolution rate, a product with a slower dissolution rate, and an intravenous dose (or perhaps an oral dose of an immediate release formulation). Using an appropriate deconvolution method, for each subject and each CR product the percent absorbed is estimated for each time post dose for which the *in vitro* percent dissolved is measured. We present a method for the assessment of whether the relationship between *in vivo* percent absorbed and *in vitro* percent dissolved is the same for the three CR products

Statistical analysis of the data should be done taking into account the dependence of the data points from the same subject. (Each subject contributes an estimate of percent absorbed for several time points for each CR product.) The use of mixed effect models is discussed. A data point consists of the *in vitro* percent dissolved at a given time for one of the CR products and the measure of percent absorbed for a given subject at the same time post dose for this CR product. The subjects of the study are viewed as a sample from a large population. The parameters of greatest interest define the relationship (e.g. a linear relationship) between the population central value for percent absorbed and the *in vitro* percent dissolved. Analyses are performed to address the question of whether the values of the parameters that define the relationship between percent absorbed central value and *in vitro* percent dissolved are the same for the three CR products (i.e. to address the question of whether the IVIVC is the same for the three CR products).

1. INTRODUCTION

Establishing a correlation between the *in vivo* plasma concentration profile and the *in vitro* dissolution profile of a controlled release (CR) formulation has been of great interest for a number of years. For a new CR formulation, it is now expected that an attempt will be made to find an *in vivo-in vitro* correlation (IVIVC), although this will not be successful for every product. Level A, B, and C correlations have been described in USP XXIII Chapter <1088>. The most useful of these is a Level A correlation, which is described as a point-to-point correlation in which the *in vivo* percent absorbed curve is compared to the *in vitro* percent dissolved curve. In the ideal case, *in vivo* percent absorbed is actually equal to *in vitro* percent dissolved. However, more generally a Level A correlation may be said to have been found if a function (a predictive model) is identified that describes well the relationship between *in vivo* percent absorbed and *in vitro* percent dissolved. In this article we are concerned only with Level A correlation. In order to be useful for predicting *in vivo* bioavailability characteristics when changes are made in the future or to assist in setting specification ranges for dissolution tests, the same correlation (same function or predictive model) should hold for a reasonably wide range of values of the formulation parameters that determine the release characteristics.

A good study design by which to address these issues is a four period cross-over design in which each subject receives the to-be-marketed CR product, a product with a faster vitro dissolution rate, a product with a slower dissolution rate, and an intravenous dose (or perhaps an oral dose of an immediate release formulation). Using an appropriate deconvolution method and the data from the intravenous dose, for each subject and each controlled release product, the percent absorbed is estimated for each time post dose for which *in vitro* percent dissolved is measured. The objective is to show that the same correlation (same predictive model) holds for all three CR products. It must also be shown that the correlation model is a good enough predictor of in vivio bioavailability for all three CR products. The purpose of this article is to present an approach for the assessment of whether the same IVIVC holds for all three CR products.

A notable feature of the data from an *in vivo* cross-over study as described above is that the relatively large number of percent absorbed determinations from the same subject are correlated (dependent). The more common methods of statistical analysis are applicable when the data points represent independent observations, but such is not the case here. A more complicated approach that takes into account the dependence of the data points from a subject while allowing for a distinction between CR products should be utilized.

2. MIXED EFFECTS MODEL

In this section, we discuss the use of mixed effects models for IVIVC investigations. We assume that the variability of *in vitro* percent dissolved measurements among individual dosing units (e.g. tablets) is very small so that the mean percent dissolved at the various time points can be assumed to define the true *in vitro* dissolution profile. The percent dissolved measurement will be labelled as X. The percent absorbed measurement will be labelled as Y. The model will account for both variability among subjects and variability within a subject.

We denote the underlying relationship between *in vivo* percent absorbed and *in vitro* percent dissolved as $Y = f(X, \theta)$ where f is some linear or nonlinear function and θ is the vector of parameters for the function. For a linear relationship, $Y = \alpha + \beta X$ for values of X such that $\alpha + \beta X$ is between 0 and 100%, with θ consisting of the two components α and β.

For the data of the cross-over study, and with variability incorporated, the model can be described as follows:

$$Y_{ikt} = f(X_{kt}, \theta_k, \varsigma_{ik}) + \epsilon_{ikt} \tag{1}$$

where Y_{ikt} denotes percent absorbed from the i-th subject and k-th product at the t-th time point, where X_{kt} is percent dissolved for the k-th product at the t-th time point, where θ_k and ς_{ik} are vectors that define the parameters of the relationship and the variability among subjects with respect to those parameters for the k-th product, and where ϵ_{ikt} represents within subject randomness. The subjects of the study are viewed as a sample from a large population. The vector θ_k consists of the population central values of the parameters of the function f that describes the IVIVC, and the vector ς_{ik} consists of the deviations from the parameter central values for the i-th subject's IVIVC. In the case of a linear relationship,

$$Y_{ikt} = [\alpha_k + \varsigma_{1ik}] + [\beta_k + \varsigma_{2ik}]X_{kt} + \epsilon_{ikt}.$$

That is, for the i-th subject, the intercept and slope of the line are $\alpha_k + \varsigma_{1ik}$ and $\beta_k + \varsigma_{2ik}$, respectively. The components of θ_1, θ_2, and θ_3 are fixed effects, and the components of ς_{i1}, ς_{i2}, and ς_{i3} are random effects. Hence, the presence of both fixed and random effects is why this is called a mixed effects model. Thus, ς_{ik} represents the variability among subjects and also accounts for the dependence of the data points from the i-th subject.

The assumptions and structure for the random components, including ϵ_{ikt}, must be specified. An important simplification is to assume that ς_{i1}, ς_{i2}, and ς_{i3} are the same. In the examples that are given in the next section, we made this assumption so that for a linear relationship (1) becomes

$$Y_{ikt} = [\alpha_k + \varsigma_{1i}] + [\beta_k + \varsigma_{2i}]X_{kt} + \epsilon_{ikt}. \tag{2}$$

Observations from different subjects are assumed to be independent. We assume that the components of ς_i, which we have denoted as ς_{1i} and ς_{2i} in the case of a linear relationship, have a multivariate normal probability distribution with each component having a mean of 0. Another way to state the random effects aspect of the model is that the parameters of the function f have a multivariate normal distribution with θ_k as the mean vecfor for the k-th CR product. If there is confidence that the parameters are independent or have low correlations, it could be assumed that the covariances of the parameters (i.e. the covariances of the components of ς_i) are all 0. Although we have assumed a multivariate normal distribution, one could adopt other distributional assumptions such as lognormal distributions. Finally, we assume that ϵ_{ikt} has a normal distribution with mean 0 and the same variance for all data points, and that all these within subject random variations are independent of each other and independent of all the random effects represented by the ςs.

Statistical analysis of Model (1) has been studied by Beal and Sheiner (1992), Vonesh and Carter (1992), Wolfinger (1992), Wolfinger and O'Connell (1993), and Davidian and Giltinan (1995). Although much of the available software is not yet to the point of being user friendly and commercially marketed, software for mixed effects modeling is becoming widely available. A description of all the products available is beyond the scope of this article. The most commonly used ones are the NONMEM package (Beal and Sheiner, 1992), which is widely used with pharmacokinetic data, and SAS Procedure MIXED (Wolfinger, 1992), which may be used if $f(X,\theta,\varsigma)$ is a linear function of θ.

A likelihood ratio test is very useful for testing a statistical hypothesis in the setting of a mixed effects model. In our case, we want to compare the full model in (1), in which separate IVIVC's are assumed for the three CR products, to a reduced model in which the same IVIVC is assumed for all three CR products. This simpler model is

$$Y_{ikt} = f(X_{kt}, \theta, \varsigma_i) + \epsilon_{ikt} \tag{3}$$

In the reduced model, the parameters of the function f that describes the IVIVC are the same for all three CR products. In the case of a linear relationship, the intercept and slope central values are the same for all three CR products under the reduced model. The hypothesis that the simpler model is true (same IVIVC for all three CR products) can be tested against the alternative that the full model is required (differences among the three IVIVC's) by using the likelihood ratio test. If the objective function of the goodness of fit of the data to a model is -2log(likelihood), subtraction of the objective function value for the full model from the objective function value for the reduced model may be used to perform the likelihood ratio test. If the number of data points is large enough, the difference in objective functions has an approximate chi-square distribution with q degrees of freedom, where q is the difference in the number of parameters of the full and reduced models. A large value of the test statistic is evidence that the IVIVC is not the same for all three CR products.

3. SIMULATION ANALYSES

3.1. Data

In vitro percent dissolved data are given in Figure 1 for three Formulations A, B and C at times 1, 3, 5, 9, 12, 18 and 24 hours. For a simple simulation illustration, a linear IVIVC is assumed. Two sets of *in vivo* simulated data are summarized here. For each simulation, there were 16 subjects. The percent dissolved Y_{ikt} for i-th subject and k-th formulation at time t after a dose is given by Equation (2). X_{kt} was percent dissolve for k-th formulation at t-th time t. The error term ϵ_{ikt} was assumed to be normally distributed with mean 0 and variance σ^2 and independent of all other error terms and independent of the random components of ς_{1i} and ς_{12}. ς_{1i} was assumed to be normally distributed with mean 0 an standard deviation 0.316 and ς_{12} was assumed to be normally distributed with mean 0 an standard deviation 0.100 and independent of ς_{1i}. It was assumed that the standard deviation of residual error σ was 14.1. Simulation I was intended to show a very similar IVIVC among three formulations. α_k was chosen to be 0.01, 0.02 and 0.03 for Formulations A, B and C respectively. β_k was 0.98, 1.00 and 1.02 for Formulations A, B and C respectively. Simulation II was designed to show a different IVIVC among three formulations. α_k was assumed to be 0.01, 5.00 and 10.00 for Formulations A, B and C respectively. β_k was 0.95, 1.00 and 1.05 for Formulations A, B and C respectively.

Simulated data were obtained using SAS software Version 6.11 (SAS Institute, 1990). Independent samples for ς_{1i}, ς_{12}, ϵ_{ikt} were obtained seperately. Simulated Y_{ikt} were obtained using Equation (2). Negative values of percent dissolved that were generated

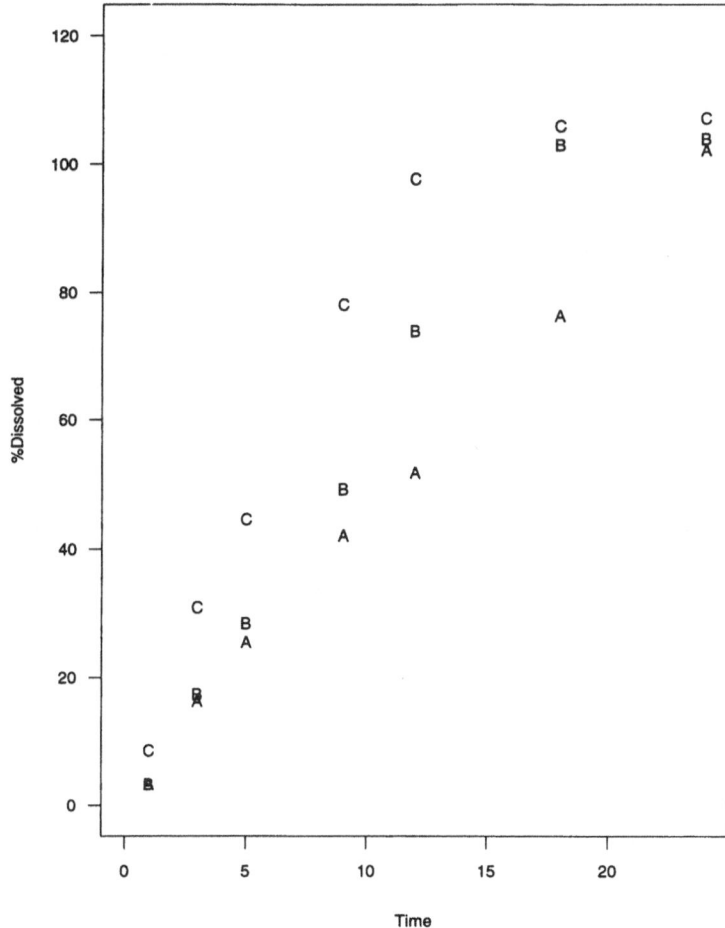

Figure 1. Mean percent dissolved versus time.

were changed to 0. Figure 2 displays the simulated mean percent dissolved data for both cases.

3.2. Analyses

For each simmulated data, the NONMEM package was used to analyze the data. Stage 1: Assuming different formulations have different IVIVC's, that is, α_k and β_k were different for each k, a full model was estalished and an objective function was obtained. Stage 2: Assuming different formulations have the same correlation, that is, α_k and β_k were the same for each k, a reduced model and its objective function were obtained.

3.3. Results

Table 1 displays the results of Simulation I and Table 2 shows the results of Simulation II. The difference of objective functions from full and reduced models in Simulation I

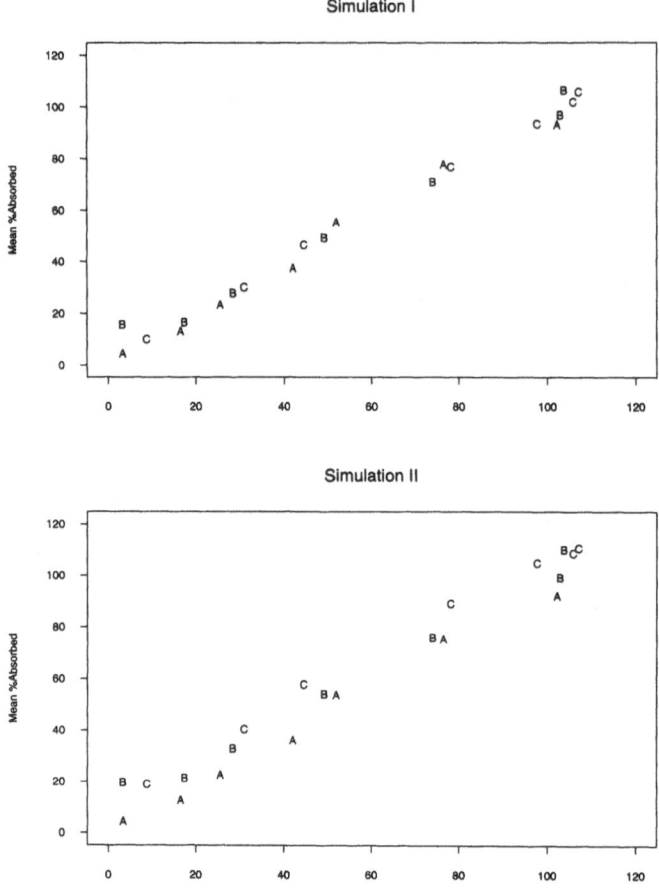

Figure 2. Mean *in vivo* percent absorbed versus *in vitro* percent dissolved.

was 3.257 which gives the p-value 0.52 for the chi-square distribution with four degree of freedom. There is no statistically significant difference among the IVIVC's from three formulations. The common slope 0.94 was slightly lower due to the big intercept 2.290. The difference of objective functions from full and reduced models in Simulation II was 43.969 which is corresponding to the p-value less than 0.001. In latter case, the correlations are statistically significantly different, so the correlation in Simulation II is not "real".

Table 1. Objective function and estimates of intercept and slope for simulation I

Model	Objective func	Formulation	Intercept (s.e.)	Slope (s.e.)
Full	1945.315	A	0.036 (1.13)	0.96 (0.04)
		B	5.090 (1.95)	0.92 (0.03)
		C	2.140 (2.40)	0.95 (0.04)
Reduced	1948.572	A, B and C	2.290 (0.75)	0.94 (0.018)

Table 2. Objective function and estimates of intercept and slope for simulation II

Model	Objective func	Formulation	Intercept (s.e.)	Slope (s.e.)
Full	1880.105	A	-0.25 (1.21)	0.943 (0.041)
		B	10.1 (2.09)	0.903 (0.036)
		C	12.6 (2.61)	0.942 (0.038)
Reduced	1924.074	A, B and C	5.83 (0.824)	0.960 (0.020)

4. TOPICS FOR FURTHER WORK

We assumed that variability in the observed *in vitro* percent dissolved means is negligible. For most, if not all, cases the variability in the *in vitro* data is small relative to that of the *in vivo* data. However, it is doubtful that the *in vitro* variability is always negligible. Therefore, it would be desirable for the methodology to account for the variability of *in vi-*

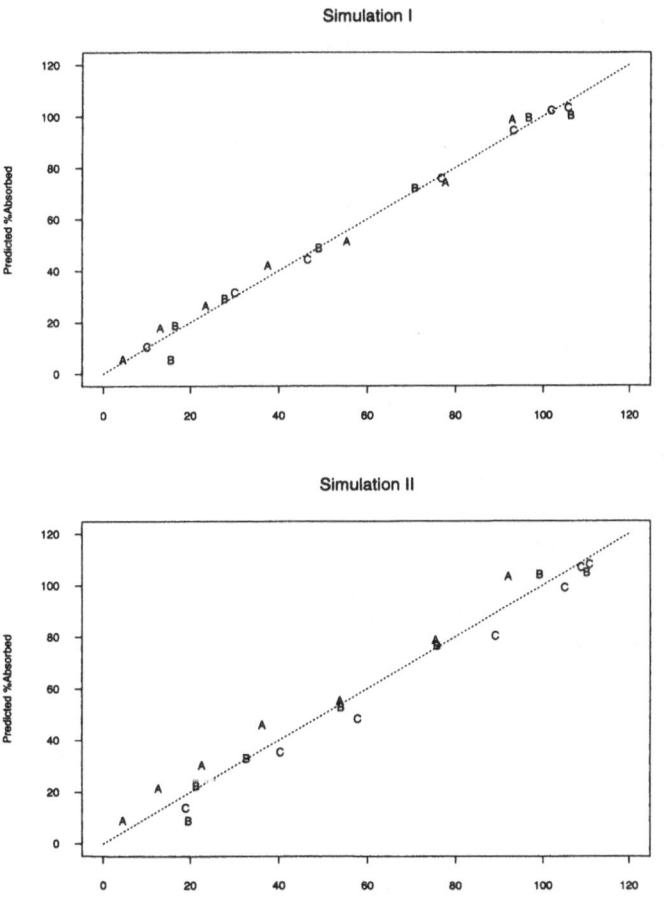

Figure 3. Predicted versus observed mean percent absorbed.

tro percent dissolved measurements. We are not aware of readily available methodology and software to handle this problem.

In some crossover trials, there are period effects. In the statistical analysis of bioavailability variables such as maximum observed concentration (Cmax) , time to maximum observed concentration (Tmax), and area under the concentration-time curve (AUC), the possibility of period effects is routinely taken into account. Period effects could be incorporated into the function that describes the IVIVC. The way this would be done would depend on the nature of the function and on the plausibility of various kinds of period effects. In the case of a linear relationship, one would have to decide whether there might be a period effect for the intercept or slope or both. The presence of a period effect for AUC would not necessarily mean that there is a period effect for the IVIVC. A period effect for Tmax or Cmax might be more likely to indicate a period effect on the IVIVC since these variables seem more likely to be associated with release characteristics of the formulation or absorption rate.

5. CONCLUSIONS

This article provides an introduction of mixed effects modeling to the area of *in vitro-in vivo* correlation. The major advantage of using mixed models is that the dependence of all data points from the same subject is taken into account. The model provides a framework for a proper assessment of whether the same IVIVC holds for the CR products that result from a reasonably wide range of values of the formulation parameters that determine the release characteristics. The model also provides opportunity for accounting for other sources of variation such as period effects.

REFERENCES

Beal, B. L. and Sheiner, L. B. (1992). NONMEM User's Guide, NONMEM Project Group. SCF, CA.
Davidian, M. and Giltinan, D. (1995). Mixed effects models for repeated measurement data, Chapman and Hall.
FDA (1996) Guidance for industry: extended release solid oral dosage forms development, evaluation and application of *in vitro/in vivo* correlations. center for drug evaluation and research (CDER).
SAS Institute (1990). SAS Procedures Guide, Version 6, 3th ed.
USP XXIII Chapter <1088>. *In vitro in vivo* evaluation of dosage forms, United States Pharmacopoeal Concention, Inc. 1927–1929
Vonesh, E. F. and Carter, R. L. (1987). Efficient inference for random coefficient growth curve models with unbalanced data. Biometrics 43, 617–628.
Weisberg, S. (1985) Applied linear regression, 2nd ed. Wiley, New York.
Wolfinger, R. (1992). A tutorial on mixed models. SAS institute Inc.
Wolfinger, R. and O'Connell, M. (1993). Generalized linear mixed models: a pseudo-likelihood approach. Journal of Statistical Computation and Simulation, 48, 233–243.

16

COMPARISON OF METHODOLOGIES FOR EVALUATING REGIONAL INTESTINAL PERMEABILITY

A. Raoof,[1] D. Moriarty,[1] D. Brayden,[1] O. I. Corrigan,[2] I. Cumming,[1]
J. Butler,[1] and J. Devane[1]

IVIVR Cooperative Working Group
[1]Elan Corp. Plc
Athlone, Ireland
[2]Trinity College
Dublin, Ireland

1. INTRODUCTION

Drugs are most frequently administered orally and any delay or losses during absorption may contribute to variability in drug response and thus to failure in drug therapy.[1] How drugs cross the cell membrane was first described by Overton in 1899 which resulted in "Overton's Law" i.e. permeability coefficients correlate well with oil/water partition coefficients.[2] Modern work, however, indicates that the ability of a drug to traverse a biological membrane is difficult to predict from a simple physicochemical measurement (solubility, lipophilicity, pKa, hydrogen bonding capacity, molecular size or weight) and that other biological factors such as stomach-emptying rate, intestinal motility, the composition (pH profile, volume, enzymes and food) of the intestinal lumen and intrinsic membrane permeability can further limit the bioavailability of drugs.[3] Recently, factors related to drug formulation such as particle size and shape, degradation and dissolution of the dosage form have been found to influence the absorption kinetics of drugs.[4, 5]

There is evidence that oral absorption is site dependent i.e. it varies from duodenum to colon. This was first reported by Nobecourt and Vitry in 1904 when they described a regional intestinal absorption of ions and fluids using isolated segments of rabbit small intestine[6]. Recently, regional differences in intestinal permeability and oral bioavailability have been reported for different compounds in animals and humans.[7–10] Regional reduction in the permeability of drugs and/or nutrients can be ascribed to morphological differences of the mucosal cells along the intestine e.g. type of the cells, absorption surface area, paracellular pore sizes and numbers, pH profiles, availability of transport systems, enzymatic capacity, etc.[11, 12]

Several model systems are now available for studying the intestinal permeability of drugs.[4, 13] Among them are intestinal perfusion techniques in both animals and humans, excised intestinal segments, everted intestinal sacs, excised intestinal rings and epithelial cell culture models. The effectiveness of any of these models depends solely upon how closely it mimics the characteristics of the *in vivo* biological barriers.[4] It is important to note that each model has its own characteristics and biological limitations and therefore a combination of several models may be necessary to assess the mechanism(s) and extent of absorption of the intact molecule.

In this study we compare the permeability of a number of drug candidates for inclusion in extended release products. The following three (different) model systems were used: the *in vitro* vascularly perfused rat gut segment, the in situ (single pass) rat gut perfusion system and the Caco-2 cell monolayer system. In the two rat gut systems, three separate sections of gut were investigated, the upper small intestine (13 cm below pyloric sphincter), lower small intestine (16 cm proximal to caecum) and the large intestine (9 cm from caecum to anus). This allowed comparison of regional variations in permeability. Caco-2 display microvilli and express transporters normally found in jejunal and ileal enterocytes, although their transepithelial electrical resistance (an indirect indication of the leakiness of the tight junctions) is more reminiscent of that of colonocytes (reviewed in [14]).

2. METHODS

2.1. *In Vitro* Vascularly Perfused Rat Gut Segments

The procedure of the *in vitro* isolated vascularly perfused rat gut segment is described in detail by Roy et al. (1991).[15] In brief: Male Sprague-Dawley rats (weighing 250–300 g) were anaesthetised with halothane followed by the I.V. injection of sodium pentobarbitone (25 mg/kg) and the following were cannulated: the trachea (for adequate respiration during anaesthesia), the right jugular vein (for administration of sodium pentobarbitone and heparin) and the common bile duct (to divert the bile from the gut lumen). Sutures were then placed around the mesenteric artery and hepatic portal vein at the points of entry into the liver. After an I.V. injection of heparin (500 N.I.H. units), both vessels were rapidly cannulated and an infusion of Krebs-Ringer bicarbonate buffer, pH 7.4 at 37 °C was delivered via the mesenteric artery. Once the effluent flow in the portal vein cannula was established (within seconds), the preparation was transferred to the perfusion apparatus and perfused at approximately 15 ml/min and 50–80 mmHg via the mesenteric artery with 120 ml of warm (37 °C), heparinised (15 N.I.H. units/ml) and oxygenated rat blood. The blood was diluted with Krebs-Ringer bicarbonate containing 6% albumin such that the packed cell volume was 15% v/v. The effluent blood from the portal vein cannula was returned to the perfusion apparatus for reoxygenation, reheating to 37 °C and recirculation through the vascular bed of the entire intestinal tract from the pyloric sphincter to the rectum.

Three segments of gut, namely the upper small intestine, lower small intestine and the large intestine were isolated by tying the gut lumen with two sutures at measured distances apart. As regards the upper small intestine, one suture was tied 1 cm below the pyloric sphincter and another 13 cm distal to the first. For the lower small intestine, one suture was tied 1 cm before the caecum and a second 16 cm proximal to the first, while for the large intestine sutures were placed immediately distal to the caecum and approximately 1 cm proximal to the anus (a length of 9 cm).

The test compounds together with the internal standard atenolol (as a poorly permeable marker) were prepared in the following buffer system: NaCl (48 mM), KCl (5.4 mM), Na_2HPO_4 (28 mM), NaH_2PO_4 (43 mM) and mannitol (35 mM). The pH and osmolarity of the buffer system were always kept at 6.4 and 290 mOsm/l respectively. Aliquot (0.75 ml) of the test solutions were injected into each gut segment as a bolus. Blood samples (3 ml) were collected pre-dose and at different time intervals up to 180 minutes and the plasma obtained by centrifugation was kept at -20 °C until analysis.

2.2. *In Situ* Rat Gut Perfusion System

This technique is based on the one described by Komiya et al. (1980).[16] The method can be described as follows: Male Spraque-Dawley rats (weighing 250–300 g) were anaesthetised with halothane followed by I.V. injection of sodium pentobarbitone (25 mg/kg) and the trachea, right jugular vein and the common bile duct were cannulated as described above (*in vitro* perfusion system).

The segments of the gut, upper small intestine, lower small intestine and the large intestine were isolated by tying as described in the *in vitro* system using the same sites and length of gut. In each of the three regions an inflow cannula and outflow cannula was inserted into the lumen approximately 0.5 cm from each of the sites and the lumen between the sites was perfused with buffer containing test compounds and internal markers (PEG 4000 as a volume marker, antipyrine as a highly permeable marker) at 37 °C and a flow rate of 0.2 ml/min. The composition of the buffer system was as described in the *in vitro* perfusion system. Effluent from the lumen was collected at timed intervals up to 180 minutes and was centrifuged to remove the debris. Changes in the water flux were measured by the differences in the concentration of ^{14}C-PEG in and out of the lumen, it being assumed that there is no absorption (negligible) of PEG 4000 through the gut lumen as described by Hirtz (1985).[17]

2.3. Caco-2 Cell Monolayer System

Caco-2 cells (passage 30–40) were grown in Costar 25 cm^2 vented tissue culture flasks in a maintenance medium of Dulbecco's Modified Eagles Medium (DMEM) with Glutamax (Gibco), 10% foetal calf serum, 1% non essential amino acid, 1% sodium pyruvate, 50 U/ml penicillin and 150 U/ml streptomycin. Cells were incubated at 37 °C with 5% CO_2 in air. At confluence Caco-2 cells were subcultured using 0.25% trypsin-EDTA solution onto tissue culture-treated Costar Transwells Snapwells (catalogue number 3407) of area 1.13 cm^2 at density of 1×10^6 cell/cm^2.

The integrity of the monolayers was routinely checked by measurements of TEER (transepithelial electrical resistance) using an EVOM chopstick epithelial voltmeter (WPI). For Caco-2 cells TEER values at day 20 ranged from 300–500 ohms cm. Experiments were carried out at between days 20 and 30, a range which the monolayers expressed differentiated properties.[14]

Transport studies were performed directly on the filter inserts mounted in 6-well plates. Incubation solution consisted of Hanks balanced salt solution with addition of 25mM HEPES and 11.1mM d-glucose at a pH of 7.4. Drugs were added to the donor (apical) side of the monolayers. Samples were withdrawn from the receiver (basolateral) side for analysis at 30 minute time intervals over 120 minutes with sink conditions being maintained by addition of fresh buffer to the receiver side. Samples were also taken from the donor side at the beginning and end of the time period. Analysis was by UV spectrophotometry for all the selected agents except for Elan 1, which was measured by Gas Chromatography linked to electron capture.

3. DATA ANALYSIS

The apparent permeability coefficient (P_{app}, cm/sec) *in vitro* models (vascularly perfused rat gut and Caco-2 cell monolayer system) was calculated according to the following equation:[18]

$$ P_{app} = \frac{\Delta Q/\Delta t}{60.C_0 A} $$

where $\Delta Q/\Delta t$ is the flux rate ($\mu g/min$), C_0 is the initial concentration ($\mu g/ml$) of the test compounds and A is the surface area of the gut segments ($2\pi rl$, r = radius of the intestine 0.138 cm (upper small intestine), 0.2 cm (lower small intestine) and 0.38 cm the large intestine),[19] l = length of the intestine 13 cm (upper small intestine), 16 cm (lower small intestine) and 9 cm (large intestine). The surface area of the Caco-2 monolayers was 1.13 cm^2.

The apparent permeability in the in situ rat gut perfusion model is calculated based on the parallel-tube model as follows:[20]

$$ P_{app} = \frac{-Q.\ln(C_{out}/C_{in})}{60.A} $$

Where Q is the perfusion flow rate (0.2 ml/min), C_{out} and C_{in} is the outlet and inlet (fluid transport corrected) concentrations ($\mu g/ml$) of the compounds and A is the surface area of the intestinal segments calculated as described above.

The recovery of ^{14}C-PEG 4000 (PEG_{rec}) was calculated as follows:

$$ PEG_{rec} = \Sigma PEG_{out}/\Sigma PEG_{in} $$

Where ΣPEG_{in} and ΣPEG_{out} were the accumulated amounts of the ^{14}C-PEG 4000 entering and leaving the intestine.

Results in the text and tables are expressed as mean ± standard deviation (SD).

4. RESULTS

The test compounds used in this study are of different chemical classes (acids, bases and non-ionizable agents). The physicochemical properties (molecular weight, partition between octanol/buffer, pH 7.4) and the concentration of the drugs used in each experiment are summarised in Table 1.

Table 1. Physicochemical properties (molecular weight MW, pKa, log D octanol/buffer pH 7.4) and initial (inlet) drug concentrations of compounds used in situ, *in vitro* and in Caco-2 cell experiments. n.i. = not identified

Compounds	pka	MW (g/mol)	log $D_{oct,7.4}$	Concentration (mM)		
				in vitro	*in situ*	Caco-2
Elan 1	-	191	-1.62	3.2	1.0	4.2
Elan 2	9.11	94	-0.77	1.0	0.2	1.0
Elan 3	5.50	254	0.70	19	0.67	1.0
Elan 4	4.39	206	0.55	14	1.0	2.0
Elan 5	9.10	484	0.54	0.2	0.02	0.2
Atenolol	9.6	266	-1.80	1.9	-	-
Antipyrine	1.5	188	0.54	-	1.05	-
PEG 4000	n.i	4000	n.i	-	2.5	-

In the *in vitro* gut model, atenolol was used as a poorly permeable marker and also as an indicator of the viability of the intestine. The permeability values were low in the upper and the lower small intestine while they were negligible in the large intestine. The viability of the intestinal mucosa, however, in the in situ model was calculated by the % of recovery of ^{14}C-PEG 4000 in the upper small intestine, lower small intestine and the large intestine. The values were 96.9 ± 6%, 109.3 ± 8% and 103.3 ± 2% respectively. The apparent permeability coefficient in this model was calculated when a physiological steady state condition (using ^{14}C-PEG out/in concentrations) was achieved i.e. after 50–70 minutes of the start of the perfusion (Figure 1).

The apparent permeability coefficients of the drugs and of the marker compounds using the three different models (i.e. *in vitro*, in situ and the Caco-2 cell monolayer system) are summarised in Table 2. All the compounds appear to have a high permeability coefficient relative to their poorly (atenolol) and highly (antipyrine) permeable markers in the first two models. In Caco-2 monolayers all of the compounds tested gave Papp values

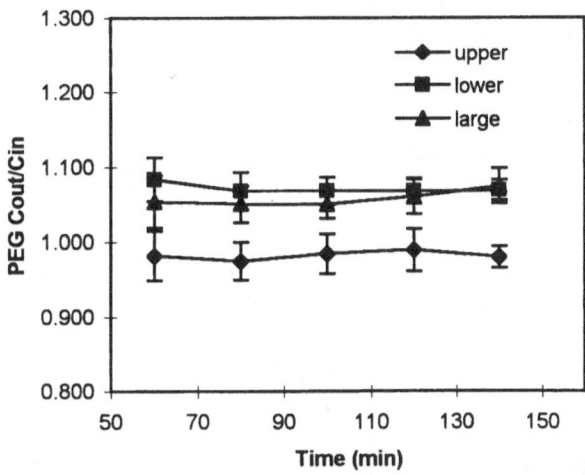

Figure 1. Steady state levels of ^{14}C-PEG in the upper small intestine, lower small intestine and the large intestine during in situ rat gut perfusion system. Results are expressed as mean (n = 16) ± SD.

Table 2. Regional apparent permeability coefficients (p_{app} x 10^{-6}, cm/sec) of the studied drugs and the marker compounds in the upper small intestine, lower small intestine and the large intestine using *in vitro*, in situ and Caco-2 cell monolayer systems. The results are expressed as mean (n = 4) ± SD with the exception of *in vitro* model where n = 2. * n = 10

Compounds	*in vitro*			in situ			Caco-2
	upper	lower	large	upper	lower	large	
Elan 1	22.6	18.2	10.4	181.0 ± 115	40.2 ± 16	25.0 ± 18	74.6 ± 19
Elan 2	22.7	9.0	2.9	-	-	-	26.1 ± 3
Elan 3	50.0	18.4	12.2	276.2 ± 61	156.4 ± 11	59.4 ± 22	21.8 ± 2
Elan 4		20.9	17.7	212.1 ± 43	130.5 ± 28	101.1 ± 19	40.2 ± 4
Elan 5	7.3	2.8	3.7	242.9 ± 30	170.1 ± 29	108.0 ± 4	27.7 ± 4
Atenolol*	1.9	0.9	0.0	-	-	-	-
Antipyrine*	-	-	-	146.9 ± 53	117.6 ± 31	59.1 ± 32	-

which were at the top end of the absorption scale (Papp > 10^{-5} cm/s, > 90% absorbed) relative to moderately absorbed agents such as loperamide (Papp = 5 x 10^{-6} cm/s, 40% absorbed) or poorly absorbed agents such as mannitol (Papp = 4 x 10^{-7} cm/s, 15% absorbed) (Figure 2). Using the *in vitro* and in situ rat gut perfusion systems, the permeability of all the compounds was found to be high in the upper small intestine and to decrease distally towards the large intestine (Figure 3). The Caco-2 cell monolayer permeability values obtained were lower per se than those obtained by the in situ model and larger than those of the *in vitro* model. No correlation was found between the apparent permeability coefficients obtained using the three different models. However, within each of the models all of the selected agents appeared to be well absorbed with respect to either the internal controls in the case of the rat gut models or in respect to the absorption of agents which are known to be moderately or poorly absorbed in the case of Caco-2.

5. DISCUSSION

In general, the compounds studied in the rat models were found to have a high permeability relative to marker compounds atenolol and antipyrine which are poor and high permeability marker compounds respectively.[18, 21] The permeability trends obtained using

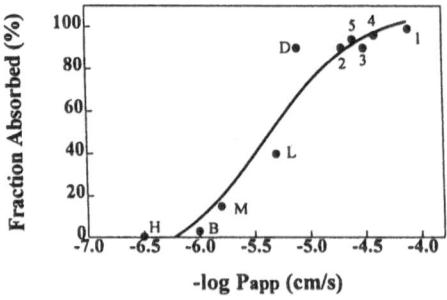

Figure 2. Plot of Papp of the 5 selected agents across Caco-2 monolayers versus fraction absorbed (%) in man. The plot was fitted by non-linear regression. For comparison data from moderately and poorly absorbed agents are also included. Code: 1–5 correspond to Elan drugs 1–5, D = diltiazem, L = loperamide, M = mannitol, B = berberine and H = heparin.

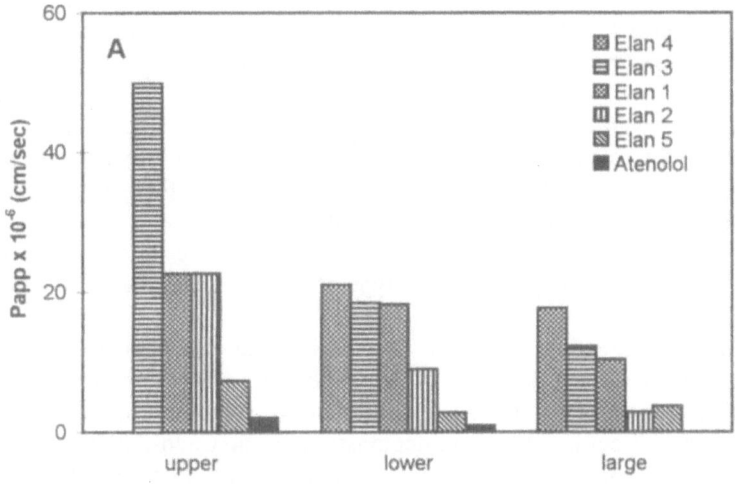

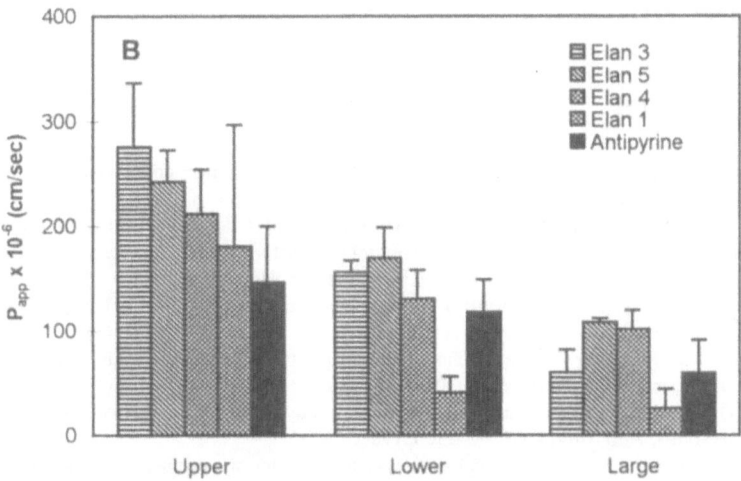

Figure 3. The apparent permeability coefficient of the test compounds and the markers in the upper small intestine, lower small intestine and the large intestine using *in vitro* (A) and in situ (B) models.

the *in vitro* model were similar to those obtained using the in situ model i.e. the both sets of values were found to decrease distally from duodenum to colon. These results are in agreement with the literature and may be due to differences in bio-membrane characteristics of the intestine. The absence of villi/microvilli and also the large diameter of the colon may play a significant role in decreasing the absorption surface area in this region by comparison to the small intestine.[22]

It has been also reported that the pore size (diameter of the tight junctions between the epithelial cells) is smaller in the colon (3 A) by comparison to the jejunum and ileum (8 A).[6] The paracellular route may be important for compounds having both a low molecular weight and a hydrophilic log D such as Elan 1 (MW 191) and Elan 2 (MW 94). However it seems that this can only be part of the explanation: the high Papp values associated with these two agents in Caco-2 are usually obtained for drugs using a predominantly passive transcellular route, because the high transepithelial resistance of Caco-2 would indicate the presence of a restricted paracellular pathway in this model. The regional differences in permeability located in the rat gut models may also be related to the metabolic capacity of the intestine.[22, 23] Metabolic intracellualr and brush border membrane activity is though to be greater in the duodenum and jejunum than in the ileum and colon and furthermore activity tends to be higher in the villous tips and decreases progressively towards the crypts.[24] Against that, bacterial enzyme levels are far higher in the colon than in the small intestine, so it depends on the type of enzymes that the selected drugs are sensitive to.

The effect of intestinal first pass metabolism has not been studied in the in situ model as no portal venous sampling was available. This effect however was considered in the *in vitro* and in the Caco-2 cell monolayer systems. It is important however to note that intestinal first pass effect is difficult to demonstrate in the Caco-2 cell system owing to the reduction of certain enzymatic activity [25]. However there are now sub-clones of Caco-2 available which appear to have adequate levels of the metabolic enzyme system, cytochrome P450 3A4 [25]. As a result of the above mentioned differences, no correlation was established between the three different models.

6. REFERENCES

1. Rowland M. and Tozer T. Absorption. In Clinical Pharmacokinetics: concepts and applications. Lea and Febiger, Philadelphia, pp. 113–147, 1989.
2. Overton E. Ueber die allgemeinen osmotischen Eigenschaften der Zelle, ihre vermutlichen Ursachen und ihre Bedeutung fur die Physiologie. Viertelijahrsschr. Naturforsch. Ges. Zuerich, 44:88–135, 1899.
3. Audus K. and Raub T. Barriers to protein delivery. Plenum, New York, 1993.
4. Hillgren K., Kato A. and Borchardt R. *In vitro* systems for studying intestinal drug absorption. Medicinal Research Reviews, 15:83–109, 1995.
5. Dressman J. and Fleisher D. Mixing-tank model for predicting dissolution rate control of oral absorption. Journal of Pharmaceutical Sciences.75:109–116, 1986.
6. Powell D. Intestinal water and electrolyte transport. In Physiology of the gastrointestinal tract. (Johnson R., ed), Raven Press, New York, pp. 1267–1306, 1987.
7. Park G. and Mitra A. Mechanism and site dependency of intestinal mucosal transport and metabolism of thymidine analogues. Pharmaceutical Research. 9:326–331, 1992.
8. Seta Y, Higuchi F., Kawahara Y., Nishimura K. and Okada R. Design and preparation of captopril sustained-release dosage forms and their biopharmaceutical properties. International Journal of Pharmaceutics. 41:245–254, 1988.
9. Staib AH., Beermann D., Harder S., Fuhr U. and Liermann D. Absorption differences of ciprofloxacine along the human gastrointestinal tract determined using a remote-control drug delivery device (HF-capsule). American Journal of Medicine. 30:66S-69S, 1989.
10. Barr W., Zola E., Candler EL., Hang S., Tendolkar A., Shamburek R., Parker B. and Hilty M. Differential absorption of amoxacillin from human small and large intestine. Clinical Pharmacology and Therapeutics. 56:279–285, 1994.
11. Johnson L. Physiology of the gastrointestinal tract. Raven Press, New York, 1994.
12. Steed K. and Wilson C. Drug delivery to the large intestine. In: Physiological pharmaceutics: biological barriers to drug absorption. (Wilson C. and Washington N., ed.), Ellis Horwood, Chichester, pp. 91–108, 1989.
13. Lennernas H. Gastrointestinal absorption mechanisms: a comparison between animal and human models. European Journal of Pharmaceutical Sciences. 2:39–43, 1994.
14. Anderberg E. and Artursson P. Cell cultures to access drug absorption enhancement. In drug absorption enhancement concepts, possibilities, limitations and treads. (De Boer A., ed), Harwood Publishers, pp. 101–118, 1994.

15. Roy A., Curtis G. and Hughes H. The uptake of oestrone from the lumen of the isolated perfused rat gut. Xenobiotica. 21:491–498, 1991.
16. Komiya I., Park JY., Kamani A., HO NHF. and Higuchi W. Quantitative mechanistic studies in simultaneous fluid flow and intestinal absorption using steroids as model solutes. International Journal of Pharmaceutics. 4:249–262, 1980.
17. Hirtz J. British Journal of Clinical Pharmacology. 19:77S-83S, 1985.
18. Artursson P. Epithelial transport of drugs in cell culture. I: a model for studying the passive diffusion of drugs over intestinal absorption (caco-2) cells. Journal of Pharmaceutical Sciences. 79:476–482, 1990.
19. Kararli T. Comparison of the gastrointestinal anatomy, physiology and biochemistry of humans and commonly used laboratory animals. Biopharmaceutics and Drug Disposition. 16:351–380, 1995.
20. Amidon G., Kou J., Elliott R. and Lightfoot E. Analysis of models for determining intestinal wall permeabilities. Journal of Pharmaceutical Sciences. 69:1369–1373, 1980.
21. Fagerholm U., Johansson M. and Lennernas H. Comparison between permeability coefficients in rat and human jejunum. Pharmaceutical Research. 13:1336–1342, 1996.
22. Fara J. Colonic drug absorption and metabolism in novel drug delivery and its therapeutic applications. (Prescott L. and Nimmo W., ed), John Wiley & Sons Ltd, 1989.
23. Ilett K., Tee L., Reeves P. and Minchin R. Metabolism of drugs and other xenobiotics in the gut lumen and wall. Pharmacology and Therapeutics. 46:67–93, 1990.
24. Krishina D. and Klotz U. Extrahepatic metabolism of drugs in humans. Clinical Pharmacokinetics. 26:144–160, 1994.
25. Hu M., Li.,Ha H., Crespi L. and Huang, S. Drug metabolism by transfected Caco-2 cells expressing CYP3A4: comparison studies. Pharmaceutical Research. 13: S237, 1996

17

IN VITRO–IN VIVO RELATIONSHIPS OF SEVERAL "IMMEDIATE" RELEASE TABLETS CONTAINING A LOW PERMEABILITY DRUG

James E. Polli

School of Pharmacy
University of Maryland at Baltimore
20 North Pine Street
Baltimore, Maryland 21201

1. ABSTRACT

The objective of this work was to gain insight into the biopharmaceutical performance of four different but bioequivalent ranitidine hydrochloride tablet formulations. This analysis employed a recently described method[1] to relate *in vitro* and *in vivo* data and aimed to facilitate an understanding of oral drug product performance. For each ranitidine formulation, dissolution was performed using the USP procedure. A four-way, single dose bioequivalence study (n = 14) was performed. The fraction of the total amount of dose absorbed at each plasma sample time was determined by the Wagner-Nelson method. Equation 1 (see below) was fitted to the *in vitro* vs. *in vivo* data. For all four formulations, this analysis suggests absorption was permeation-rate limited, where ranitidine exhibited a low permeation rate constant of 0.01/min.

2. INTRODUCTION

The elucidation and quantification of factors which contribute to the success or failure of dosage forms has been an unifying interest of pharmaceutical scientist. Ever since the evolution of pharmaceutics as a discipline, pharmaceutical scientists have attempted to link physicochemical properties of a drug and dosage form with its resulting biological performance. In particular for orally-administered drugs that need to be absorbed to impart pharmacologic action, the effects of excipients, dosage form fabrication processing, and drug crystallinity and particle size have been identified as possibly important determinants of pharmacokinetic or pharmacodynamic performance. In light of important possible consequences of drug and drug product "*in vitro*" properties on "*in vivo*" drug product charac-

In Vitro–in Vivo Correlations, edited by Young *et al.*
Plenum Press, New York, 1997

ter, the propensity to relate or "correlate" *in vitro* data to *in vivo* data has been a natural one.

In a sense, the general scope of this work was to perform "*in vitro-in vivo* correlation" on a set of ranitidine tablets. The term "*in vitro-in vivo* correlation" is now commonly used and, in fact, the USP identifies three categories of *in vitro-in vivo* correlation[2]. USP Level A correlation, for example, highlights the fraction of drug dissolved (F_d) as the *in vitro* parameter and the fraction of the total amount of drug absorbed (F_a) as the *in vivo* parameter.

However, as suggested in this manuscript's title, the phrase "*in vitro-in vivo* correlation" and specifically the word "correlation" are purposely avoided here. The term correlation generally means "the degree of relationship between two random variables."[3] That is, correlation concerns itself with the extent to which two variables obey a particular mathematical relationship (or rank-order relationship); correlation is generally not concerned with the nature of the relationship between the two variables. Although this notion of correlation can be applied to two variables which are associated with one another in a nonlinear way, the term is often used in a more limited sense to indicate a linear relationship between the two variables (or agreement between two sets of ranks). USP Level A correlation appears to follow this common application of "correlation" analysis since its description makes reference to attaining a linear or linearized relationship between F_a and F_d.

Here, F_a and F_d data from four ranitidine tablet formulations are analyzed. Neither a linear nor a linearized relation between F_a and F_d is assumed. In this sense, a USP Level A approach and "*in vitro-in vivo* correlation" were not undertaken, although the USP Level A approach does highlight the use of F_a and F_d data. Rather, the analysis here can be better described as the characterization of "*in vitro-in vivo* relationships" for several ranitidine tablets. Moreover, the main intent of this work was to make use of the *in vitro-in vivo* relationship for each tablet formulation in order to elucidate the relative roles of dissolution and intestinal permeation in overall ranitidine absorption from each dosage form.

3. THEORY

Analysis of *in vitro-in vivo* relationships was of the form[1]:

$$F_a = \frac{1}{f_a}\left(1 - \frac{\alpha}{\alpha-1}(1-F_d) + \frac{1}{\alpha-1}(1-F_d)^\alpha\right) \tag{1}$$

where F_a is the fraction of the total amount of drug absorbed at time t, f_a is the fraction of the dose absorbed at $t = \infty$, α is the ratio of the first-order permeation rate constant (k_p) to the first-order dissolution rate constant (k_d), and F_d is the fraction of drug dose dissolved at time t.

4. MATERIALS AND METHODS

Eq. 1 was fitted to F_a vs. F_d data from Zantac[R] and three "immediate" release ranitidine hydrochloride tablet formulations. Each product contained the equivalent of 300 mg of ranitidine in each tablet. The three test formulations (fast, medium, and slow) were development at the University of Maryland at Baltimore (UMAB) School of Pharmacy and were formulated[4] to span a range of dissolution. Dissolution was performed on six tablets of each formulation according to the USP monograph method. Dissolution employed the paddle method at 50 rpm. The medium was 900 ml of purified water. Dissolution samples were collected at 10, 20, 30, and 45 min. As shown in Fig. 1, the dissolution profiles of the four formulations visually differed from one another (fast > Zantac[R] > medium > slow). Dissolution profiles were fit to a first-order release model to yield k_d.

Figure 2 illustrates the plasma concentration-time profiles for the four products from a four-way, single 300 mg dose bioequivalency study (n = 14)[5]. Each formulation was bioequivalent to the other three. F_a was determined by the Wagner-Nelson method from the plasma profiles. Eq. 1 was fitted to the F_a vs. F_d data to yield an estimate of α. In eq. 1, individual values for f_a were estimated from relative $AUC_{0-\infty}$ and from requiring mean f_a for Zantac[R] to be 0.52[6]. k_p was calculated from $k_p = \alpha\, k_d$.

All regressions employed nonlinear least squares by the Quasi-Newton procedure in SYSTAT 5.03 (SYSTAT Inc., Evanston, IL).

5. RESULTS AND DISCUSSION

5.1. Wagner-Nelson Profiles

Figure 3 illustrates the Wagner-Nelson plots of the plasma data. Drug absorption occurred over a period of about 2.5 hr from each formulation. In general agreement with the observed bioequivalence, the cumulative drug input from each product was similar to the input profile from each of the other products.

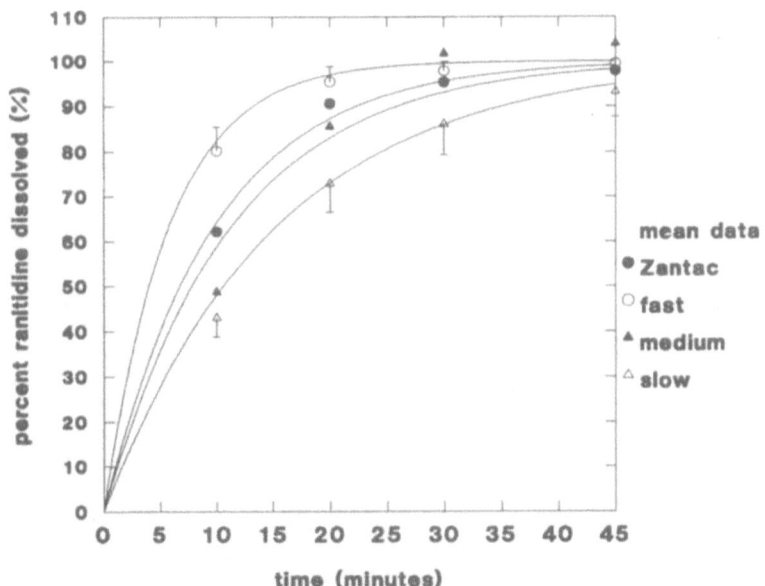

Figure 1. Dissolution profiles of the four ranitidine hydrochloride formulations. The rate of release of ranitidine spanned about a three-fold range. Solid lines are the mean first-order fits.

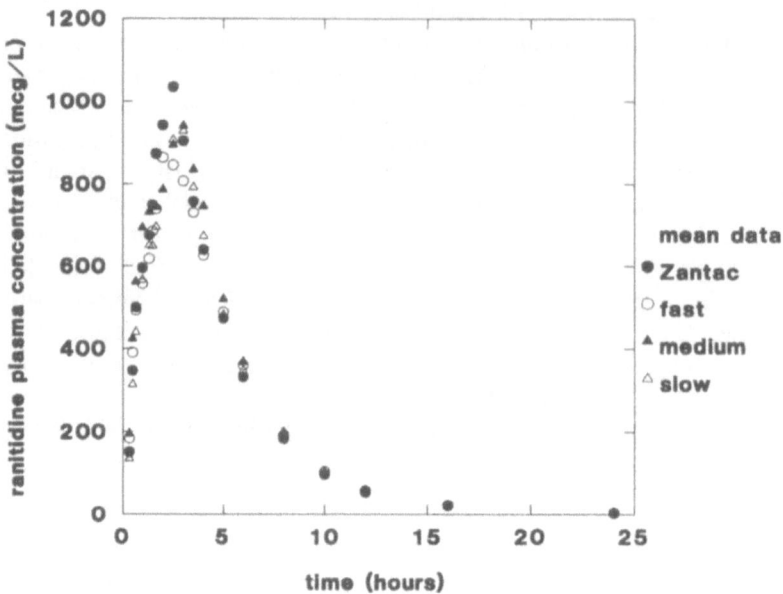

Figure 2. Ranitidine plasma profiles. In spite of differing *in vitro* dissolution profiles, the four ranitidine formulations were bioequivalent to one another.

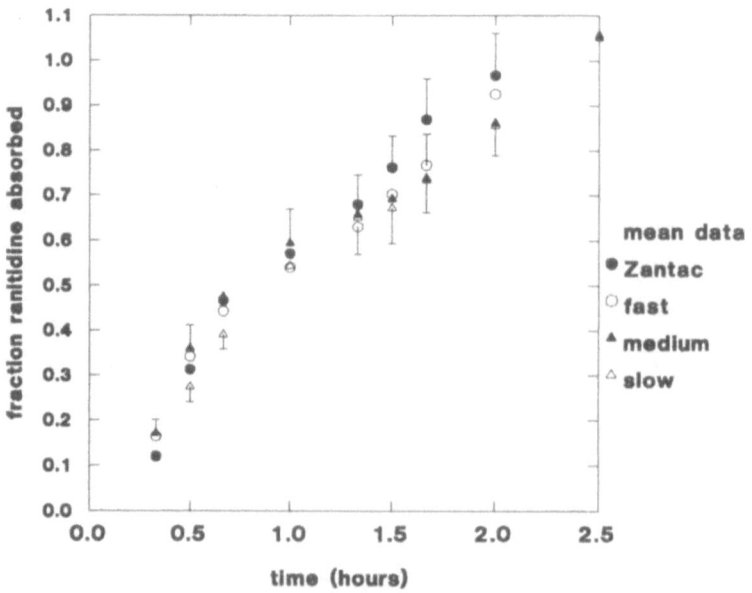

Figure 3. Wagner-Nelson plot for the four ranitidine hydrochloride formulations.

5.2. *In Vitro-in Vivo* Relationships: Semi-Quantitative Inspection

Figure 4 plots the fraction absorbed versus the fraction dissolved and graphs the mean fit of eq. 1 to each of the four formulations. Each profile constitutes the *in vitro-in vivo* relationship for each respective product. More specifically, since each F_a and F_d are both dependent on the same variable (i.e. time) and are plotted against one another, Fig 4 can be considered the phase plane[7] for the *in vitro-in vivo* relationship problem. Plotted in the phase plane is each product's trajectory.

As the name implies, each trajectory describes the time-dependent "path" of each product's simultaneous dissolution and absorption performance. For example, the first three plotted mean points for fast are ($F_a = 0.164$, $F_d = 0.954$), ($F_a = 0.342$, $F_d = 0.978$), and ($F_a = 0.443$, $F_d = 0.988$) and correspond to $t = 20$ min, 30 min, and 40 min, respectively. Hence, the trajectory characterizing the dissolution and absorption of ranitidine from the fast test product rapidly moves from ($F_a = 0$, $F_d = 0$) in the lower left corner of the phase plane at $t = 0$ min to the 20 min data point in the lower right corner of the phase plane. This portion of the trajectory reflects the rapid dissolution of ranitidine during the first 20 min and the relatively low fraction of absorbed drug at 20 min. Over the following 40 min, the trajectory curves "upward" due to the relatively large fraction in ranitidine absorption while only the small remaining fraction of ranitidine dissolves during that time frame. For the next one to two hours, essentially only ranitidine absorption occurs since dissolution had been complete. Hence, this semi-quantitative analysis of the F_a vs. F_d plot indicates intestinal permeation, rather than dissolution, is the rate-limiting step in the overall absorption of ranitidine hydrochloride from the fast test formulation. This analysis of the fast test formulation is in agreement with idealized cases[1] which indicated that highly permeation rated-limited absorption results in a "reverse L" appearance of a trajectory.

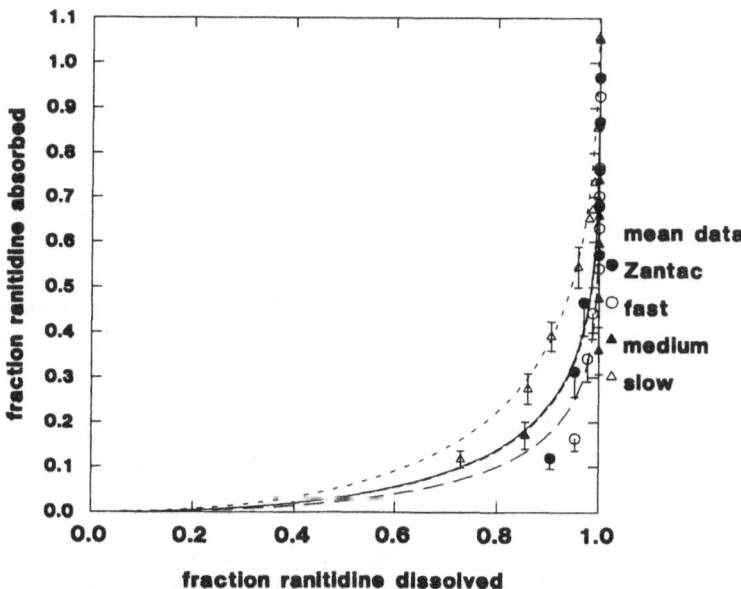

Figure 4. Plot of the fraction absorbed *in vivo* versus fraction dissolved *in vitro* for each ranitidine hydrochloride formulation. For each formulation, the curve for eq. 1 using the mean values for f_a and a is drawn.

Like for fast, the relationship between F_a and F_d for the slow test formulation is substantially nonlinear and possesses appreciable "reverse L" character. Hence, even this slowly dissolving "immediate" release formulation of ranitidine appears to be substantially permeation-rate limited. However, compared to fast, the trajectory of slow is less "reverse L" in appearance and more "hockey stick" in appearance. In agreement with idealized trajectories[1] where a product with "hockey stick" character relative to a product with "reverse L" character is less permeation rate controlled, Fig 4 suggests slow to be less permeation rate-limited than is fast.

Situated between the trajectories of fast and slow in Fig. 4 are the trajectories of Zantac[R] and medium. Again, semi-quantitative inspection of their trajectories suggests that intestinal permeation limits the rate of overall ranitidine absorption. In order to quantifiably assess the degree of permeation rate-controlled absorption, numerical fitting of eq. 1 to the F_a vs. F_d data was undertaken.

5.3. *In Vitro-in Vivo* Relationships: Quantitative Assessment

Numerical fits of eq. 1 to F_a vs. F_d data yielded quantitative values for α. Since, $\alpha = k_p/k_d$ can be interpreted to be a metric identifying either dissolution or intestinal permeation as the rate-limiting step for absorption from a drug product. Table 1 summarizes the mean (±SE) values for α. Values for k_d and f_a are also noted and were determined according to the procedures described in the Materials and Methods section. For Zantac[R], fast, medium, and slow, $\alpha \ll 1$ and indicates that $k_p \ll k_d$. Hence, for each formulation, overall absorption was intestinal permeation rate-limited, as concluded above for Zantac[R], fast, medium, and slow through semi-quantitatively inspection.

For fast, regression suggests that dissolution was 16-fold faster than permeation (i.e. $1/\alpha \cong 16$). For slow, dissolution was only 6-fold faster than permeation, yet still was considerably faster than the permeation step. This appreciable degree of permeation rate-controlled absorption is evident in the nonlinear trajectory of the slow formulation in Fig. 4. In comparing slow test product performance to fast test product performance, this nearly 3-fold difference in the extent to which permeation kinetically "lagged" behind dissolution is presumably attributable to the dissolution rate differences between fast and slow. The nearly 3-fold difference between fast k_d and slow k_d appears to account for this difference in biopharmaceutical performance.

5.4. Model Consistency and *in Vivo* Permeation Rate Constant

Although eq. 1 is based upon a simplified view of drug absorption where drug in a dosage form is subject to first-order dissolution and subsequent first-order permeation, the values in Table 1 suggest a degree of model consistency and applicability of eq. 1 when applied to

Table 1. Absolute and relative contributions of dissolution and permeation to overall ranitidine absorption kinetics

Formulation	f_a (±SE)	alpha (±SE)	k_d (±SE)(min^{-1})	k_p (±SE) (min^{-1})
Zantac[R]	0.520(±0.016)	0.0943 (±0.0181)	0.103(±0.005)	0.00971 (±0.00187)
fast	0.502(±0.018)	0.0646 (±0.0095)	0.174(±0.023)	0.0113 (±0.0016)
medium	0.541(±0.016)	0.0964 (±0.0194)	0.0889 (±0.0049)	0.00857 (±0.00173)
slow	0.517(±0.021)	0.156(±0.020)	0.0656 (±0.0106)	0.0102 (±0.0013)

these ranitidine formulations. Since the rank order of k_d are fast > Zantac^R > medium > slow, the anticipated rank order of a would be the reverse (i.e. slow > medium > Zantac^R > fast). Table 1 shows this rank order of values for α, a one in 24 chance occurrence.

Moreover, eq. 1 can be used to estimate an *in vivo* permeation rate constant (k_p) which is uncoupled from dissolution. Table 1 lists the estimated k_p ($\pm$SE) of ranitidine from each formulation. k_p values are about 0.01/min for each formulation. Although eq. 1 does not take into account any regional or dosage form-dependent permeability differences, similar values for k_p would be expected for ranitidine across all formulations. ANOVA testing suggests the values for k_p to be the same across all four formulations (p = 0.67). Hence, in spite of the simplified model underpinning eq. 1, the consistence of eq. 1 in estimating an *in vivo* ranitidine permeation rate constant suggest the model's validity and biopharmacentic identifiability in analyizing the oral absoption performance of ranitidine from these drug products.

5.5. Applications to a Biopharmaceutics Classification System (BCS)

In the last several years, drug regulatory agencies throughout the world have moved towards the using *in vitro* data as a means to maintain product quality and performance characteristics. For example, in the recent U.S. Food and Drug Administration "SUPAC IR" guidance for scale up and post-approval changes, *in vitro* dissolution testing serves, in many cases, as an important tool to assure drug product bioequivalence. Included in SUPAC IR is a Biopharmaceutics Classification System (BCS) which is used to set bioequivalence standards. One attribute of the BCS is drug permeability. Hence, a sponsor interested in invoking SUPAC IR would need to classify their drug product according to the BCS and, thus, would need a metric for drug permeability. Application of eq. 1 to the four ranitidine formulations indicate k_p for ranitidine from across all four formulation is about 0.01/min, which indicates ranitidine to be a low permeable drug. The observed low intestinal permeability here through the quantitative analysis of *in vitro-in vivo* relationships is in agreement with Caco-2 permeability studies. While this single example of one drug suggests the possible utility of *in vitro-in vivo* relationships to assist in classifying drug products according to a BCS, many more drug products need to be analyzed before the merits of this method to classify drug products can be determined.

6. CONCLUSIONS

The general objective of this work was to gain insight into the biopharmaceutical performance of four ranitidine hydrochloride tablet formulations. This analysis employed a recently described method[1] to relate *in vitro* and *in vivo* data and aimed to facilitate an understanding of oral drug product performance by elucidating the absolute and relative rates of dissolution and intestinal permeation in overall ranitidine absorption from each dosage form. For all four formulations, this analysis suggests absorption was permeation-rate limited, even for slow, where ranitidine exhibited a low permeation rate constant of 0.01/min across all formulations.

7. REFERENCES

1. Polli, J.E.; Crison, J.R.; Amidon, G.L. *J. Pharm. Sci.* **1996**, *85*, 753–760.
2. *USP 23-NF 18*; United States Pharmacopeial Convention, Inc.: Rockville, MD, 1994.

3. Kachigan, S.K. *Multivariate Statistical Analysis*; Radius Press, New York, 1991.
4. Goskonda, S.; Propst, C.; Augsburger, L.; Schwartz, P.; Lesko, L. *Pharm. Res.* **1994**, *11*, S-163.
5. Piscitelli, D.A.; McGlone Dalby, J.; Augsburger, L.; Shah, V.P.; Lesko, L.J.; Young, D. *Pharm. Res.*, **1995**, *12*, S-417.
6. Grant, S.M.; Langtry, H.D.; Brogden, R.N. *Drugs*, **1989**, *37*, 801–870.
7. Boyce, W.E.; DiPrima, R.C. *Elementary Differential Equations*; John Wiley & Sons: New York, 1986.

18

NONLINEAR *IN VITRO–IN VIVO* CORRELATIONS

Jeanne Mendell-Harary, James Dowell, Sian Bigora, Deborah Piscitelli, Jackie Butler, Colm Farrell, John Devane, and David Young

IVIVR Cooperative Working Group
University of Maryland at Baltimore
Pharmacokinetics and Biopharmaceutics Laboratory
Baltimore, Maryland
Elan Corporation PLC
Anthlone, Ireland

1. INTRODUCTION

A Level A *in vitro-in vivo* correlation (IVIVC) has been defined as a predictive mathematical model for the relationship between the entire *in vitro* dissolution/ release time course and the entire *in vivo* response time course (e.g. the time course of the plasma drug concentration or amount of drug absorbed.)[1]. A definite, reproducible model across individuals would have great advantages in drug development and manufacturing[2]. For example, a Level A correlation can be used to establish dissolution specifications required for quality control. More recently, a Level A correlation has been proposed as a surrogate marker for human bioequivalence studies. The goal has been to obtain a linear correlation in which the profiles of *in vitro* and *in vivo* percent released versus time are parallel. Rather than achieve linearity by iteratively altering *in vitro* dissolution tests to match the *in vivo* release data or by employing other methods such as time scaling, nonlinear functions could be used to adequately predict *in vivo* response time course. Application of nonlinear IVIVC has been suggested by several authors[3,4,5]. There are a number of examples of *in vitro-in vivo* profiles in the literature which appear to indicate curvature and where use of a nonlinear function may be more appropriate than linear regression analysis[6,7,8]. This chapter outlines several nonlinear functions which could be used to characterize nonlinear IVIVC in lieu of linear regression.

2. DATA AND FUNCTIONS

The nonlinear functions listed in Table 1 were considered as empirical equations to describe the IVIVC and compared with the linear model. In vitro mean percent released was calculated from the dissolution of 6 capsules sampled at 7 time points (0 - 12 hours). The *in vivo* percent released for each of 9 subjects was determined by deconvolution of the modified release (MR) beaded capsule formulation with an immediate release (IR) formulation using PCDCON[10]. This method provides an estimate of the *in vivo* percent released as each subject received both formulations in a cross-over fashion. It was assumed that the absorption window was similar for both products. As consistent with standard two stage analysis, parameter estimates for each individual were statistically summarized for each function.

The mean *in vitro* percent released was related to the *in vivo* percent released for each individual using the least squares options in ADAPT II[11]. Each function was independently reviewed for goodness of fit which included minimizing weighted residual sum of squares (WRSS), coefficient of variation (CV) of each parameter estimate and bias. Model discrimination was assessed by comparing Akaike information criteria (AIC)[9] of each function with those values obtained for the linear function calculated as delta AIC ($\Delta AIC = AIC_{linear} - AIC_{nonlinear}$). The criteria for the best model were based on: 1) number of $\Delta AIC > 7$, 2) ability to converge and 3) total number of significant ΔAIC (i.e. $\Delta AIC > 4$).

3. COMPARISON

The profiles for the *in vitro* and *in vivo* mean percent released are presented in Figure 1. The *in vivo* dissolution profile is not parallel to the *in vitro* dissolution profile. When *in vitro* mean percent released was plotted against *in vivo* percent released for each individual, a nonlinear shape was apparent as shown in Figures 2 - 9 for a typical subject. From visual inspection alone, the nonlinear functions would appear to fit these data better than the linear model. Similarly, curvature was seen for all the individual subject plots, and most of the nonlinear functions fit these data better than the linear model.

The parameter estimates obtained from fitting the individual IVIVC curves are presented in Table 2 as the means and standard deviations (SD). High SD values were observed with the Sigmoid model. In addition, high variability was also associated with the parameter estimates (CV% > 100) indicating that this model was overparameterized for the number of data points. Reducing the number of parameters in the Sigmoid model such as excluding the intercept (parameter D) can help alleviate this problem. High variability was also observed with the Gompertz model (parameter A). Removal of subject 3 from the summary analysis reduces the SD from 2015.77 to 5.27.

Quantitative assessment of the nonlinear functions included comparison of AIC for the nonlinear models with the linear model (ΔAIC) as summarized in Table 3. A positive number indicates the nonlinear function fits better than the linear one; most of the values

Table 1. Functions used to fit *in vitro-in vivo* data

Model name	Function	Model name	Function
Linear	$y = Ax + B$*	Weibull	$y = A - Bexp(-Cx^D)$
Sigmoid	$y = D + (Ax^B)/(C^B + x^B)$	Higuchi	$y = (Ax)^{0.5}$
Hixson-Crowell	$y = A - (A^{0.3333} - Bx)^3$	Mitcherlich	$y = A - Bexp(-Cx)$
Gompertz	$y = Aexp(-Bexp(-Cx))$	Logistic	$y = A/(1 + B exp(-Cx))$

*y = in vivo data (percent released), x = in vitro data (percent released)

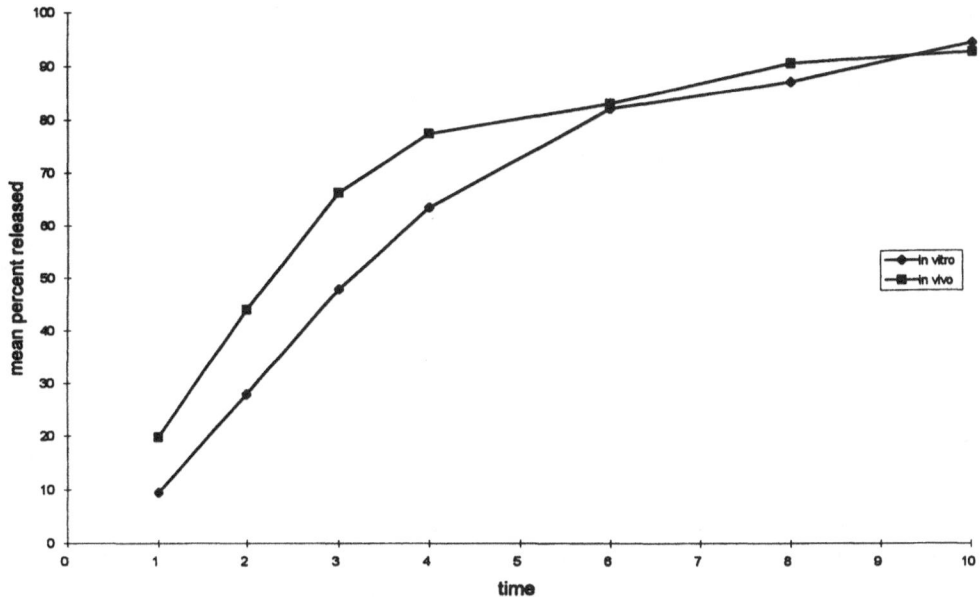

Figure 1. *In vitro* and *in vivo* dissolution profiles.

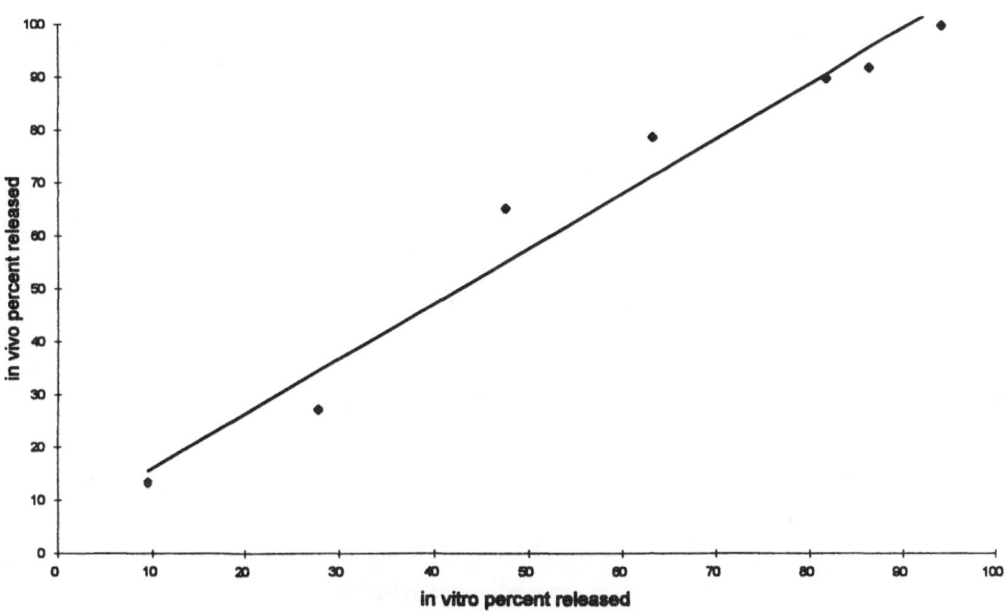

Figure 2. Linear (typical subject).

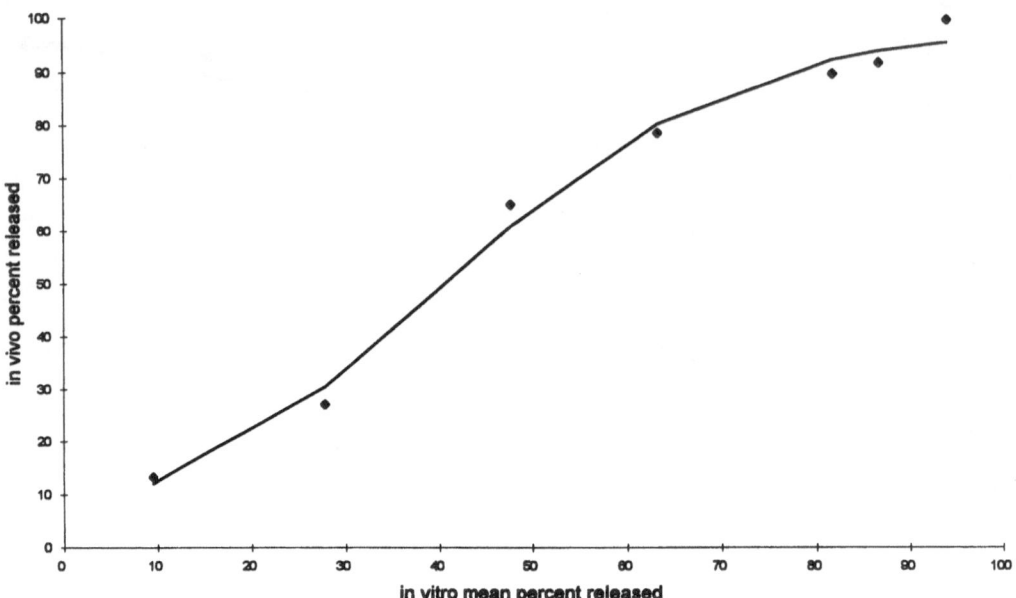

Figure 3. Weibull (typical subject).

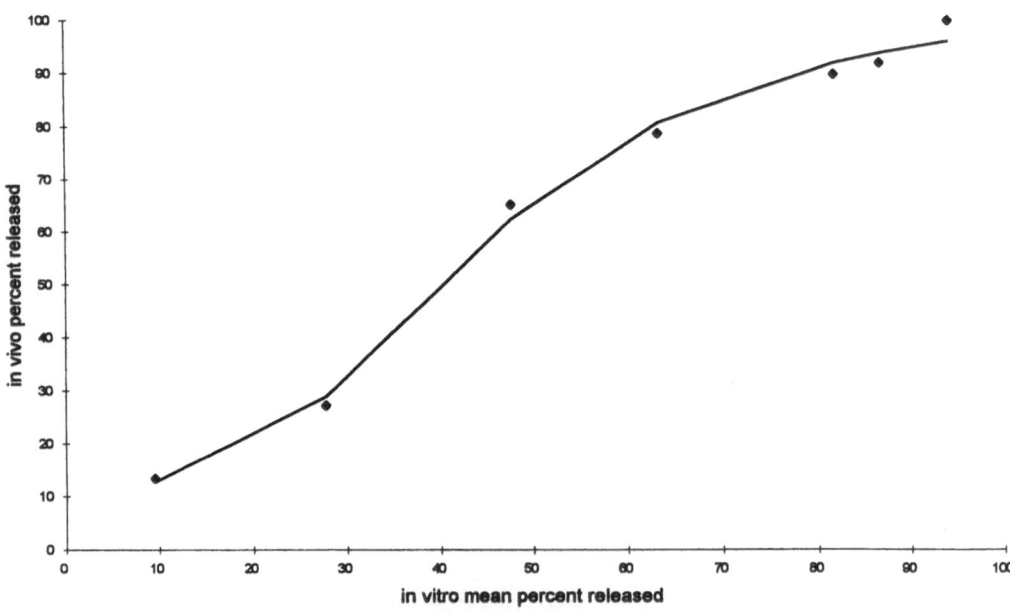

Figure 4. Sigmoid (typical subject).

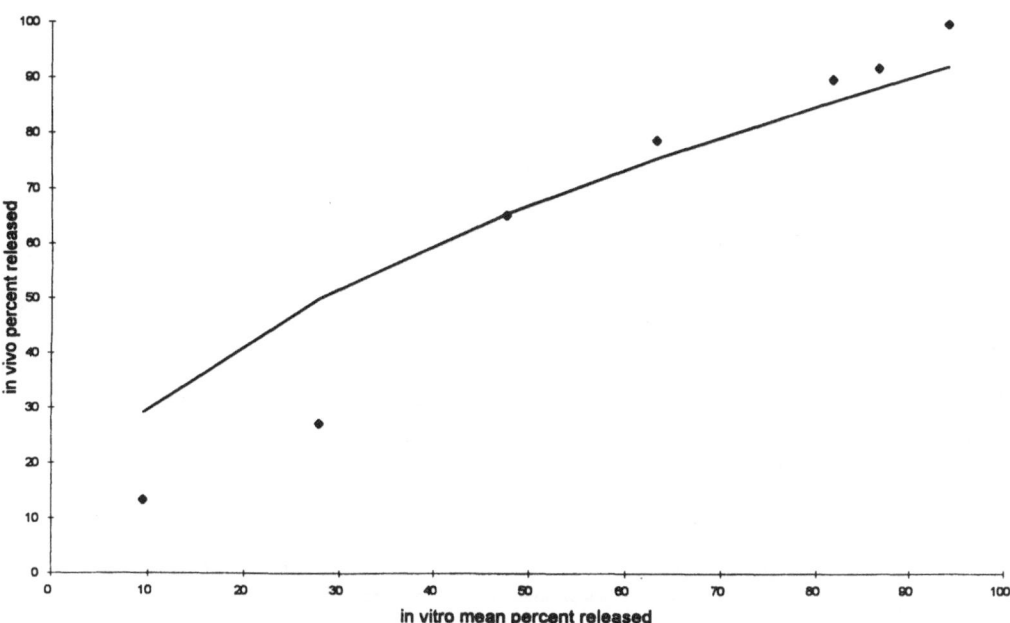

Figure 5. Higuchi (typical subject).

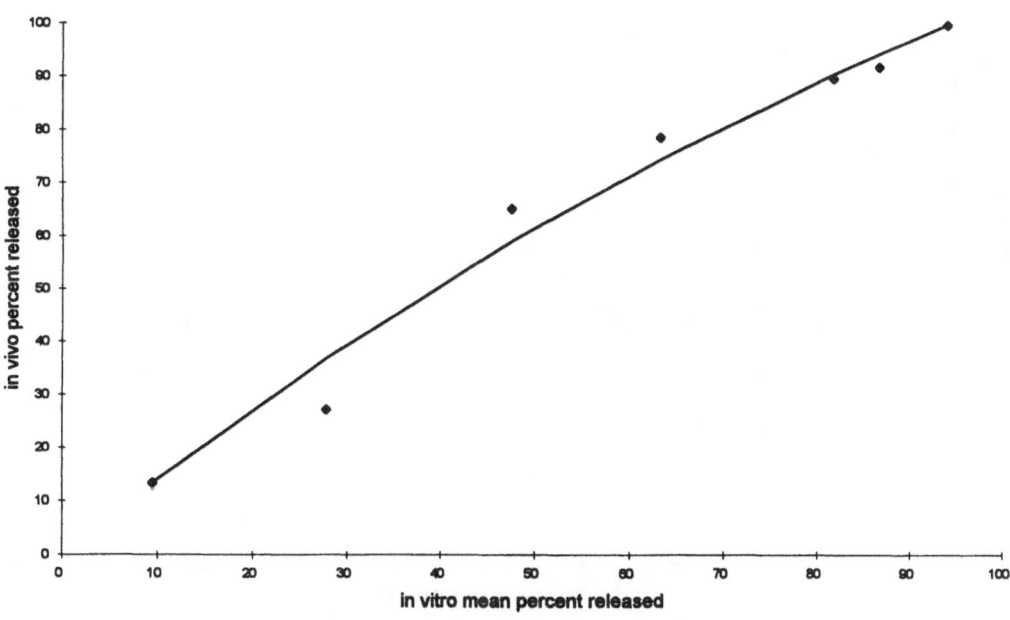

Figure 6. Hixson-Crowell (typical subject).

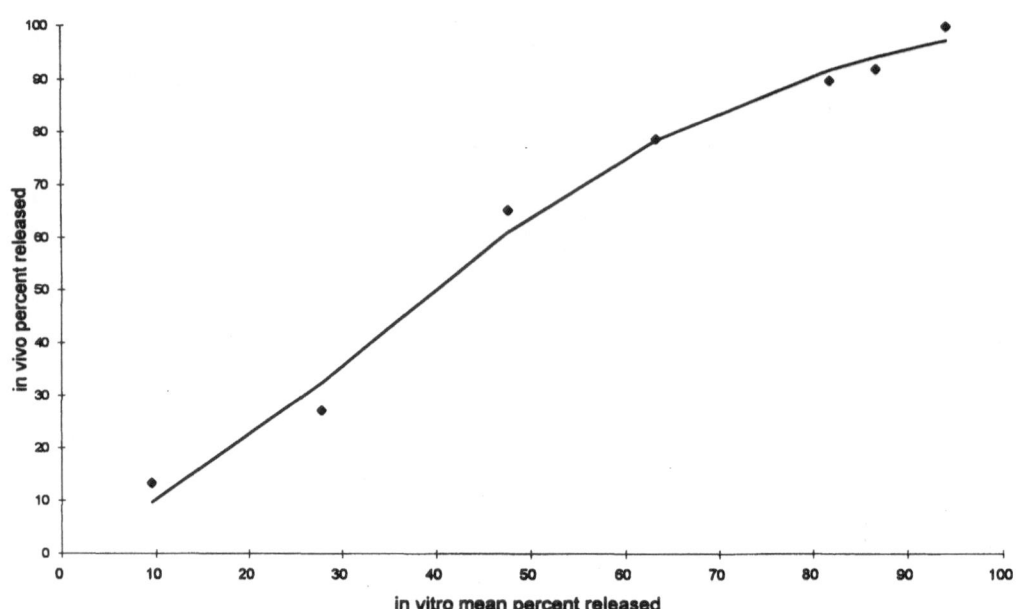

Figure 7. Gompertz (typical subject).

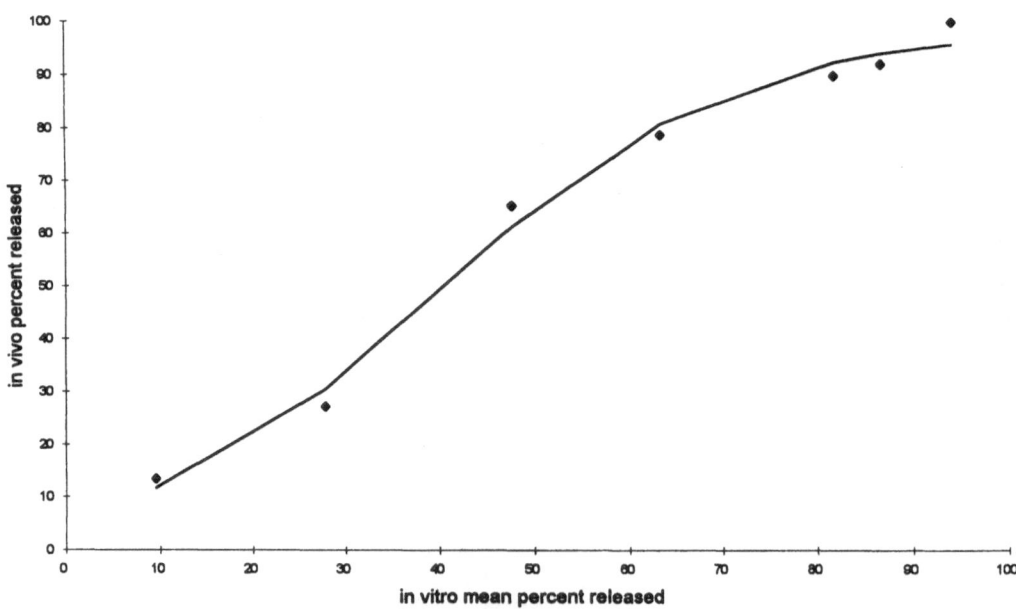

Figure 8. Logistic (typical subject).

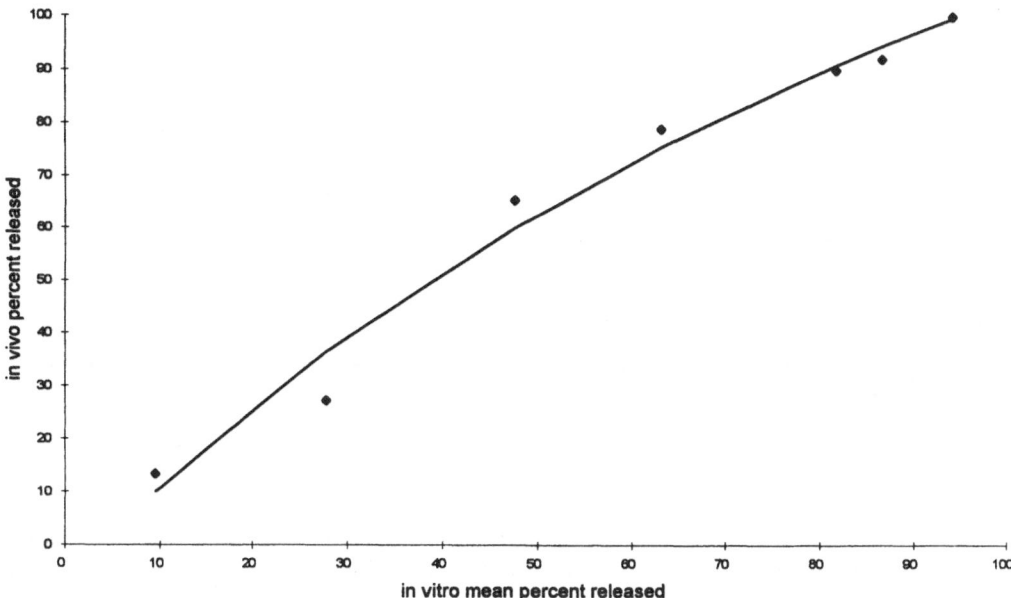

Figure 9. Mitcherlich (typical subject).

were positive. A statistically significant difference in AIC was considered as $\Delta AIC \geq 7$. On this basis, the best models were the Sigmoid, Hixson-Crowell, Gompertz and Mitcherlich. The Logistic and Weibull functions could be classified as providing moderate fits while the Higuchi resulted in the worst fit.

4. DISCUSSION

Although this is an original application of these models to IVIVC, some of these equations have been used previously to fit dissolution data. For the purpose of establishing IVIVC the equations were fit empirically, and the value of the parameters are not necessarily meaningful to the interpretation of the relationship. The number of parameters included in a model is also a consideration since usually few timepoints are collected for *in*

Table 2. Mean parameter estimates and variability

Function	A	SD	B	SD	C	SD	D	SD
Linear	0.82	0.16	19.13	15.72				
Weibull	209.98	279.22	232.71	294.74	0.0914	0.22	1.52	1.12
Sigmoid	169.67	153.88	2.51	2.68	71.66	71.66	7.2	18.33
Higuchi	88.41	17.99						
Hixson-Crowell	103.03	20.41	0.04	0.02				
Gompertz	769.63[1]	2015.77	2.97	1.47	0.04	0.02		
Logistic	106.99	40.97	8.32	4.94	0.06	0.02		
Mitcherlich	115.86	22.67	118.53	25.17	0.02	0.01		

[1]Excluding Subject 3 from the summary analysis, the mean value is 97.71 with an SD of 5.27.

Table 3. Comparison of linear with nonlinear functions (ΔAIC)

Subject	Weibull	Sigmoid	Higuchi	Hixson-Crowell	Gompertz	Logistic	Mitcherlich
1	9.11	9.1	4.15	12.77	9.59	8.61	10.4
2	12.41	12.2	1.16	6.8	9.63	5.97	14.34
3	3.06	1.73	-5.71	-0.54	2.64	2.9	NF[b]
4	5.82	5.56	1.24	13.39	5.92	6.38	5.45
6	5.78	8.93	-6.9	4.7	6.31	7.56	1.75
7	3.36	2.88	2.94	6.17	3.28	1.73	4.67
8	0.28	7.03	-1.49	9.14	7.92	5.07	8.44
9	-0.3	-0.1	0.72	9.36	1.71	1.66	1.3
10	6.82	4.86	-0.28	6.36	2.03	-0.38	5.85
No. > 4	5	6	1	8	5	5	6
No. > 7	2	4	0	4	3	2	3

[a] $\text{AIC}_{\text{linear}} - \text{AIC}_{\text{nonlinear function}}$. Positive numbers indicate a better fit than the linear model.
[b] NF indicates no fit was obtained for this subject.

vitro dissolution data (e.g. FDA guidance requires a minimum of 3 timepoints) which can quickly lead to overparameterization.

In summary, for these *in vitro-in vivo* data, the Hixson-Crowell, Sigmoid, and Gompertz functions fit best, and these functions should be considered when linear regression does not adequately characterize the data. However, these nonlinear functions should be validated with other data for different drugs and drug products and assessed for predictive performance to complete the evaluation of these functions.

REFERENCES

1. Guidance for Industry: Extended release solid oral dosage forms development, evaluation and application of *in vitro-in vivo* correlations. FDA, Center for Drug Evaluation and Research (CEDER). July 1996
2. Guidance for Industry: Immediate release solid oral dosage forms; scale-up and post-approval changes: chemistry, manufacturing and controls, In vitro dissolution testing, and *in vivo* bioequivalence documentation. Center for Drug Evaluation and Research. November 1995
3. Leeson LJ. *In vitro/In vivo* correlations. Drug Infor J 1995; 29:903–915
4. Skoug JW, Borin MT, Fleishaker JC and Cooper AM. *In vitro* and *in vivo* evaluations of whole and half tablets of sustained-release adinazolam mesylate. Pharm Research 1991; 8:1482–1488
5. Dietrich BR, Brausse R, Benedikt G and Steinijans VW. Feasibility of *in vitro/in vivo* correlation in the case of a new sustained-release theophylline pellet formulation. Arz Forsch 1988; 38:1229–1237
6. Hayashi T, Ogura T and Takagishi Y. New evaluation method for *in vitro/in vivo* correlation of enteric-coated multiple unit dosage forms. Pharm Research 1995; 12:1333–1337
7. Mendes RW, Masih SZ and Kanumuri RR. Effect of formulation and process variables on bioequivalency of nitrofurantoin II: In vitro-*in vivo* correlation. J Pharm Sci 1978; 67:1616–1619
8. Benedikt G, Steinijans VW and Dietrich R. Galenical development of a new sustained-release theophylline pellet formulation for once-daily administration. Arz Forsch 1988; 38:1203–9
9. Yamaoka A, Nakagawa T, and Uno T. Application of the Akaike Information Criteria (AIC) in the evaluation of linear pharmacokinetic equations. J Pharmacokinet Biopharm 1978; 6:165–175
10. Gillespsie WR. PCDCON: Deconvolution for pharmacokinetic applications. University of Texas at Austin. Austin, TX 1992
11. ADAPT II. David D'Argenio. Biomedical Simulations Resource. University of Southern California. Los Angeles, CA 1992

19

USE OF NONLINEAR MIXED EFFECTS MODELLING IN THE DEVELOPMENT OF *IN VITRO–IN VIVO* CORRELATIONS

Sian Bigora,[1] Deborah Piscitelli,[1] James Dowell,[1] Jackie Butler,[2] Colm Farrell,[2] John Devane,[2] and David Young[1]

IVIVR Cooperative Working Group
[1]Pharmacokinetics-Biopharmaceutics Laboratory
University of Maryland at Baltimore
Baltimore, Maryland
[2]Elan Corporation PLC
Athlone, Ireland

1. INTRODUCTION

The variability associated with the estimation of the pharmacokinetic and pharmacodynamic parameters of a population has traditionally been described using simple statistical terms such as the mean and standard deviation. Other sources of variability exist within the population, such as the quantitative relationship of the parameter to individual physiology (such as weight, age, kidney function), the magnitude of the intersubject variability across the population and the magnitude of the residual deviations between the predicted and observed drug concentrations within a subject[1,2].

Using the naive averaging of data approach (NAD), the average value of the data for each observation time is calculated. A pharmacokinetic model is then fit to the mean data. This results in the derivation of mean parameters for a mean response. The disadvantage of this approach, is it removes any reference to the individual and requires that the observation times be identical for each individual.

The standard two-stage approach first estimates the individual subject's parameters and then traditionally, reports the mean and standard deviation associated with the parameters. In order for the parameters to be estimated for each individual, sufficient data or samples have to be available.

In population analysis, three parameters are used to characterize the variability associated with the population. They are the fixed effect mean parameter values, the random intersubject effect and the random residual error (intrasubject variance) of the parameter[3]. These effects can be calculated using the two-stage approach and more advanced techniques such as iterative two-stage method (IT2S) and nonlinear mixed effects modelling (NONMEM).

In Vitro–in Vivo Correlations, edited by Young *et al.*
Plenum Press, New York, 1997

Nonlinear mixed effects modelling (NONMEM) was developed to address some of the limitations of the two stage approach used in population analysis[4,5]. NONMEM uses extended least squares regression to estimate the fixed effect, random intersubject effect and the random residual error parameters. The major advantages of this type of modelling are the simultaneous estimation of all population parameters, the ability to estimate the confidence intervals for the parameters, a statistical evaluation of the model adequacy and the ability to use sparse data where traditional methods would not be able to characterize the individual pharmacokinetic parameters.

The classical approach for the development of an *in vitro-in vivo* correlation (IVIVC) is to use linear regression to compare a measure of *in vitro* dissolution or release to a measure of *in vivo* absorption or release[6]. The most desirable relationship is a Level A correlation, which directly compares point-to-point from the *in vitro* and *in vivo* profiles[7]. Based on this relationship, *in vitro* and *in vivo* time profiles are parallel to each other. The slope of the linear regression line is tested for any significant difference from one. The intercept is assumed to be zero unless a time-lag is present.

In many cases, observation of the IVIVC may suggest a nonlinear relationship rather than the usual linear relationship[8-10]. However, the shape of the correlation is generally ignored and a "best fit" line is used to describe the correlation, which can result in significant bias. Other functions can be used to describe the shape of such a correlation, e.g. Sigmoid, Mitcherlich, Gompertz functions. These functions may more accurately describe the IVIVC resulting in a more reproducible model across subjects which can be used in drug development and manufacturing.

This chapter outlines the development of an *in vitro-in vivo* correlation using nonlinear mixed effect modelling (NONMEM). Using this modelling technique, the ability of a linear and 4 nonlinear functions to relate the *in vitro* measure of dissolution for one formulation of a drug to an *in vivo* measure of the absorption can be compared. From the evaluation of the models, the intersubject variability of the *in vitro-in vivo* correlation could also be estimated and compared to the standard two-stage method.

2. METHODOLOGY

2.1. Data

The data for this analysis were obtained from a phase I, 9 subject crossover study for an immediate and an extended release formulation. The mean *in vitro* cumulative amount dissolved was calculated from the dissolution of 6 capsules sampled at each of the time-points 1, 2, 3, 4, 6, 8, and 10 hours. The *in vivo* cumulative amount of drug absorbed at each time-point was estimated by performing deconvolution of the individual plasma concentration-time profiles for the extended release product using PCDCON [11,12]. The unit impulse parameters for each individual were obtained by fitting a one-compartment oral absorption model to the immediate release formulation *in vivo* data using ADAPT II[13]. The relationship between *the in vitro* and *in vivo* cumulative amount dissolved and absorbed is presented in Figure 1.

2.2. *In Vitro-in Vivo* Correlations

Four nonlinear functions were evaluated in this population analysis. The nonlinear functions were Sigmoid, Gompertz, Mitcherlich and Weibull. The equations for each of these functions are presented in Table 1.

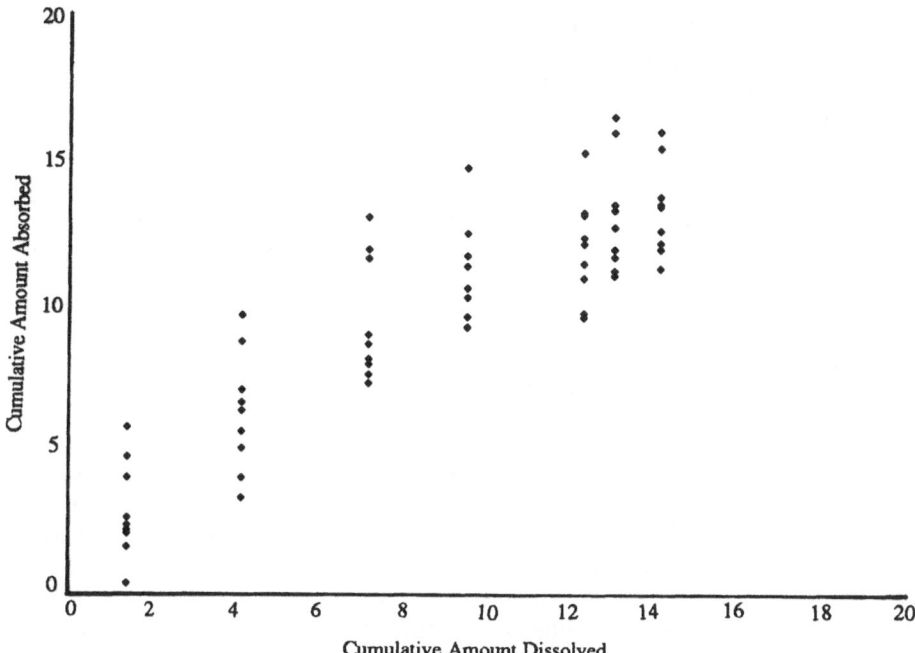

Figure 1. *In vitro-in vivo* correlation between the cumulative amounts dissolved and absorbed.

The functions were evaluated in NONMEM using the follow routines and structural model features:

- first order estimation (posthoc)
- $pred subroutine
- exponential intersubject error (η)
- proportional random residual error (ε)
- intersubject covariance matrix (omega block)

The criterion for determining the best model were:

- difference in the maximum likelihood function(MOF) defined by the use of Akaike Information Criteria (AIC):

$(MOF_A - MOF_B)*(2*(p_A - p_B))$

where p=number of parameters evaluated in the function

Table 1. *In vitro-in vivo correlation* equations

Model	Equation
Linear	$Y = A*Diss + B$
Gompertz	$Y = A*exp(-B*exp(-C*Diss))$
Sigmoid	$Y = D+(A*Diss^B)/(C^B+Diss^B)$
Weibull	$Y=A-(B*exp(-C*Diss^D)$
Mitcherlich	$Y = A-(B*exp(-C*Diss)$

Diss = Cumulative Amount Dissolved at Time(t)
Y= Cumulative Amount Absorbed at Time (t)

- decrease in the inter-(η) and intrasubject(ε) variability
- visual inspection of the observed versus individual predicted cumulative amount absorbed plots
- visual inspection of the weighted residuals versus predicted cumulative amount absorbed plots

3. RESULTS

3.1. Nonlinear Mixed Effects Model Results

The differences in the maximum likelihood function for each model with regard to the linear model are described in Table 2. Based on the AIC comparison of the maximum likelihood function, all four nonlinear IVIVCs produced a better fit than the linear IVIVC. There does not appear to be significant differences between the Weibull and Sigmoid functions, but based on the difference in the maximum likelihood function, these two functions did perform better than the Gompertz and Mitcherlich functions. The parameter estimates, standard errors and intersubject variability for all the models are presented in Table 3. The coefficient of variation (CV%) of the parameter estimates was relatively small for each of the models.

The predictive performance of each of the models are illustrated in the observed vs. individual predicted cumulative amount absorbed profiles (Figures 2–6). Based on visual inspection of these profiles and the weighted residuals vs. predicted cumulative amount absorbed profiles, there did not appear to be any major differences in bias between any of the models.

3.2. Standard Two-Stage Results

More detailed information on this approach can be found in the Nonlinear IVIVC chapter. A comparison of the intersubject variability associated with the parameter estimates from each of the four nonlinear and linear models, using the NONMEM and standard two-stage methods can be found in Table 4. For all the models, except the Mitcherlich model, the intersubject variability was lower from the NONMEM method compared to the standard two-stage method.

Table 2. Comparison of linear and nonlinear IVIVCs

Function	MOF[†]	# Parameters	Δ AIC[‡]
Linear	143.3	2	
Gompertz	94.6	3	-97.4
Sigmoid	89.4	4	-215.6
Weibull	89.9	4	-213.6
Mitcherlich	91.5	3	-103.6

[†] If Δ AIC < 0 the nonlinear model performs best.

[‡] MOF - Minimum Objective Function

Table 3. Vector estimates from NONMEM IVIVC models

Model and vector	Vector estimate	Intersubject variability (coefficient of variation)
Linear		
A	0.83 (5.7 %)[†]	16 %
B	2.80 (25.3 %)	65 %
Sigma	0.0217 (12.5 %)	
Gompertz		
A	13.7 (5.5 %)	20 %
B	2.08 (11.2 %)	37 %
C	0.25 (13.1 %)	58 %
Sigma	0.0061 (13.3 %)	
Weibull		
A	14.4 (12.3 %)	29 %
B	13.1 (19.2 %)	39 %
C	0.09 (47.1 %)	121 %
D	1.24 (21.5 %)	37 %
Sigma	0.005 (20.3 %)	
Sigmoid		
A	14.9 (24.3 %)	61 %
B	1.60 (25.9 %)	53 %
	6.91 (28.6 %)	84 %
D	1.78 (56.2 %)	141 %
Sigma	0.005 (22.0 %)	
Mitcherlich		
A	15.5 (9.2 %)	41 %
B	15.1 (7.9 %)	35 %
C	0.13 (21.4 %)	103 %
Sigma	0.006 (18.1 %)	

[†] Mean (Coefficient of Variation (CV %) of parameter estimate)
Sigma - intrasubject variabilty

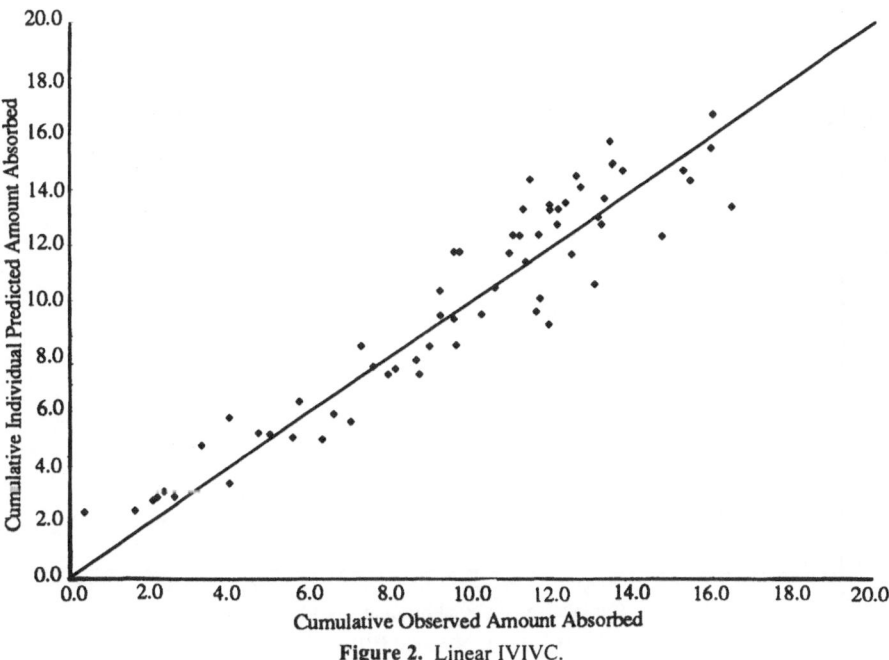

Figure 2. Linear IVIVC.

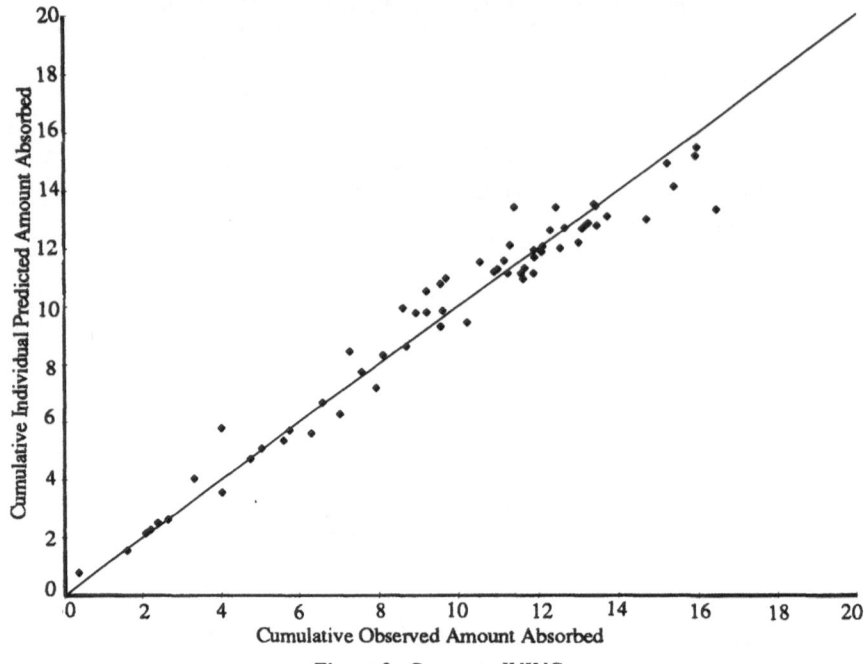

Figure 3. Gompertz IVIVC.

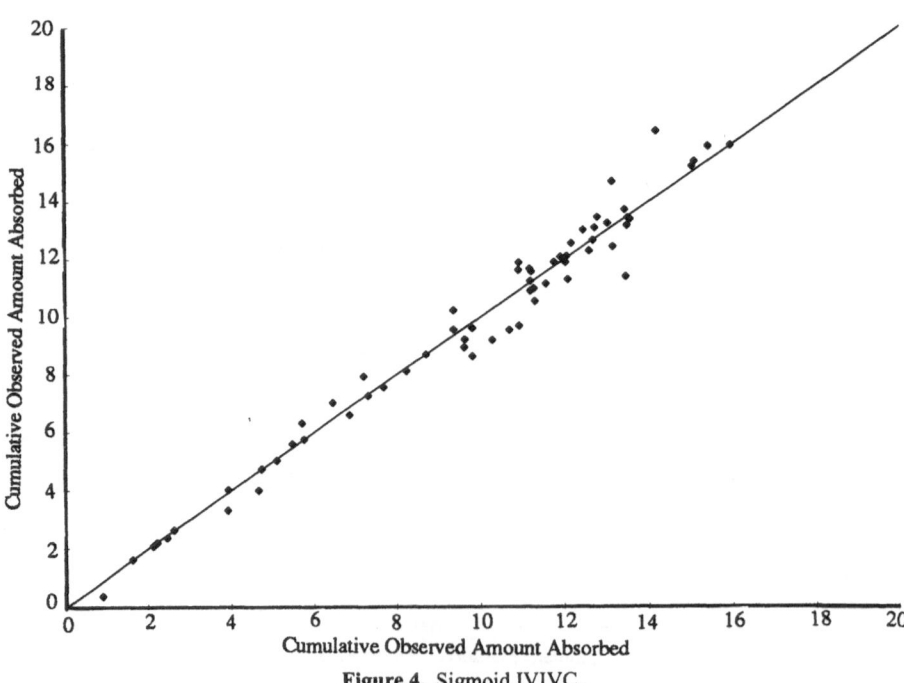

Figure 4. Sigmoid IVIVC.

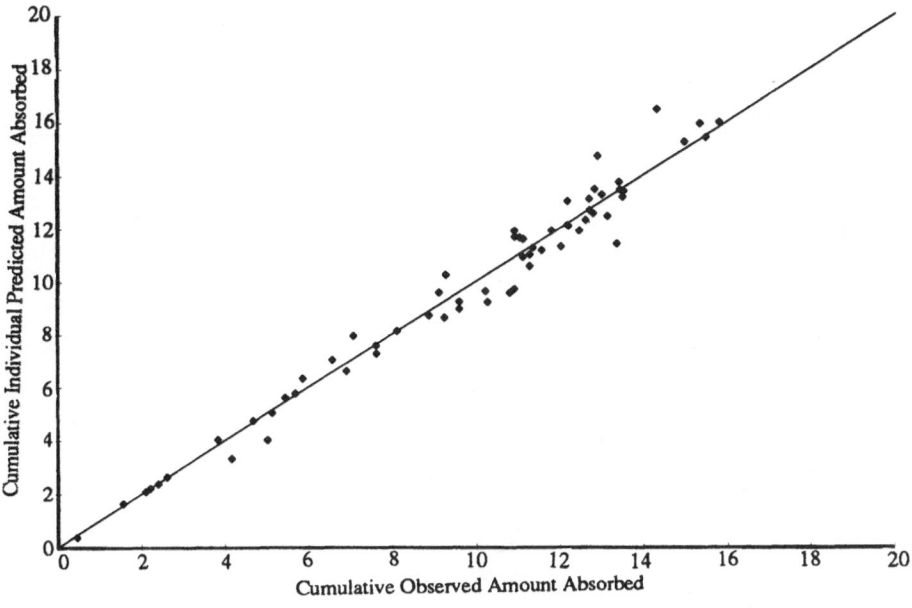

Figure 5. Weibull IVIVC.

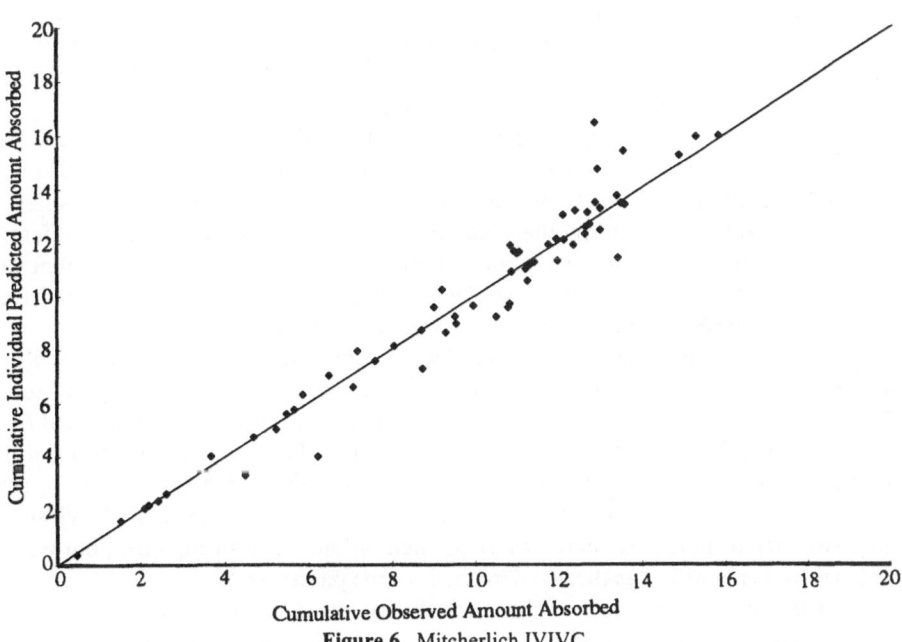

Figure 6. Mitcherlich IVIVC.

Table 4. Comparison of intersubject variability from the NONMEM
and standard two-stage methods

Model and vector	NONMEM intersubject variability	Standard two-stage intersubject variability
Linear		
A	16 %	20 %
B	65 %	82%
Gompertz		
A	20 %	262 %
B	37 %	49 %
C	58 %	50 %
Weibull		
A	29 %	111 %
B	39 %	127 %
C	121 %	241 %
D	37 %	74 %
Sigmoid		
A	61 %	91 %
B	53 %	107 %
C	84 %	72 %
D	141 %	254 %
Mitcherlich		
A	41 %	20 %
B	35 %	21 %
C	103 %	50 %

4. DISCUSSION

The objective of this research was to evaluate a new methodology for developing *in vitro-in vivo* correlations. The advantage of population analysis techniques in the development of IVIVCs is the additional information that can be gained in the estimation of the intersubject variability, random residual variability (intrasubject variability) and the intersubject covariance between the parameters of the model.

Nonlinear mixed effects modeling is a feasible method for developing new IVIVCs. One major advantage of the NONMEM approach in the development of IVIVCs, is the simultaneous modelling of all the subjects. Additionally, the estimation of parameters was possible for all subjects. During the standard two-stage approach, some subjects were eliminated from the estimation of the mean parameter estimates. This was either because of an inability to fit the model to the subject's data, or the parameters obtained from the individual's fit caused a large increase in the intersubject variability.

In this analysis, the parameter estimates were similar from both methods. However, the coefficient of variation associated with the parameter estimates, was lower with the NONMEM method. This may be related to the overparameterization with the individual fits. Generally, the individual IVIVCs were based on 7 data points. For many of the nonlinear models, the number of parameters being estimated was 3 or 4. With the NONMEM method, this is no longer the case, as the data from all the subject are simultaneously evaluated. Within the parameter estimates from the standard two stage method there is a fitting error, but with NONMEM this error is identified as the standard error of the parameter[1,2].

The estimates of the intersubject variability are also lower with the NONMEM method compared to the standard two-stage approach. This may be because of the limited data or variability in an individual's *in vivo* data. The estimate of intersubject variability is more easily biased with the two-stage method by one subject[1,2].

Based on the change in maximum likelihood function, the nonlinear IVIVCs resulted in a better fit of the data then the linear correlation. Although some of the nonlinear models examined appear to more accurately describe the data than others, the objective of this research was not to determine the best nonlinear function in comparison to other nonlinear functions. Our intention was to illustrate the concept of a nonlinear IVIVC instead of the classical linear correlation. Other nonlinear functions also could have been used to describe this data, such as a polynomial, logistic or Hixson-Cromwell function.

In summary, population analysis can be used to develop a IVIVC. Nonlinear mixed effects modelling was able to efficiently determine the parameter estimates for either a nonlinear or linear function. The intersubject variability was also decreased using this method in comparison to the standard-two stage method. This technique could also be expanded to include multiple formulations and to evaluate the effect of different excipients on parameters within the IVIVC.

5. REFERENCES

1. Sheiner LB, Grasela TH. An introduction to mixed effects modeling: concepts, definitions and justification. J Pharmacokinetic Biopharm 1991;19:11S-23S.
2. Sheiner LB. The population approach to pharmacokinetic data analysis: rationale and standard data analysis methods. Drug Metab Rev 1984;15:153–71.
3. Sheiner LB, Beal SL. Evaluation of methods for estimating population pharmacokinetic parameters II. Biexponential model and experimental pharmacokinetic data. J Pharmacokinet Biopharm 1981;9:635–651.
4. Beal SL, Sheiner LB. Population pharmacokinetic data and parameter estimation based on their first two statistical moments. Drug Metab Rev 1984;15:173–93.
5. Beal SL and Sheiner LB. NONMEM Users guides, NONMEM Project Group, UCSF, San Francisco, CA, 1992.
6. In vitro-*in vivo* correlations. Pharmacopeial Forum USP 1993;19:5374–9.
7. 'Draft guidance for industry: modified release solid dosage forms; scale-up and post-approval changes: chemistry, manufacturing and controls, *in vitro* dissolution testing and *in vivo* bioequivalence documentation. U.S. Department of Health and Human Services, Food and Drug Administration, July 1996.
8. Leeson LJ. In vitro/*in vivo* correlations. Drug Info J 1995;29:903–15.
9. Hussein Z, Freidman M. Release and absorption characteristics of novel theophylline sustained-release formulations: *in vitro-in vivo* correlation. Pharm Res 1990;7:1167–71.
10. Hyashi T, Ogura T, Takagishi Y. New evaluation method for *in vitro/in vivo* correlation of enteric coated multiple unit dosage forms. Pharm Res 1995;12:1333–7.
11. Gillespie WR, Veng-Pedersen P. A polyexponential deconvolution method. Evaluation of the "gastrointestinal bioavailability" and mean *in vivo* dissolution time of some ibuprofen dosage forms. J Pharmacokinet Biopharm 1985;13:289–307.
12. Gillespie WR. PCDCON: Deconvolution for pharmacokinetic applications. University of Texas at Austin, Austin TX. 1992
13. ADAPT II. D'Argenio D. Biomedical Simulations Resource, USC, CA.

20

THE DEVELOPMENT OF A NOVEL *IN VITRO* DISCRIMINATORY DISSOLUTION METHOD FOR A CLASS I DRUG IN A MATRIX TABLET FORMULATION

C. Farrell, J. Butler, P. Stark, H. Madden, and J. Devane

IVIVR Cooperative Working Group
Elan Corporation
Athlone, Co. Westmeath, Ireland

1. INTRODUCTION

A controlled release hydrophilic matrix tablet formulation of a Class I (Amidon et al., 1995) drug containing hydroxypropylmethylcellulose (HPMC) has been developed. The level of HPMC has previously been shown to influence the *in vivo* release of the drug from this formulation. However, a standard dissolution test utilising USP II aqueous methodology failed to discriminate sufficiently between formulations. Therefore, work was undertaken to develop a novel discriminatory dissolution method based on retrospective *in vivo* data.

2. METHODS

2.1. *In Vitro* Dissolution Methods

USP II (paddles) and USP III (BioDis) apparatus was used to develop the novel discriminatory test. Both distilled water (dH_2O) and phosphate buffer (KH_2PO_4) was examined as media for this method. The speed for the USP II apparatus ranged from 50 to 250 rpm while the speed for the USP III apparatus was 20 to 30 dpm.

2.2. *In Vivo* Methods

To verify the novel discriminatory dissolution method, a prospective *in vitro in vivo* study was designed. Four formulations, varying only in the level of HPMC, were evaluated in a ten-subject, randomised crossover biostudy. An immediate release formulation of the drug was also included to assist in the deconvolution of the plasma concentration versus time data.

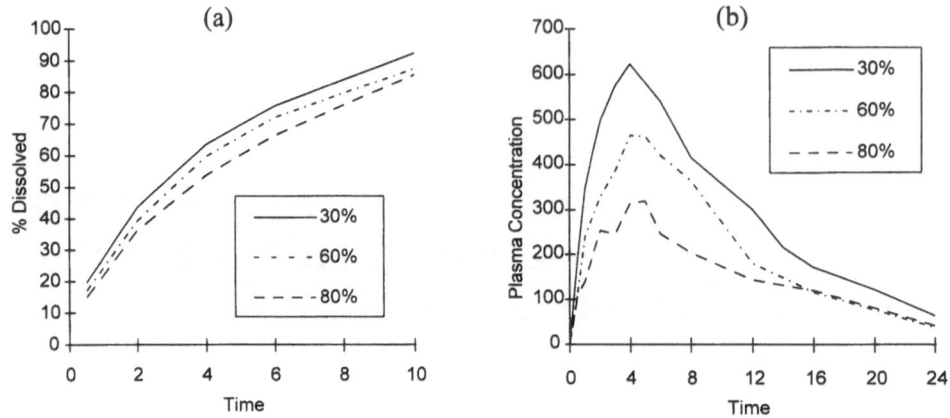

Figure 1. (a) *In vitro* dissolution profiles of formulations with varying HPMC (30, 60 and 80%) content using standard USP II test (dH$_2$O, 100 rpm). (b) Retrospective *in vivo* pharmacokinetic profiles of these formulations.

3. RESULTS

3.1. Standard USP II apparatus

Using dH$_2$O as a medium, a standard dissolution test using USP II methodology failed to discriminate sufficiently between formulations of varying HPMC levels (Figure 1).

Altering the agitation (50 - 250 rpm) in this USP II system did not change the dissolution profiles of the three formulations. Changing the dissolution apparatus to USP III methodology also did not give sufficient discrimination between the *in vitro* profiles of the formulations (Figure 2).

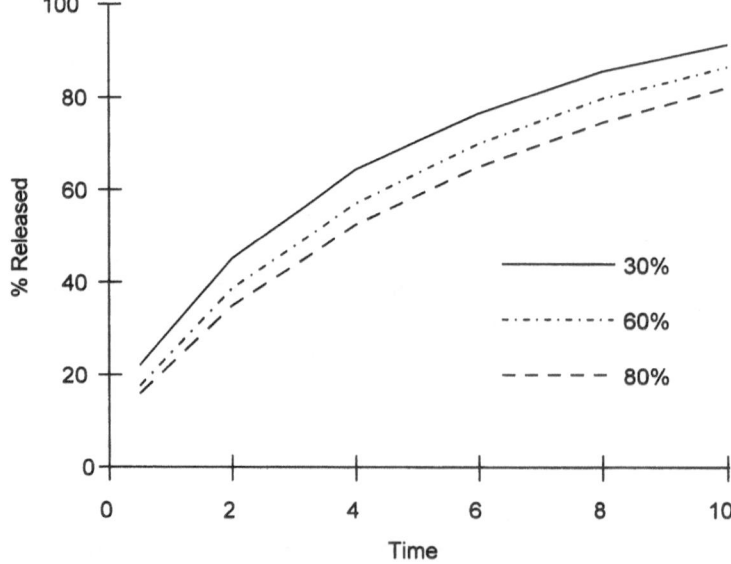

Figure 2. *In vitro* dissolution profiles using USP III apparatus (dH$_2$O, 20 dpm).

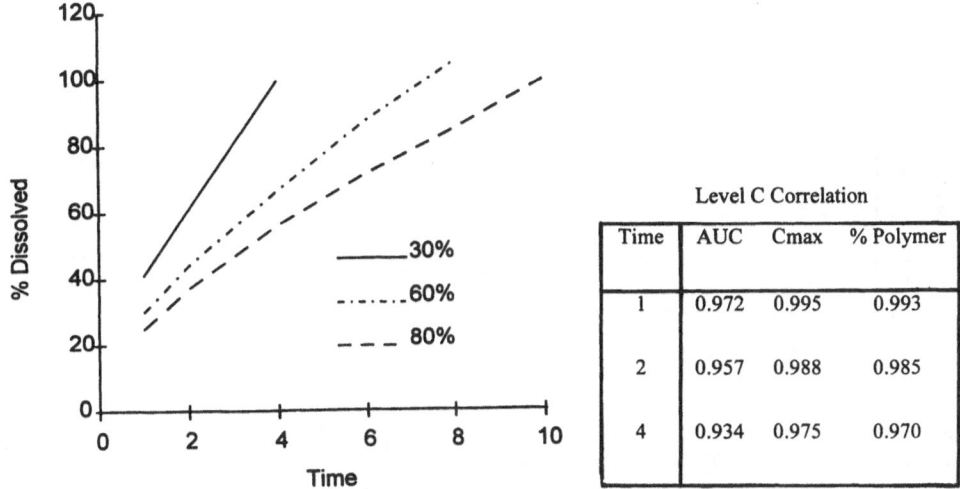

Figure 3. *In vitro* dissolution profiles using USP III with 0.35M KH₂PO₄ and 20 dpm.

The introduction of phosphate buffer to the USP II system resulted in a better discrimination between the formulations. A combination of USP III apparatus and 0.35M KH₂PO₄ gave good discrimination between the *in vitro* profiles and also demonstrated good correlation between the *in vitro* dissolution and the pharmacokinetic parameters, Cmax and AUC, and the % polymer composition, particularly in the range of 40 to 80% HPMC range (Figure 3).

However, further studies revealed that this test was highly sensitive to other variables such as slight variation in the dissolution media temperature and to changes in tablet

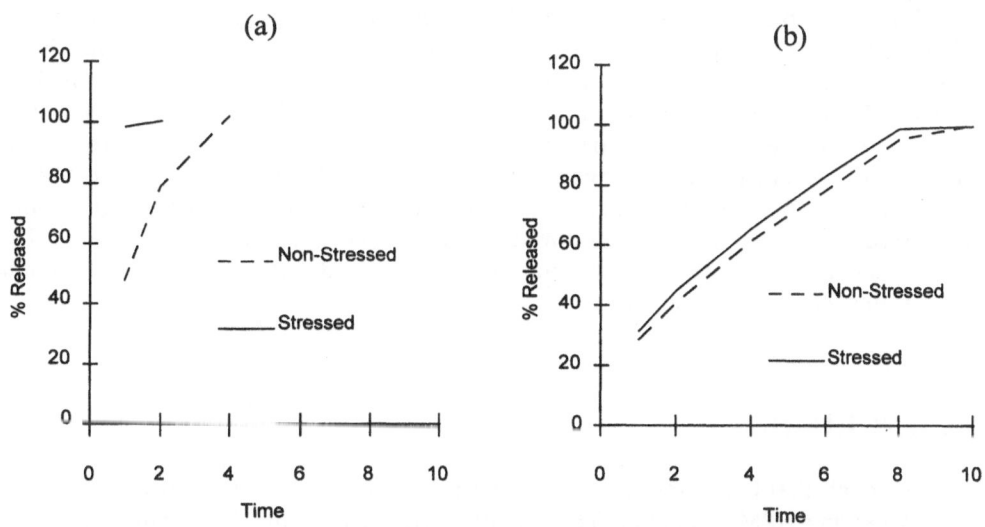

Figure 4. *In vitro* dissolution profiles of non-stressed (RT) and stressed (40°C, 75% RH) products in USP III apparatus with (a) 0.35M KH₂PO₄, 20 dpm and (b) 0.32M KH₂PO₄, 30 dpm.

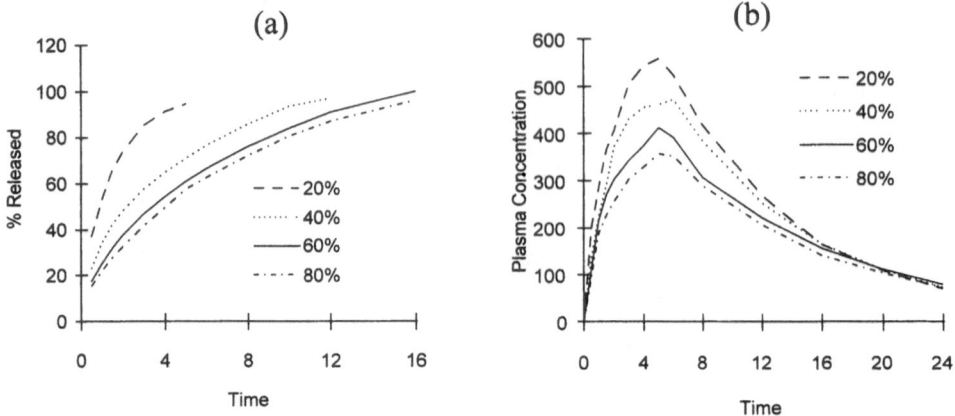

Figure 5. (a) *In vitro* dissolution profile of formulations with varying HPMC levels (20 - 80%) using USP III apparatus with 0.32M KH_2PO_4, 30 dpm. (b) *In vivo* pharmacokinetic profiles of these formulations.

hardness over time for stressed (40^0C, 75% RH) stability lots, which were found not to differ *in vivo*. Reducing the ionic strength to 0.32 M and increasing the dipping rate of the USP III apparatus eliminated these findings while retaining discrimination between % polymer content (Figures 4 and 5).

A further biostudy showed that, despite reducing the ionic strength, this method still retained some sensitivity to variables such as MeO/HOPrO ratio for which no *in vivo* sensitivity was found.

4. DISCUSSION

For a controlled release HPMC-containing hydrophilic matrix tablet formulation of a Class I drug, standard USP II aqueous methodology failed to sufficiently discriminate between formulations for which significant difference in *in vivo* performance had been shown. The use of a novel discriminatory dissolution method (USP III (BioDis), 0.35M KH_2PO_4), developed using retrospective *in vivo* data, could discriminate between these formulations. Reducing the molar concentration of the medium to 0.32M eliminated *in vitro* sensitivities to factors for which no *in vivo* sensitivity had been found while retaining the required discrimination between formulations of varying polymer composition. However, this method did retain some sensitivity to other manufacturing variables for which *in vivo* differences were not shown.

5. CONCLUSION

A novel dissolution method has been developed to discriminate between formulations of varying HPMC content. The role of *in vivo* data in the development of *in vitro* dissolution methods has been highlighted.

21

IN VIVO–IN VITRO EVALUATION OF THE IMPACT OF ACCELERATED STABILITY CONDITIONS ON A HYDROPHILIC MATRIX TABLET

P. Stark, A. Kinahan, S. Cunningham, C. Farrell, J. Butler, M. Reilly, and J. Devane

IVIVR Co-operative Working Group
Elan Corporation plc
Athlone, Co. Westmeath, Ireland

1. INTRODUCTION

Changes occur in the tensile strength of compacts produced from powdered Hydroxypropyl methylcellulose (HPMC) after storage at different relative humidities (S. Malamataris et al., Int. J. Pharm., 1994). Studies in this laboratory have noted that subjecting a hydrophilic matrix formulation of a Class I drug (Amidon et al., 1995) to accelerated stability conditions resulted in a reduction in tablet crushing strength. An *in vitro* dissolution method developed in this laboratory indicated that such physical changes in this hydrophilic matrix formulation could potentially result in an altered *in vivo* profile.

This study is concerned with the use of the *in vitro* discriminatory dissolution method, developed in this laboratory (C. Farrell et al., The development of a novel *in vitro* discriminatory dissolution method for a class I drug in a matrix tablet formulation), to predict the *in vivo* performance of a hydrophilic matrix tablet formulation.

2. MATERIALS AND METIIODS

2.1. Formulation Details

The formulation of the Hydrophilic Matrix Tablet is described in Table 1.

In Vitro–in Vivo Correlations, edited by Young *et al.*
Plenum Press, New York, 1997

Table 1. Formulation details for the hydrophilic matrix tablet

Component	Function	% w/w
Class I drug	Active	-
Hydroxypropyl methylcellulose	CR polymer	60
Diluent 1	Filler	q.s
Diluent 2	Filler	q.s
Glidant	Glidant	<1%
Lubricant	Lubricant	<1%

2.2. Packaging

100 Tablets were packaged in 100ml white high density polyethylene wide-round bottles with a 38/400 mm finish (Treitler-Owens) with 12g of polyester coil and one 2g sorb-it desiccant canister. The bottles were sealed with 38/400mm white polypropylene fine ribbed closure with plain HS-130 heat induction liner.

2.3. Accelerated Stability

Containers were stored in stability cabinets at accelerated stability conditions of 40°C + 75% RH and at room conditions of 25°C + 60% RH. The resultant products were regarded as 'stressed' and 'non-stressed' respectively. Tablet crushing strength was recorded at appropriate intervals during storage using a Schleuniger Tablet Tester 6D.

2.4. *In-Vitro* Dissolution Method

In-Vitro Dissolution Testing was carried out using a BioDis Apparatus (USP III) with a 20 mesh polypropylene screen, at 30 dips/minute in 250ml potassium di-hydrogen phosphate buffer (KH_2PO_4, 0.32M, pH 4.3) maintained at 37°C. UV analysis of the dissolution samples was performed using a Waters 480 HPLC system at 220nm.

2.5. *In-Vivo* Evaluation

Tablets stored at 40°C + 75% RH and at 25°C + 60% RH for three months were administered in an eight subject randomised cross-over study.

3. RESULTS AND DISCUSSION

3.1. Tablet Crushing Strength

After storage at 40°C + 75% RH for three months, the crushing strength of the resultant stressed product was 37% lower than the non-stressed product, which had been stored at 25°C + 60% RH. No significant change in moisture content or in the potency of the tablets was noted. Within three months at accelerated stability conditions, the tablet crushing strength of the stressed product was recorded to drop from 269N (Month 0) to 173N (Month 3), with the greater reduction occurring during the first two months. The stressed product did not undergo any further decrease in tablet crushing strength after the third

Table 2. Tablet crushing strength of non-stressed and stressed
hydrophilic matrix tablets

Tablets conditions	Non-stressed 25°c/60%RH		Stressed 40°c/75%RH	
Storage (months)	Range	Mean	Range	Mean
Month 0	249-283N	269N	249-283N	269N
Month 1	N/E	N/E	208-241N	226N
Month 2	N/E	N/E	166-199N	186N
Month 3	260-288N	276N	159-186N	173N
Month 4	N/E	N/E	163-180N	173N
Month 6	256-277N	268N	158-183N	172N

N/E = Not examined

month. Table 2 summarises the reduction in tablet crushing strength over six months storage for both the non-stressed and stressed tablets.

3.2. *In-Vitro* Evaluation

The *in vitro* discriminatory method produced similar *in vitro* profiles for the stressed and corresponding non-stressed products and thus suggested that there would be no *in vivo* difference between the tablets. The *in vitro* dissolution profiles for each of these products are illustrated in Figure 1.

3.3. *In-Vivo* Evaluation

In vivo evaluation demonstrated that both products are bioequivalent in terms of extent and rate of absorption in this group of volunteers. No significant difference in the Cmax or AUC parameters were reported between the treatments and thus confirmed the results forecasted by the *in vitro* dissolution testing. Table 3 summarises the relevant pharmacokinetic parameters, while Figure 2 illustrates the mean plasma profiles after administration.

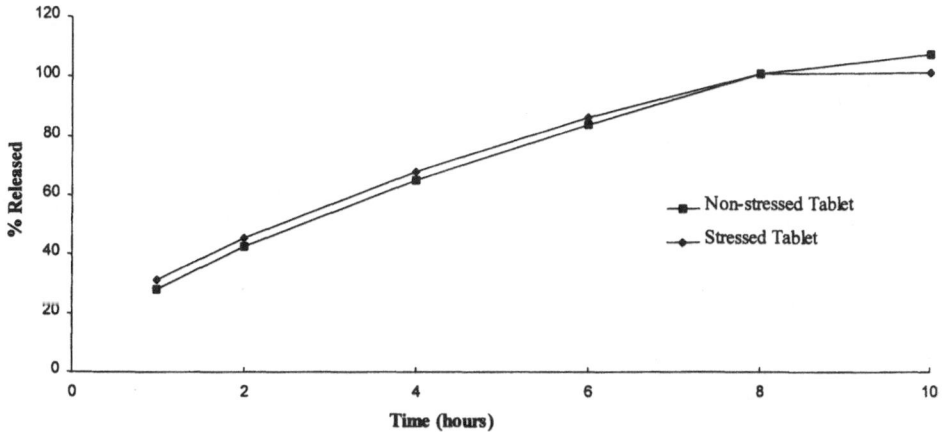

Figure 1. *In vitro* dissolution profiles of non-stressed and stressed hydrophilic matrix tablets

Table 3. *In-vivo* evaluation of non-stressed and stressed hydrophilic matrix tablets

Parameter	Non-stressed geometric mean (GSD)	Non-stressed geometric mean (GSD)	90% Confidence interval (stressed/non-stressed)log transformed
AUC $_{(0-infinity)}$	6008.75(1.30)	5573.80(1.22)	83 - 104
C_{max}	423.62 (1.26)	431.01(1.13)	94 - 110

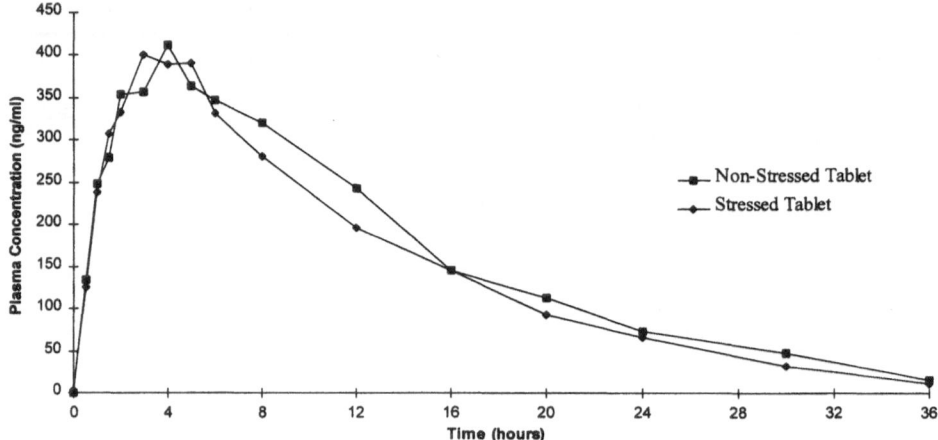

Figure 2. Mean plasma profiles of non-stressed and stressed hydrophilic matrix tables.

4. CONCLUSIONS

The *in vivo* performance of the hydrophilic matrix formulation is not affected by storage at accelerated stability conditions. The *in vitro* discriminatory dissolution method has successfully indicated that the stressed and non-stressed products are bioequivalent and that storage at accelerated stability conditions does not impact on the *in vivo* performance of the resultant product.

22

DEVELOPMENT OF *IN VITRO–IN VIVO* CORRELATIONS USING VARIOUS ARTIFICIAL NEURAL NETWORK CONFIGURATIONS

James A. Dowell,[1] Ajaz S. Hussain,[2] Paul Stark,[3] John Devane,[3] and David Young[1]

[1]University of Maryland at Baltimore
Baltimore, Maryland
[2]Food and Drug Administration/CDER
Rockville, Maryland
[3]Elan Corporation, PLC
Athlone, Ireland

1. INTRODUCTION

It is desirable to have a predictive tool to determine the *in vivo* pharmacokinetics based on the *in vitro* dissolution and other important variables. We can see the *in vitro - in vivo* correlation (IVIVC) as an input-output relationship, and often are not interested in the internal structure of this model as long as we have a good, validated, predictive tool. This may be important, for example, in product development or to establish dissolution specifications. Many of the previous examples in this book use parametric models to define an IVIVC. For example, simple linear models are often used to relate a parameter or a time point descriptive of the dissolution to a parameter or a time point descriptive of the pharmacokinetic absorption[1-3]. These models, however, can be unsuccessful in completely describing the IVIVC, and sometimes no correlation can be determined. The number of possible variables, the model unable to account for some physiological rate determining process, and the possible amount of variability intrinsic to the parameters of these modeled relationships are some examples of these difficulties [4-6].

It is an aim of our working group to extend the development of IVIVC, using newer modeling tools, such as those in the field of artificial intelligence. The self-organizational properties of these methods and their ability to incorporate a large number of possible variables and relationships without a predefined model structure, encourage the evaluation of artificial neural networks (ANN) in determining an IVIVC.

The term ANN refers to a group of algorithms used for pattern recognition and data modeling. As its name implies, ANN systems are loosely based on neural physiology, using the concept of a highly interconnected system of parallel processing units. It is the in-

In Vitro–in Vivo Correlations, edited by Young *et al.*
Plenum Press, New York, 1997

tent of this chapter to illustrate the application of ANN in developing an IVIVC and not to intensively cover ANN methodology; a review of the development of ANN can be found elsewhere [7], as well as a complete description of the theory involved and tested in this research [7–9].

The application of neural network concepts is relatively new to the field of pharmacokinetics and pharmacodynamics. An introduction to ANN as applied to the field of pharmacokinetics was given by Erb, describing the common backpropagation learning algorithm[10] and demonstrated its ability to be used as a bayesian classifier using simulated data [11]. The application of real pharmacokinetic data for the task of learning interspecies scaling, using different input-output data formats and neural network configurations, has been described by Hussain *et al* [12]. They also described the problem of the lack of a structured set of rules or guidelines in determining network configuration variables, such as the number of hidden nodes, the necessary number of training iterations and the proper data format.

It is our eventual aim to develop a methical approach to ANN - IVIVC, and the intent of this chapter to discuss ANN-IVIVC and show its feasibility by presenting the results from some common ANN configurations.

2. OVERVIEW OF ARTIFICIAL NEURAL NETWORKS AND IVIVC

2.1. A Brief Description of Terms

There is a great deal of diversity in terminology within the literature concerning this new field and it is necessary to define some important terms before we proceed. This is not intended to be a complete glossary of terms in the area of artificial intelligence programming, but rather a necessary beginning in establishing a common foundation. Some of these terms are described in a previous chapter of this book. Also, Table I presents some common terms and a brief description that is consistent with their use in this chapter.

2.2. The Artificial Neural Network - *in Vitro-in Vivo* Correlation Problem

The approach when we apply ANN to any type of problem is very similar. The output variable, or variables, are the dependent expressions we would like to predict, such as the pharmacokinetic profile itself. Other output variables in an ANN-IVIVC model that may be considered include the absorption rate, lag time, or parameters descriptive of the drug's input function.

The success of an ANN-IVIVC is often dependent on the type of input variables selected. Obviously, the dissolution kinetic profile, or parameters descriptive of this profile may be used, but often additional parameters are necessary. As we know, the pharmacokinetics of a drug are often influenced by covariates such as patient demographics. Additionally, the influence of the dissolution profile on the *in vivo* pharmacokinetics is usually only observed through the absorption and distribution phases after administration of the drug, but contributes little information about the elimination. The pharmacokinetic profile of the drug's impulse response, or parameters descriptive of this profile, may be used as additional inputs to the ANN-IVIVC model to better characterize drug elimination. This is analogous to performing a convolution operation with the pharmacokinetics of the associated immediate release product and the drug input function. We will also see later by ex-

Table 1. A description of artificial neural network terms

Artificial neural network	A group of algorithms used for pattern recognition and data modeling, which are loosely based on neural physiology. These algorithms use the general concept of a highly interconnected system of parallel processing units.
Neuron	The basic building block of artificial neural networks. Each single neuron sums the input values and applies a function to this sum to produce an output.
Layer	A grouping of neurons in a network. The first layer is sometimes referred to as the "input layer" and consists of a number of neurons equal to the number of inputs. The last layer is referred to as the "output layer" and consists of a number of neurons equal to the number of outputs. The layers between the input and output layer are referred to as hidden layers.
Transfer function (f)	The functional attribute of each neuron.
Association	The paired input and output in a data file.
Training pattern file	The data file used to train the artificial neural network.
Validation pattern file	The data file used to validate the trained network. This data file is not involved in training, and is used as a predictive measure of the trained artificial neural network.
Test pattern file	A portion of the training file, which is not used for training, but as a periodic measure to guard against the artificial neural network mimicking or "memorizing" the training pattern file.
Backpropagation	This is the most widely used learning algorithm employed in training neural networks. In its simplest form, it is an iterative gradient descent procedure that minimizes error.

ample, that the way by which we relate the inputs and outputs is also important in obtaining a successful ANN-IVIVC.

2.3. Artificial Neural Network Architectures and Learning Paradigms

One aim of our early studies was to determine acceptable network configurations for different sets of *in vitro - in vivo* data. Part of choosing a network configuration is the choice of an ANN architecture. When we discuss ANN architecture, we are usually referring to how the individual nodes are related. Sometimes, mistakenly, the way an ANN learns is also taken to be considered architecture. Here we will make an emphasis on the separation of the two, ANN architecture and ANN learning paradigms.

The selection of the proper ANN structure is influenced by the type of input-output relationship we are trying to model. Figures 1a-g shows a number of the more common architectures that we have employed in the example discussed later. In these diagrams, the nodes are grouped together by their functional layers. Displaying an ANN in this manner allows us to identify two distinct groups of architectures, feed forward neural networks (FFNNs) and recurrent neural networks (RNNs). The FFNNs are usually used to relate two functions, or establish an input-output relationship that is not dependent on a previous or sequential input-output relationship. Any type of nonlinearity in the system must be within each input-output association. The RNNs, however, allow us to model nonlinearity across input-output associations, by allowing an input-output association to be dependent on the association that came before it. For example, if we try to model the entire dissolution profile of each formulation to the entire pharmacokinetic output as one input-output association, then the FFNN architecture should suffice. The input-output association in this case does not depend on the previous input-output association. If we model each dissolution time point to each pharmacokinetic time point as an input-output association, then

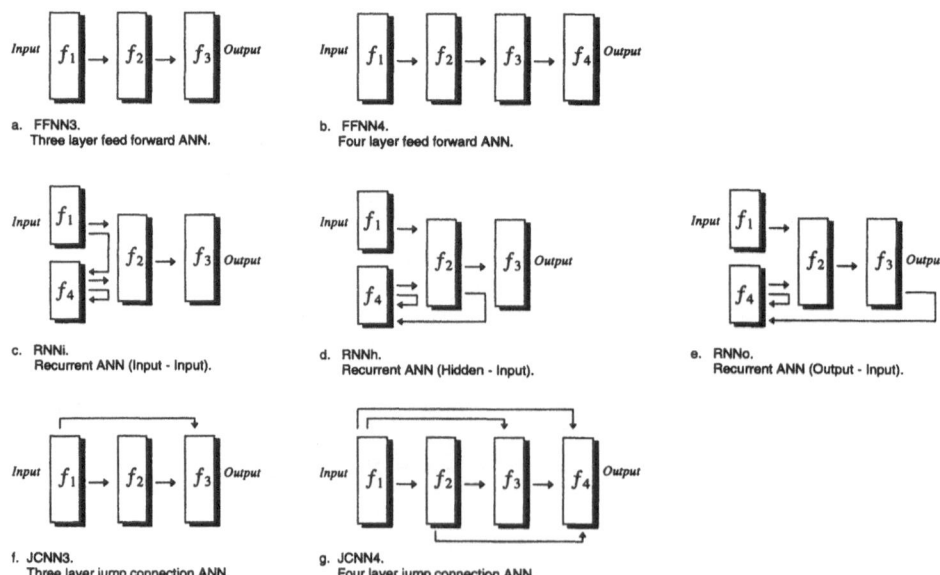

Figure 1. Block diagrams of the ANN architectures used in the study. All of the architectures employ some type of backpropagation learning.

in order to model any nonlinear properties, recurrent connections must be used. The recurrent connections allow previous history to influence the current output. This is usually important when the data follows a time series.

The most common, and most widely employed learning paradigm is backpropagation. This is an iterative gradient descent procedure that minimizes the error [13]. In previous years backpropagation would have been the only method mentioned in learning, however, newer paradigms are now being developed and used that offer promise to problems like ANN-IVIVC. Radial basis function networks, polynomial networks, probabilistic neural networks, generalized regression neural networks, and many other types of networks are examples of ANNs with different learning paradigms, and sometimes, different nodal architectures [9]. It is too complicated to detail the algorithms for these types of ANNs, but it needs to be pointed out that some of the problems we encounter with ANN-IVIVC, such a limited amount of data, may be solved by applying a non-backpropagation type of ANN.

2.4. The Input-Output Relationship

As we mentioned above, we can present IVIVC data as a functional relationship by having the complete dissolution profile as a set of inputs related to the pharmacokinetic profile per input-output association. We can also present the data as a sequential time series, with the individual measurements at each time point related in each input-output association. These are general examples of the four different input-output associations that we have used to construct ANN-IVIVC pattern files that will be shown in the example that follows.

2.5. Artificial Neural Network Training Criteria

Methods in artificial intelligence, as well as traditional methods, require a parameter or function to optimize against. This, sometimes referred to as a fitness function, is usually a measure of prediction or bias, such as the R^2, mean prediction error (MPE), and mean absolute error (MAE), defined as,

$$R^2 = \frac{\sum (y - \hat{y})^2}{\sum (y - \bar{y})^2}$$

$$MPE = \frac{1}{N} \sum (\hat{y} - y)$$

$$MAE = \frac{1}{N} \sum |y - \hat{y}|$$

where,

$y = actual\ observation$

$\hat{y} = ANN\ prediction$

$\bar{y} = average\ observation$

$N = number\ of\ observations$

Training proceeds, iteratively if backpropagation is applied, until some predetermined measure of a good model is met. If training is too short, then predictions may be poor. If training proceeds for too long a period, then memorization may become a problem. When memorization occurs, the ANN will mimic the data used for training, but will fail to be predictive when applying other data. To prevent memorization, we often use a separate set of data, referred to as a test pattern file, that is periodically applied during training. The error is tested and used as a criteria to stop training. In the example given below, training was stopped following 20,000 applied input-output associations following a minimum in the error of the test pattern file.

3. AN ANN - IVIVC EXAMPLE

3.1. The IVIVC Data

This example includes *in vitro* inputs measured as percent dissolved and *in vivo* outputs consisting of the plasma concentrations. Inputs in the training pattern files consisted of the dissolution values from two extended release formulations with 7 dissolution time points each. For each formulation, 6 tablets were tested. The two formulations represent the upper and lower dissolution profiles of this study. Each formulation was administered to 9 individuals in a crossover trial. The drug plasma concentrations were sampled at 15 time points following oral tablet administration. A third extended release formulation with the same experimental setup, and part of the same crossover study, was used as a validation set. The success of the ANNs were based on the prediction of the middle validation profile.

The pharmacokinetics of the product were known to follow a "flip-flop" model, where the absorption of this drug was relatively slow compared to its elimination. This situation, as a trial for ANN-IVIVC, gave a reasonable assurance that the dissolution kinetics could be considered an influential variable throughout the pharmacokinetic profile.

3.2. ANN-IVIVC Configurations and Training

ANN training and application of the data were performed using Ward Systems' software package, NeuroShell 2®[14]. Three basic types of ANN architectures contained within the NeuroShell 2® software were used: traditional feed forward neural networks, recurrent neural networks, and jump connection neural networks. Diagrams of these network structures, with the nodes represented collectively as functional blocks, are shown in Figures 1a-g. A detailed description for each architectural setup is given in Table 2. Including the type of network architecture and the number of hidden layers, we tested a total of seven types of network architectures.

Two of these ANN architectures are the common three layer (FFNN3) and four layer feed forward neural networks (FFNN4) shown in Figures 1a and 1b, which have one and two hidden layers, respectively. To give the network functional flexibility, linear functions ($f_1(x) = x$) were used for the nodes in the input layer and a logistic function ($f_2(x)$, $f_3(x)$, $f_4(x) = 1/(1+\exp(-x))$) was used for each node in the hidden and output layers.

The recurrent networks defined as RNNi, RNNh, and RNNo (Figures 1c-e) had recurrent connections to the input, hidden, and output layers, respectively. The transfer function of each node in the hidden and output layers was set as a logistic function, while the input layer nodes were set to linear functions. The fourth layer can be called the network's "long term memory", and has no node functionality. It contains the contents of the connected layer as it was in the previous training. These types of networks have been shown to work well with time series data that depend on history [15].

The following two network architectures are a type of ANNs known as jump connections. In this type of backpropagation network, every layer is connected in a feed forward manner. Three and four layer jump connection ANN architectures, designated JCNN3 and JCNN4 (Figures 1f and 1g), were used as they may be possible alternatives to

Table 2. Summary of the seven types of ANN architectures tested

	Architecture	Data presentation	Test set	Node configuration (input-hidden-output)
FFNN3	Feed Forward 3 Layers	Random	≈10% Randomly Selected	Linear - Logistic - Logistic
FFNN4	Feed Forward 4 Layers	Random	≈10% Randomly Selected	Linear - Logistic - Logistic - Logistic
RNNi	Recurrent Input - Input	Rotational	Individual Subject with Single Formulation with all 6 Dissolution Sets	Linear - Logistic - Logistic
RNNh	Recurrent Hidden - Input	Rotational	Individual Subject with Single Formulation with all 6 Dissolution Sets	Linear - Logistic - Logistic
RNNo	Recurrent Output - Input	Rotational	Individual Subject with Single Formulationwith all 6 Dissolution Sets	Linear - Logistic - Logistic
JCNN3	Jump Connections 3 Layers	Random	≈10% Randomly Selected	Linear - Logistic - Logistic
JCNN4	Jump Connections 4 Layers	Random	≈10% Randomly Selected	Linear - Logistic - Logistic - Logistic

the traditional feed forward structures. They were given the same node functions as FFNN3 and FFNN4.

The pattern files were constructed from different input-output associations from the same data and were named ASSOCIATION 1 through 4. Each training pattern file had corresponding validation and test pattern files constructed with the same type of input - output association. A diagram of each type of input - output association is shown in Figure 2 and a summary of the constructed pattern file is shown in Table 3.

ASSOCIATION 1 was a pattern file in which each input-output association is a functional relationship. The pattern file then contained each pharmacokinetic observation set associated with each of the tablet dissolution profiles. The dissolution mean was not used in this type of pattern file, nor in any of the other pattern files. Like ASSOCIATION 1, the input - output associations in the ASSOCIATION 2 pattern file included the complete kinetic set of dissolution values for each tested tablet, but each was associated with a single respective pharmacokinetic output. Collectively, the input - output association lines of the pattern file formed a pharmacokinetic time sequence. The pharmacokinetic time point was included as an input for this pattern file, and all other time sequential pattern files. The input - output associations of ASSOCIATION 3 consisted of each *in vitro* value as an input associated with each *in vivo* output. Pharmacokinetic observations with no directly associated dissolution observations were not used in the training. ASSOCIATION 4 attempted to unite some of the more desirable features of the previous pattern files which included presenting the entire dissolution profile per tested tablet as inputs (ASSOCIATION 1 and 2), presenting the data as a time sequence (ASSOCIATION 1, 2, and 3), and utilizing all of the *in vitro* data (ASSOCIATION 1, 2, and 4). Pattern file ASSOCIATION 4 was a sequential time series, and included previous dissolution values as inputs. This type of pattern file can be termed a *memorative association* and was a type of time progressive synthesis neural network configuration described by Veng-Pedersen [16]. The output consisted of the pharmacokinetic concentration value, while the inputs were the pharmacokinetic time point and all the dissolution values that preceded that point in time.

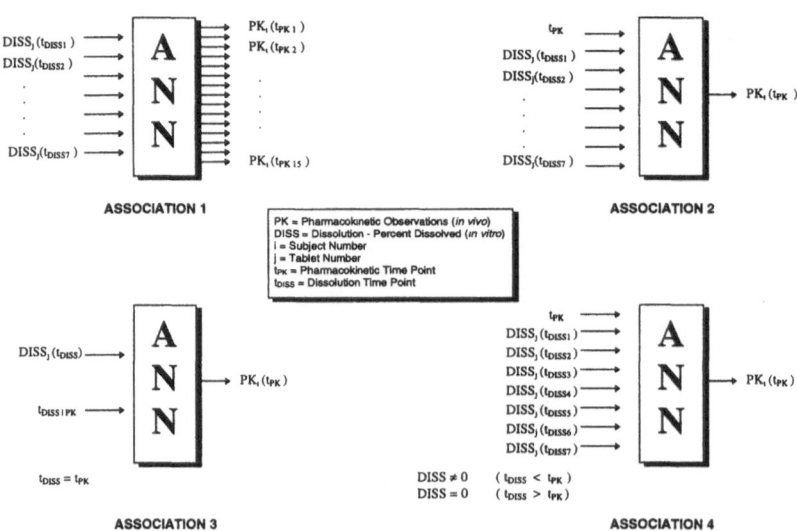

Figure 2. Diagrams of the input - output associations used in pattern files ASSOCIATION 1 through 4.

Table 3. Pattern files constructed from the different input - outp

#	Association type	Association input(s)	Association output(
1	Functional	(7) Dissolution set $_j$ (t_{DISS1} : t_{DISS7})	(15) PK $_i$ ($t_{PK\ 1}$: t_{PK15})
2	Time series	(8) t_{PK} , dissolution set $_j$ (t_{DISS1} : t_{DISS7})	(1) PK $_i$ (t_{PK})
3	Time series	(2) $t_{DISSIPK}$, dissolution set $_j$ (t_{DISS}),	(1) PK $_i$ (t_{PK})only those outputs where t_{PK} = $t_{DIS:}$
4	Time series memorative association	(1-8) t_{PK} , dissolution set $_j$ (if t_{DISS} < t_{PK})	(1) PK $_i$ (t_{PK})

PK = Pharmacokinetic observations (in vivo); DISS = % dissolved (in vitro); i = subject number; j = tablet number; t_{PK} = pharmacok

Dissolution values that occurred after that pharmacokinetic time point were set to zero in the pattern file, and were interpreted as null inputs by the software.

In all non-recurrent networks, 10% of the training pattern file was randomly selected and placed aside as a test pattern file during training. Recurrent neural networks, however, rely on previous history, which required the data to be presented as a time series across input - output associations. This also included the application of any test or validation pattern file. For this reason, the test pattern file applied to any recurrent architecture consisted of the associations from the 9th subject / 2nd formulation, kept in time sequence.

The test pattern file was not included in the training, but was used as a periodic measure of the network's ability to successfully predict while being trained. The test pattern file was applied to the network after every 200 input - output associations (training events), using the NET-PERFECT™ feature in the NeuroShell 2® software. The prediction of the outputs in the test pattern file was used as a stop criterion. In each case, the network was directed to stop training after 20,000 training events following a minimum error, and the weights corresponding to that minimum were saved as the trained ANN.

3.3. Results and Discussion

A total of 25 network configurations, which included the seven different types of ANN architectures and four types of input-output associations, were tested. The three recurrent architectures were not used with ASSOCIATION 1, as this type of relationship did not have a sequential format across associations.

Inputs from the training and validation pattern files were applied to the trained networks and the respective ANN outputs were compared to the actual observations. The R^2, MPE, and MAE are shown in Table 4 for both the training and validation data sets. Also shown is the ratio of R^2 between the predictions and training pattern files, as an indicator of possible network memorization.

In all cases, the ANN attempted to determine a mean concentration curve based on the information contained in the dissolution kinetics, and in some configurations, attempted to account for the variability in the pharmacokinetics due to variability in the dissolution kinetics. More than half of these ANN configuration could be considered successful in predicting the pharmacokinetic data from the dissolution kinetics. The better network architectures, based on this IVIVC data set, seem to be the feed forward architectures, based on their ability to give good model predictions with all four pattern files. The more successful pattern files included formatting the data as a functional relationship (ASSOCIATION 1) and as a memorative pattern file (ASSOCIATION 4).

An example of a model prediction from one of these network configurations is shown in Figure 3 and Figure 4. In this example, the ASSOCIATION 4 pattern file was used to train the FFNN3. Following training, the dissolution values from the training pattern files were used as inputs to predict the pharmacokinetic data. Comparisons of the actual observations with these ANN outputs are shown in Figure 3. The dissolution values from the validation pattern file were then presented to this trained ANN, interpolating the pharmacokinetic predictions shown in a comparison with the actual pharmacokinetic observations in Figure 4.

The common and relatively simple FFNN architecture worked as well as other types of IVIVC models, especially when the data was presented as a functional relationship (ASSOCIATION 1). For some IVIVC data, however, these types of architectures may not work as well as time series predictors. Some IVIVC correlations tend to be more non-

Table 4. Statistical results for the 25 network configurations applied to ANN-IVIVR

| | Training set | | | Validation set | | | R^2 ratio prediction/ |
	R^2	MPE	MAE	R^2	MPE	MAE	training
Association 1							
FFNN3	0.878	-0.229	3.110	0.803	-1.431	3.992	0.915
FFNN4	0.880	-0.196	3.109	0.790	-1.435	4.089	0.897
RNNi	N/A	N/A	N/A	N/A	N/A	N/A	N/A
RNNh	N/A	N/A	N/A	N/A	N/A	N/A	N/A
RNNo	N/A	N/A	N/A	N/A	N/A	N/A	N/A
JCNN3	0.875	0.454	3.207	0.819	-0.415	3.957	0.937
JCNN4	0.872	0.574	3.238	0.815	-0.526	4.007	0.935
Association 2							
FFNN3	0.136	-1.371	9.486	0.142	-2.180	9.588	1.041
FFNN4	0.865	0.493	3.316	0.792	-0.910	4.179	0.915
RNNi	0.732	3.642	5.080	0.742	3.061	5.422	1.013
RNNh	0.728	4.351	5.101	0.692	3.731	5.749	0.950
RNNo	0.741	4.273	5.268	0.763	2.312	5.336	1.029
JCNN3	0.101	-2.234	9.520	0.114	-3.062	9.612	1.134
JCNN4	0.078	-1.365	9.653	0.149	-1.396	9.592	1.910
Association 3							
FFNN3	0.749	-0.825	4.184	0.608	-2.931	5.352	0.812
FFNN4	0.752	-0.514	4.167	0.620	-2.623	5.296	0.824
RNNi	0.539	4.583	6.237	0.594	2.642	6.088	1.102
RNNh	0.420	6.089	7.171	0.559	3.493	6.381	1.333
RNNo	0.478	4.347	6.668	0.578	1.404	6.093	1.210
JCNN3	0.745	-0.073	4.267	0.614	-2.259	5.327	0.825
JCNN4	0.746	-0.235	4.238	0.617	-2.391	5.265	0.827
Association 4							
FFNN3	0.846	-0.076	3.540	0.771	-1.573	4.465	0.911
FFNN4	0.854	-0.538	3.389	0.770	-1.683	4.280	0.901
RNNi	0.658	4.650	5.626	0.571	5.540	7.196	0.868
RNNh	0.696	0.744	5.187	0.580	5.174	7.135	0.834
RNNo	0.596	1.252	6.111	0.628	3.404	6.303	1.054
JCNN3	0.841	-0.192	3.607	0.789	-0.798	4.230	0.938
JCNN4	0.856	-0.286	3.357	0.787	-1.618	4.098	0.919

linear, requiring the ANN to incorporate past history. The feed forward structure can not incorporate history, but this may be accounted for if the data is arranged as a memorative association (ASSOCIATION 4), which also proved successful with this IVIVC data.

An interesting example of network performance as a function of configuration variables is seen in comparing the ASSOCIATION 2 - FFNN3 trial with the ASSOCIATION 2 - FFNN4 trial, where the additional hidden layer improved prediction dramatically. Practical ANN experience has shown that a majority of problems can be solved with a three-layered design, and that a four-layered ANN may be prone to fall into a local minima [17]. However, with this IVIVC data set formatted as the ASSOCIATION 2 pattern file, a four-layered feed forward structure predicted well, while the three-layered ANN appeared to fail to converge on a solution.

The JCNN architectures, which are structurally very similar to the FFNN architectures, also compared equally as well. The lack of any significant improvement in describing this data, however, suggests that the additional jump connections were not necessary.

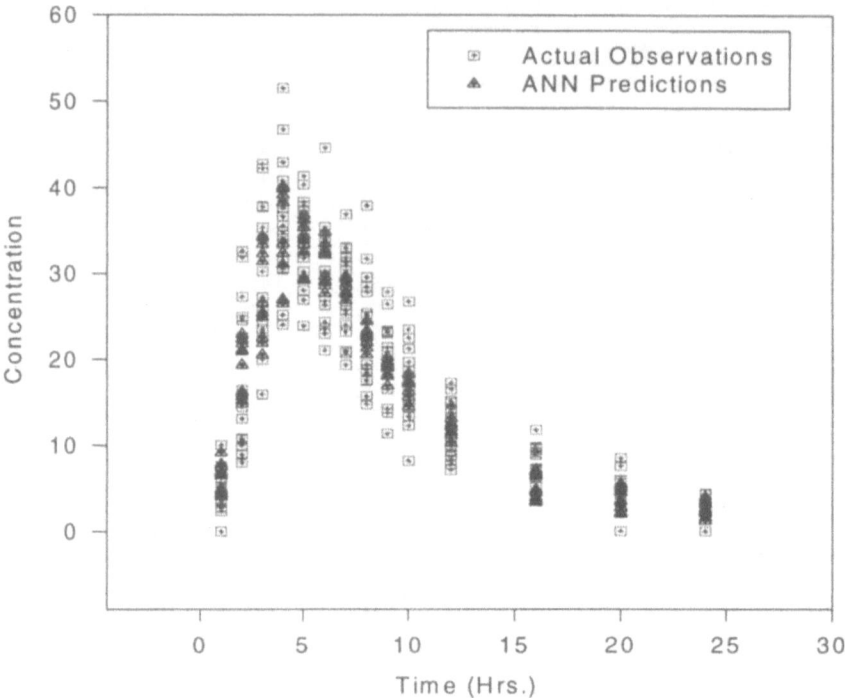

Figure 3. Actual PK observations from the training data set are compared to ANN PK predictions using *in vitro* inputs from the training data set. The FFNN3 was trained with the training pattern file ASSOCIATION 4.

A problem in implementing the recurrent network structures, as reflected in the results presented here, was the determination of a stopping criteria. As with all backpropagation network training performed in these trials, the test pattern file was applied periodically throughout training and was a measure used to indicate the completion of training before the onset of memorization. Because of the importance in keeping a sequential structure with recurrent networks, rather than the test set randomly constructed from the training data, the ninth subject receiving the second formulation was used as the selected test pattern file. This proved to be a good measure that prevents memorization, but biased the trained network in the favor of this test set. Once a minimum is found, network training oscillates across subjects, until training is stopped based on the number of iterations following a minimum in the test set. A trained network based on a minimum corresponding to the test pattern file is saved, biasing the results to the data used as the test set. So, although the recurrent ANN structures produced fair results, these results were biased to the selected test pattern file. Better results may be expected if a better "average" or unbiased test pattern can be found, the intersubject variability can be better described using additional inputs, or another method to protect against network memorization can be found.

In most ANN structures, the number of inputs and outputs dictate the number of input and output nodes, respectively. Hence, more inputs and outputs lead to a more complicated network structure. Although relatively successful here, the ASSOCIATION 1 pattern file had a total of 15 outputs, which must be considered in the evaluation of these types of input - output associations for ANN-IVIVC. When the level of complexity of the structure increases, the likelihood of obtaining a good solution decreases. As this research

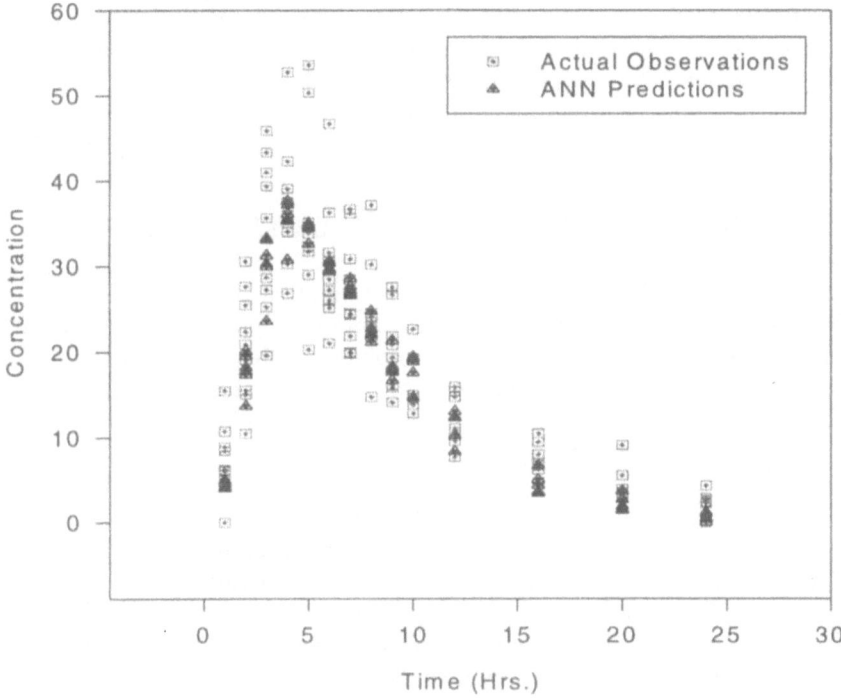

Figure 4. Actual PK observations from the validation data set are compared to ANN PK predictions using *in vitro* inputs from the validation data set. The FFNN3 was trained with the training pattern file ASSOCIATION 4.

progresses, it is expected that the number of independent variables, or inputs, will be increased, adding to the complexity of the models, and making input-output associations, like that found in ASSOCIATION 1, undesirable. The memorative type of input - output association used to construct the ASSOCIATION 4 pattern files worked well with all seven ANN architectures. This type of pattern file had the advantage of being a generalized format with a single output, which also allows the network to incorporate relationships from the previous inputs.

The number of network configurations can be immense when considering some of the variables examined here, such as network architecture, data formats, and number of hidden layers, and considering some of the other possible network configuration variables not addressed: number of hidden nodes, additional network structures, learning algorithms, and the different types of node transfer functions. The number of hidden nodes in these trials were set to the software defaults, which were allowed to be conservatively large, since the periodic application of the test pattern file helps to prevent memorization. Some of the more common network architectures were examined in this study, but there were many more that may prove to be as good, or better in ANN-IVIVC. Other possible structures that may prove to be applicable are the newer multilayer networks that include a lag in the data between the dependent and independent variables [9]. The node transfer functions were limited to the linear and logistic transfer function in this study, but, many other different types of functions, such as a limit, competitive, hyperbolic tangent, sine, or gaussian, may be used.

All of the network architectures used backpropagation as the learning paradigm. Current research suggests other learning paradigms may work well for the ANN-IVIVR problem. General regression neural networks are ANNs that involve a statistical technique known as kernel regression and require the data to only be iterated once through the network during training [14, 18]. This type of ANN has the distinct advantage in that it can converge to a linear or nonlinear regression surface, even with relatively little data.

4. FUTURE DIRECTIONS: NETWORK OPTIMIZATION

The importance of the type of network configuration and the large number of possible ANN configurations that can be examined should be apparent. For example, given an ANN with 20 inputs, even constraining the problem to the simplest of backpropagation structures, we can have millions of possible ANN configurations. Luckily, many of these structures are generalized such that most configurations will give us good solutions, but many will still fail to either converge or predict well.

Future research directions in ANN-IVIVC, and in ANN in general, include pruning and optimization algorithms. Pruning algorithms begin with a relatively large ANN structure, and proceed to reduce, or prune, the structure to an optimal form. Optimizing algorithms, on the other hand, hope to determine a set of network configurations that work well for a specified problem. Currently, we are using another artificial intelligence method, genetic algorithms, in neural network optimization. Genetic algorithms are used as a search strategy, similar to the way a simplex method is used in regression, to search the network configuration space for the best set of ANN configurations. This type of search strategy has allowed us to evaluate millions of permutations, such as the 25 examples shown above, with only a few hundred evaluations, in an automated manner.

5. CONCLUSIONS

We have demonstrated a number of possible network configurations, many of which successfully predicted a mean *in vivo* plasma concentration profile using the dissolution kinetics. This work has shown the feasibility of ANN-IVIVC by giving a number of potential ANN configurations that can be considered successful with this data set, but has illustrated a need for a methodical approach in applying ANN to problems. A basis, however, for ANN-IVIVC is now established that is now being improved upon with current work. Additional input variables including subject demographics, dissolution method variables, and formulation variables are currently being introduced to attempt to account for the non-random error associated in the relationship. ANN-IVIVC has the potential to establish complex relationships and may also posses the ability to interpolate pharmacokinetic parameters and profiles given formulation specifications. Also, algorithms and software currently exist to reverse map, possibly predicting a range of formulation variables based on desired pharmacokinetics.

ACKNOWLEDGMENTS

The work reported here was supported by the Elan Corporation, plc., as part of their overall sponsorship of the In Vitro - In Vivo Cooperative Working Group.

REFERENCES

1. T. J. Sullivan, E. Sakmar, and J. G. Wagner. Comparative bioavailability: a new type of *in vitro-in vivo* correlation exemplified by prednisone. *J. Pharmacokin. Biopharm.* **4**:173–181 (1976).
2. C. Graffner, M. Nicklasson, and J.-E. Lindgren. Correlations between *in vitro* dissolution rate and bioavailability of alaproclate tablets. *J. Pharmacokin. Biopharm.* **12**:367–380 (1984).
3. C. Caramella, F. Ferrari, M. C. Bonferoni, M. E. Sangalli, M. De Bernardi Di Valserra, F. Feletti, and M. R. Galmozzi. *In vitro / in vivo* correlation of prolonged release dosage forms containing diltiazem HCl. *Biopharm. Drug Disp.* **14**:143–160, (1993).
4. R. W. Wood, L. Martis, A. W. Gillum, T. J. Roseman, L. Lin, and P. Bernardo. *In vitro* dissolution and *in vivo* bioavailability of commercial levothyroxine sodium tablets in the hypothyroid dog model. *J. Pharm. Sci.* **79**:124–127 (1990).
5. W. H. Barr, E. M. Zola, E. L. Candler, S.-M. Hwang, A. V. Tendolkar, R. Shamburek, B. Parker, and M. D. Hilty. Differential absorption of amoxicillin from the human small and large intestine. *Clin. Pharmacol. Ther.* **56**:279–285 (1994).
6. G. Levy and L. E. Hollister. Inter- and intra subject variations in drug absorption kinetics. *J. Pharm. Sci.* **53**:1446–1452 (1964).
7. J. A. Anderson. *An introduction to neural networks*, MIT Press, Cambridge, 1995.
8. S. I. Gallant. *Neural Network Learning and Expert Systems*, MIT press, Cambridge, 1993.
9. M. T. Hagan, H. B. Demuth, and M. Beale. *Neural Network Design*, PWS Publishing Company, Boston, 1996.
10. R. Erb. Introduction to backpropagation neural network computation. *Pharm. Res.* **10**:165–170 (1993).
11. R. Erb. The backpropagation neural network - A bayesian classifier. Introduction and applicability to pharmacokinetics. *Clin. Pharmacokinet.* **29**:69–79 (1995).
12. A. S. Hussain, R. D. Johnson, N. N. Vachharajani, and W. A. Ritschel. Feasibility of developing a neural network for prediction of human pharmacokinetic parameters from animal data. *Pharm.Res.* **10**:466–469 (1993).
13. S. Haykin. *Neural networks: a comprehensive foundation*, Macmillan, New York, 1994.
14. NeuroShell® 2 Manual Third Edition, Ward Systems Group, Inc. Executive Park West, 5 Hillcrest Drive, Frederick, MD 21702. (1995).
15. J. L. Elman. Finding structure in time. *Cognitive Science.* **14**:179–211 (1990).
16. P. Veng-Pedersen and N. B. Modi. Application of neural networks to pharmacodynamics. *J. Pharm. Sci.* **82**:918–926 (1991).
17. J. De Villiers and E. Barnard. Backpropagation neural nets with one and two hidden layers. *IEEE Trans. Neural Networks.* **4**:136–141 (1992).
18. D. F. Specht. A general regression neural network. *IEEE Transactions on Neural Networks.* **2**: 568–576 (1991).

23

IMPACT OF IVIVR ON PRODUCT DEVELOPMENT

John Devane

IVIVR Co-operative Working Group
Elan Corporation

This chapter deals with a more general view of the role and impact of *in-vivo/in-vitro* relationships (IVIVR) on Product Development particularly as it pertains to extended release products.

The major areas of impact that IVIVR has on the Product Development process are summarised in **SLIDE 1**. Understandably the emphasis over the last number of years has focused on the last point, qualifying change in the context of scale-up and post approval change both for IR and ER products but in fact IVIVR is a dynamic element of Product Development and is a major thread that runs through the various critical stages of the Product Development process. To some extent qualifying change is nothing more than the logical culmination of the work done in developing, establishing and validating an IVIVR as part of the original Product Development process itself. In practice we use, IVIVR from the very early concept stages of an ER development program through the stages of defining a prototype formulation, optimising the formulation process, scaling up to and manufacture of a pivotal batch, right through the registration and approval stages and into the post approval SUPAC phase (**SLIDE 2**). Because IVIVR is not some static element of a Product Development process that is developed in isolation but rather is a dynamic aspect of a program that is developed as part and parcel of the process itself, it is useful to introduce the following terminology (**SLIDE 3**). Thus, during Stage 1 and with a particular product concept in mind one sets out to establish appropriate *in-vitro* targets to meet the desired *in-vivo* specification. This essentially is an **ASSUMED** *in-vivo/in-vitro* relationship that provides the initial guidance and direction for the early formulation development activity. In turn this assumed model will be the subject of revision as prototype formulations are developed, characterised *in-vivo* and the results often leading to a further cycle of prototype formulation and *in-vivo* characterisation. Out of this cycle of formulation development and *in-vivo* characterisation and of course extensive *in-vitro* testing is often developed what I have termed a **RETROSPECTIVE** IVIVR (**SLIDE 4**). With a defined formulation that meets the *in-vivo* specification, one then progresses through Stage 2 and at this stage now based on a greater understanding and appreciation of the defined formulation and its characteristics a **PROSPECTIVE** *in-vivo/in-vitro* relationship is established

In Vitro–in Vivo Correlations, edited by Young *et al.*
Plenum Press, New York, 1997

- DOSAGE FORM DEVELOPMENT
- METHOD DEVELOPMENT
- FORMULATION/PROCESS OPTIMISATION
- IN-PROCESS AND FINISHED PRODUCT
  CONTROLS
- QUALIFYING CHANGE

Slide 1. Major areas of impact.

through a well defined **PROSPECTIVE** IVIVR study. **(SLIDE 5)** Once the IVIVR is established and defined it can then be used to guide the final cycle of formulation and process optimisation leading into Stage 3 **(SLIDE 6)** activity of scale-up, pivotal batch
manufacture and process validation, leading to registration, approval and subsequent post-
approval scale-up and other changes. Thus, rather than viewing IVIVR as a single exercise
at a given point in a development program it should be viewed as a parallel development
in itself starting at the initial assumed level and being built on and modified through experience and leading ultimately to a prospectively defined IVIVR.

    (SLIDE 7) For an extended release product how does one go about setting these in-
itial Stage 1 targets? Clearly, from the pioneering work of Amidon in relation to the original biopharmaceutic classification and as described by Professor Corrigan in another
chapter in this book for extended release products, establishing the permeability properties
of a drug substance is a key element in setting the appropriate targets. This information
has traditionally been lacking for the majority of compounds. Yet of course, this is of
great importance both in establishing the initial feasibility of any formulation program and
also subsequently in the interpretation of the observed *in-vivo* absorption characteristics of
a given dosage form. One needs to look at the physico/chemical characteristics of the drug
substance itself, both in the context of how these impact on the formulation approach but
also in the context of relevance to dissolution at distal sites in the G.I. tract. PK/PD modelling has established what the target *in-vivo* specification should be and based on this in-

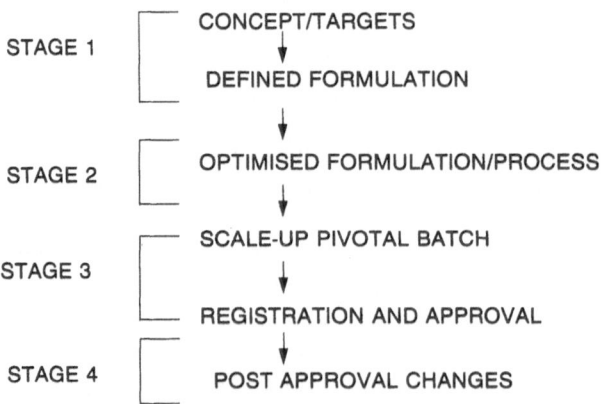

Slide 2. The product development process-extended release products.

- ASSUMED IVIVR
- RETROSPECTIVE IVIVR
- PROSPECTIVE IVIVR

Slide 3. IVIVR-product development.

formation a priori *in-vitro* methods are usually then established and we have a theoretical *in-vitro* model which based on all of the available information, physiochemical, permeability and the desired *in-vivo* targets one believes should achieve the desired absorption profile. Essentially at this stage one is assuming a level A type correlation and the formulation strategy is initiated at this point with the objective of achieving a target *in-vitro* profile. **(SLIDE 8)** The prototype formulation program itself is normally initiated with some knowledge or expectation of what technologies and/or mechanisms of release are particularly suited to meet the desired targets. This work is usually done on a laboratory level of manufacture in most cases with the simplest dissolution methodology that seems appropriate and of course starting from a base of minimal product stability. Prototypes are then selected which meet the target *in-vitro* profile, usually involving one or very often more than one technology or formulation approach. One tests one but more commonly more than one prototype within each technology or formulation approach. More extended *in-vitro* characterisation looking at the robustness of these prototypes across dissolution conditions such as pH, medium, agitation speed, apparatus type are routine at this point. Certainly, Stage 1 activity should always culminate in a pilot pk study, this is typically a 4/5 arm cross over study. The size of these pilot pk studies obviously will vary depending on the inherently variability of the drug itself but typically ranges from 6–15 subjects. The results of this pilot pk study provide the basis for establishing what has been referred to as a **RETROSPECTIVE** IVIVR. In other words a number of different prototypes with some level of variation in release rate but more often significant variation in terms of technology, formulation approach and mechanism of release have now been characterised both *in-vitro/in-vivo*. This information first of all allows a reality check on both the *in-vivo* and **ASSUMED** IVIVR either matching expectation or of course very often causing some fundamental shift in the **ASSUMED** IVIVR model. Subsequent to the results of the *in-vivo*

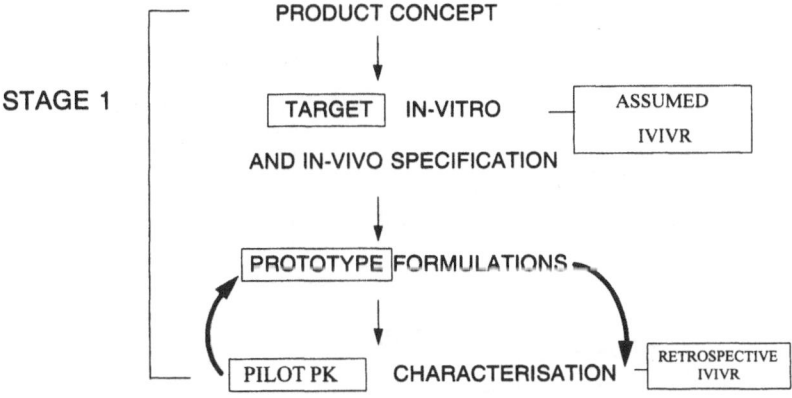

Slide 4. The product development process-extended release products.

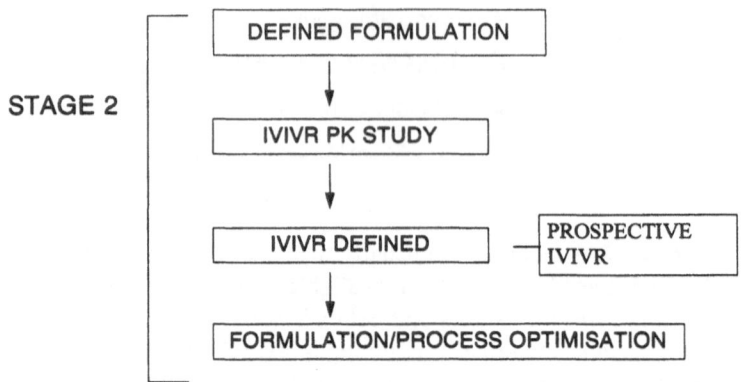

Slide 5. The product development process-extended release products.

study there is often a phase of significant revision of the *in-vitro* methods very often driven by a need to detect an *in-vitro* difference that was observed *in-vivo* but that had not been detected using the original *in-vitro* methods. Out of this work sometimes will result revised *in-vitro* targets and reformulation strategy and the same cycle of activity again.

To illustrate this phase of the Product Development process 3 examples are presented that illustrate how important and how valuable the information of the early stages of development can be in the development of a rational, well considered IVIVR. These examples are also designed to show the complexity of issues that can arise in the development process. Where the assumed IVIVR meets expectation then one can move forward very quickly. In the first example **(SLIDE 9)** there was an initial *in-vivo* target of a 6–10 hour absorption phase. Physicochemically the drug had high aqueous solubility characteristics and did not show any pH dependence in solubility. The compound had high permeability at all sites in the G.I. tract and therefore, a simple straightforward water/USP apparatus dissolution method was expected to be adequate and we expected an IVIVR to both exist and be simple and straightforward. **(SLIDE 10)** The formulation strategy involved a matrix tablet technology approach involving 4 prototypes with 3 different grades of polymer system with one grade at 2 levels and the other 2 grades at a single level. The *in-vitro* profiles **(SLIDE 11)** were as shown on this slide and demonstrated release rates over the target 6–10 hour time period. The observed *in-vivo* mean plasma concentration

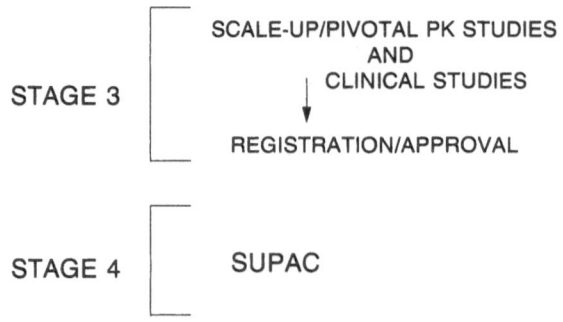

Slide 6. The product development process-extended release products.

- DRUG PERMEABILITY
- PK/PD MODELLING
- TARGET IN-VIVO PROFILE

- DRUG PHYSICO/CHEMICAL PROPERTIES
- IN-VITRO METHODS
- TARGET IN-VITRO PROFILE

- CONVOLUTION MODELLING
- FORMULATION STRATEGY BASED ON ASSUMED
  IVIVR

Slide 7. Targets.

curves are also shown on this slide and clearly what was observed *in-vivo* was a marked difference in Cmax and bioavailability across the range of formulations tested and which did rank order with the release rate shown in the earlier slide and did show a level C IVIVR. The deconvoluted absorption curves roughly rank ordered with the release rates previously shown but a level A IVIVR was not established. Therefore, from a retrospective IVIVR basis, certainly a level of *in-vivo/in-vitro* relationship had now been established but unexpectedly based on the physicochemical and more interestingly the permeability characteristics there appeared to be a rate related bioavailability loss. From a Product Development perspective what this resulted in was a revised target where the target absorption phase was narrowed to the lower end of the 6–10 hour period, in other words the 6 hour period, where acceptable bioavailability was observed. This also obviously led to the interesting question; if the permeability characteristics were acceptable why was it that a rate related loss in bioavailability was seen, given the relative short target absorption phase where G.I. transit itself would not have expected to be limiting? In retrospect in the case of this particular compound the explanation may have to do with some evidence about metabolism at distal sites in the large intestine and again it is worth pointing out that this would not have been detected in the standard permeability studies that had been performed.

The second example is as follows **(SLIDE 12)**. In this case the objective was an absorption phase of a 12–16 hour time period. Physicochemically this compound showed poor aqueous solubility in general and more particularly showed a very strong pH depend-

- # FORMULATIONS/MECHANISM OF RELEASE
- # PROTOTYPES
- IN-VITRO ROBUSTNESS
- PILOT PK STUDY
- RETROSPECTIVE IVIVR

  * REALITY CHECK ON IN-VIVO AND
    ASSUMED IVIVR
  * REVISED IN-VITRO METHODS/TARGETS
  * RE-FORMULATION STRATEGY

Slide 8. Prototypes.

IN-VIVO	:	6-10 HOUR ABSORPTION PHASE
PHYSIO/CHEMICAL	:	SOLUBILITY (AQ): HIGH NOT PH DEPENDANT
PERMEABILITY	:	HIGH NOT SITE DEPENDENT
IN-VITRO METHOD	:	SIMPLE
THEORETICAL IVIVR	:	YES/SIMPLE

Slide 9. Example.

ence. The permeability of the compound was good. From an *in-vitro* methodology perspective and from first principles the dissolution method would require control of pH and should be consistent with achieving appropriate sink conditions. Clearly from the pH solubility profile one could also anticipate such a pH condition not necessarily being consistent or appropriate to the local pH of distal sites of the G.I. tract. Based on the above characteristics from a theoretical perspective one expected it to be possible to establish an IVIVR although recognising in particular the *in-vitro* methodology would be likely to be significantly pH sensitive. From a technology perspective, a formulation approach was adopted that was designed to address both solubility and pH dependence by incorporation of appropriate pH modifying agents into the formulation and from a controlled release perspective involved a matrix tablet approach. Four different prototypes were chosen which showed different *in-vitro* dissolution profiles and associated deconvoluted absorption profiles **(SLIDE 13)**. It was very clear that the absorption profiles were significantly faster than what had been characterised as the *in-vitro* profiles. While there was perhaps a very gross rank order in that the slowest releasing formulation was also the one showing the slowest absorption rate, this did not match up for the intermediate releasing formulation, which in fact had one of the fastest absorption phases. To further understand these results we did augment the standard *in-vitro* testing by also looking at the release characteristics of the pH modifier agent that made up part of the formulation. One might speculate that part of the shift between the observed *in-vitro* release for the drug and the faster *in-vivo* absorption curves might relate to the faster pH modifier release **(SLIDE 14&15)**. Therefore, from a retrospective IVIVR basis we did not believe we had any well established relationships yet and in terms of revised targets we decided to focus on one of

PROTOTYPES

TECHNOLOGY	:	MATRIX TABLET
PROTOTYPES	:	3 GRADES A, B & C 2 LEVELS GRADE A (30%, 40%) 1 LEVEL GRADES B & C (30%) **TOTAL -** 4 PROTOTYPES

Slide 10. Example.

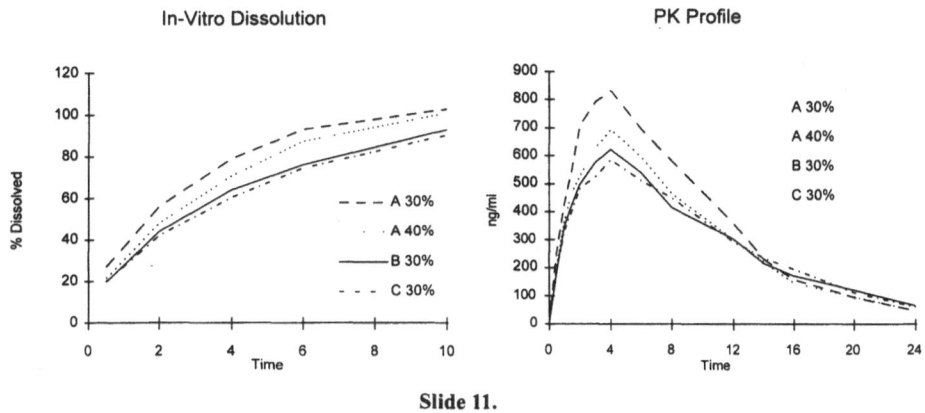

In-Vitro Dissolution   PK Profile

Slide 11.

the particular prototypes (because of its higher relative bioequivalency). Because of the shift between *in-vitro* and *in-vivo* we also looked to revised dissolution methods that might better match retrospectively the *in-vivo* profile. It was also an objective of the next phase of this particular development program that we would match both drug and pH modifier release to a greater extent.

A third example of this early phase of the Product Development process describes a project where we had a target *in-vivo* absorption phase of 24 hours. This was a short half life compound yet we wanted to achieve a fully extended once a day profile hence the need for a significantly longer absorption phase than has been described in the earlier examples. Physicochemically the drug showed very poor aqueous solubility although contrary to the last example this was not pH dependent. Permeability was good and from an *in-vitro* methodology perspective clearly the limited aqueous solubility was a major constraint and as a first step we focused on a commonly used surfactant based SLS medium with a target of sink conditions. From an IVIVR expectation point of view we expected that we should be able to develop an IVIVR but clearly the *in-vitro* methodology would potentially be problematic **(SLIDE 16)**. On **SLIDE 17**, is shown the deconvoluted absorption profile for an early prototype tested in a Stage 1 pilot pk study. The initial target *in-*

TARGETS

IN-VIVO	:	12-16 HOUR ABSORPTION PHASE
PHYSIO/CHEMICAL	:	SOLUBILITY: LOW PH DEPENDENT
PERMEABILITY	:	HIGH SITE INDEPENDENT
IN-VITRO METHOD	:	PH/SINK
THEORETICAL IVIVR	:	YES/PH SENSITIVE

Slide 12. Example.

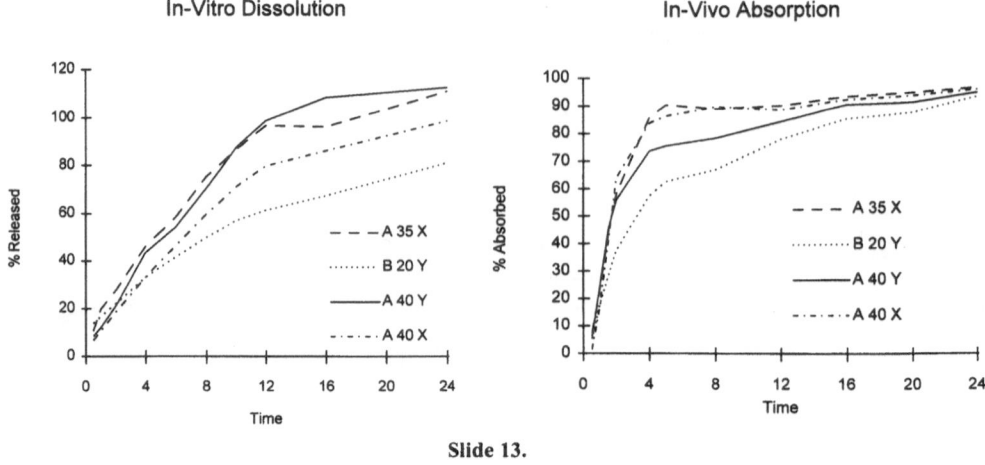

Slide 13.

vitro profile was based on a 1.25% SLS medium in USP II type of apparatus which showed quite a nice linear *in-vitro* release over 24 hours. Clearly there was no basis to a level A correlation for this method. Retrospectively we have explored a whole series of different media, conditions and apparatus and in this particular case we determined that a 0.4% CTAB surfactant system in USP II provided the best *in-vitro* to *in-vivo* match. I would point out that we explored a broad range of different media and apparatus including USP apparatus IV; the flow through system. The particular technology adopted in this project involved both solubility enhancement of the formulation and again used the matrix tablet controlled release system. From a retrospective IVIVR point of view we established a tentative level A correlation that was the basis for subsequent formulation/modification and adaptation and with a revised target based on the new *in-vitro* method the resulted in revised *in-vitro* targets for those next stages of the development program.

These 3 examples were all examples of moving from a theoretical, initial concept of what is possible in terms of an IVIVR through to actual data generated as part of the in-

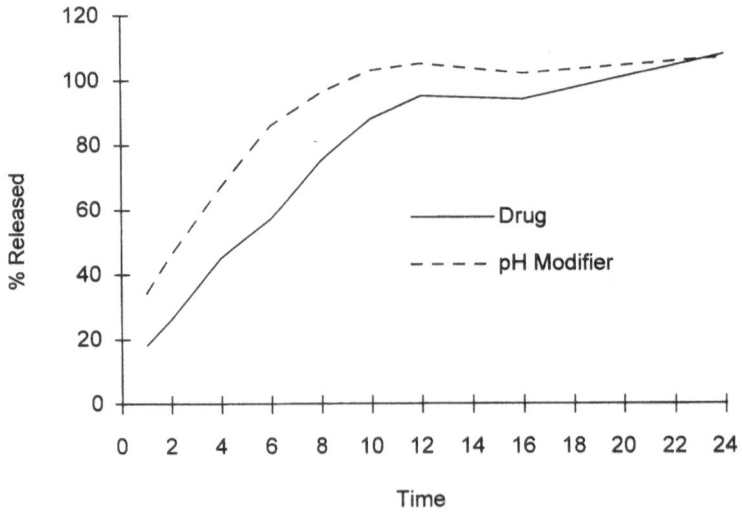

Slide 14. Prototype A35X.

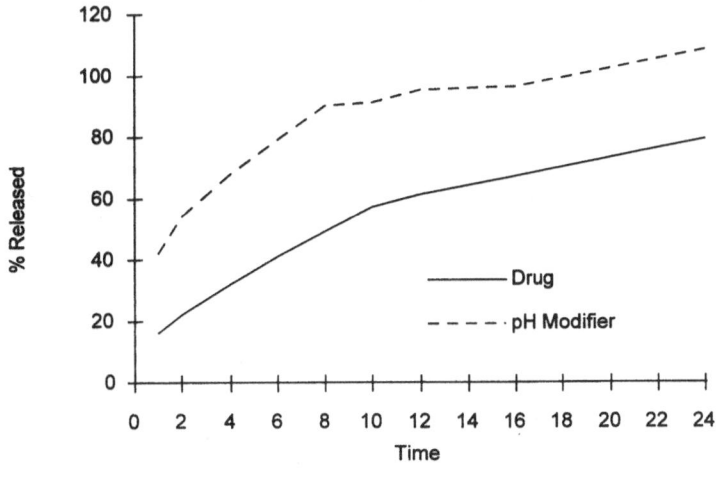

Slide 15. Prototype B20Y.

itial product formulation development process that added significant levels of information that retrospectively allowed us to come up with revised methods and revised targets, to progress onto the next phases of the development cycle.

We now move onto the next stage of the development process, in which we have a defined formulation that meets the *in-vivo* targets. One now wants to progress through the normal formulation process optimisation steps ultimately into scale-up, registration and approval. In Stage 2 one has a defined formulation and ideally one has a good understanding of the mechanism of release. Based on this understanding you have an a priori basis and from the sort of retrospective data that was shown from Stage 1, an empirical basis for determining what is the primary formulation related rate controlling variables. For extended release products this a priori understanding is usually more obvious than might be the case for immediate release products. Based on this a priori understanding we would then normally manufacture a number of products with different release rates, vary-

TARGETS

IN-VIVO	:	24 HOUR ABSORPTION PHASE
PHYSIO/CHEMICAL	:	SOLUBILITY (AQ): LOW
		NOT PH DEPENDANT
PERMEABILITY	:	HIGH
		NOT SITE DEPENDENT
IN-VITRO METHOD	:	COMPLEX/SINK
IVIVR EXPECTATION	:	YES/SOLUBILITY SENSITIVE

PROTOTYPES

TECHNOLOGY	:	SOLUBILITY ENHANCEMENT
		MATRIX TABLET

Slide 16. Example.

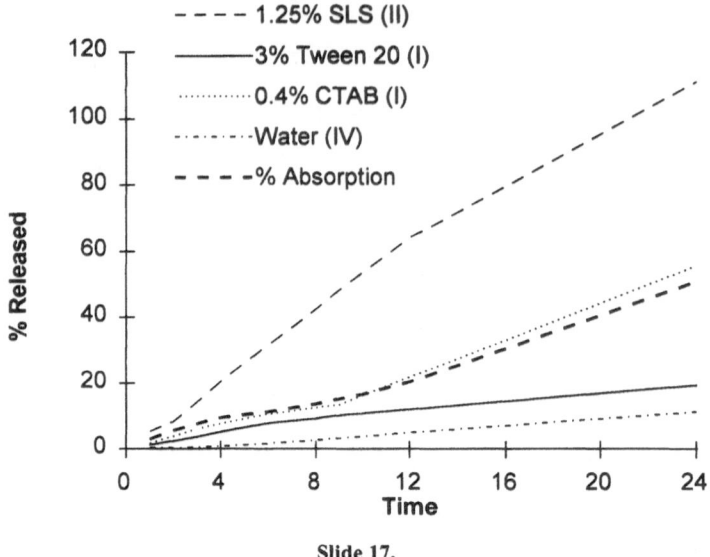

Slide 17.

ing the primary rate controlling variable but within the same qualitative formulation. We perform extensive *in-vitro* characterisation again, across pH, media and apparatus but also obviously learning from the Stage 1 work. This would lead to the execution of a **PRO-SPECTIVE** IVIVR study. The development and definition of the IVIVR is subsequent to an analysis of the results of that prospective *in-vivo* study and can often involve further *in-vitro* method development in the context of the observed results but clearly with the objective of establishing a definitive IVIVR. This ideally is a level A IVIVR but clearly level B and level C continue to be both acceptable and useful IVIVR's. Out of this work should also come the definitive *in-vitro* method that has been shown to be correlated with *in-vivo* performance and sensitive to the specific formulation variables. Once the IVIVR is established then we would routinely use this in the completion of the formulation/process optimisation program using statistically based experimental design studies looking at critical formulation and process variables and their interactions. By now having a correlated *in-vitro* method we can establish the robustness of the formulation and process and can also use this information to establish appropriate in-process and finished product specifications

USING THE ESTABLISHED IVIVR

- STATISTICALLY BASED EXPERIMENTAL DESIGN STUDIES

- CRITICAL FORMULATION/PROCESS VARIABLES AND
  INTERACTIONS (PRIMARILY ON RELEASE RATE)

- ESTABLISH "ROBUSTNESS"/RANGES/RESPONSE
  SURFACES

- IN-PROCESS AND FINISHED PRODUCT SPECIFICATIONS

- TARGETS FOR SCALE-UP

Slide 18. Formulation/process optimisation.

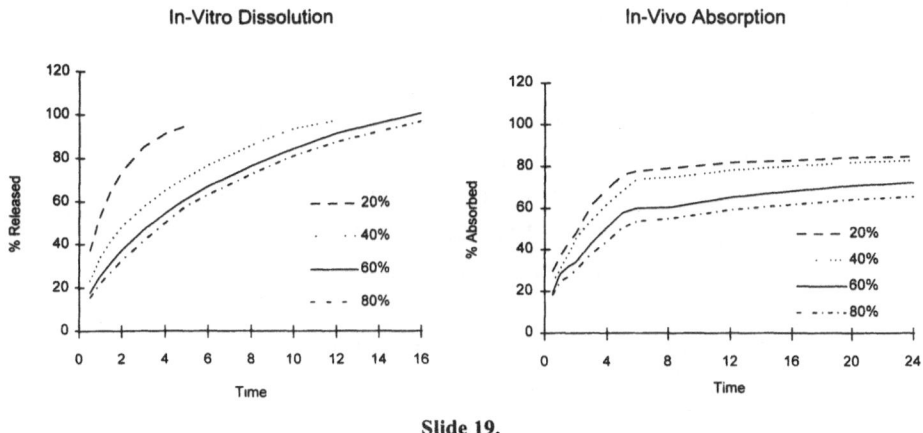

Slide 19.

and of course the appropriate targets for scale-up **(SLIDE 18)**. The next example illustrates a prospective IVIVR study. It involved a series of *in-vitro* release rates for the same qualitative formulation but varying from 20% through to 80% of rate controlling polymer. The corresponding *in-vivo* absorption rates are shown **(SLIDE 19)** and in terms of a level A correlation **(SLIDE 20)** this has been well established for the 40%, 60% and 80% polymer system. Note that the *in-vitro* method is not a standard media but was arrived at retrospectively from Stage 1 and early Stage 2 work. As a matter of interest our target formulation in terms of matching the desired *in-vivo* performance is the 60% formulation so it has been well bracketed by 20% polymer variation at either side. The 20% polymer system does not show a good level A correlation but clearly this is well outside the working range of the target formulation and also for this particular system will be viewed as marginal in terms of representing the same mechanism of release.

One should not suggest that all development programs run smoothly and that it is simply a question of gathering the right information at early stages of the development program and logically building up a well defined and easily determined level A correla-

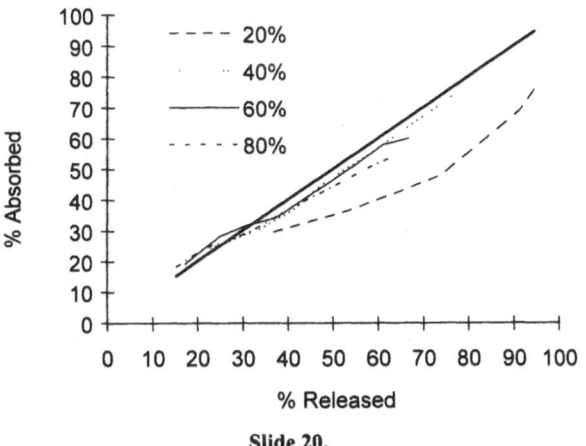

Slide 20.

- RETROSPECTIVE IVIVR BASED ON ABILITY OF IN-VITRO METHOD TO DIFFERENTIATE OBSERVED IN-VIVO STAGE 1 PROTOTYPES
- IN-VITRO METHOD VERY SENSITIVE TO PRODUCT VARIATION AND IN-VITRO TEST CONDITIONS
- RETROSPECTIVE IVIVR PREDICTED NEXT STAGE SCALE-UP PRODUCT WOULD NOT MATCH IN-VIVO TARGET
- IN-VIVO RESULTS INDICATED SCALE-UP SATISFACTORY (MODEL NOT VALID)
- REVISED IN-VITRO METHOD DEVELOPED - LESS SENSITIVE
- PROSPECTIVE IVIVR USED TO VERIFY NEW IN-VITRO METHOD

Slide 21. Example false positive.

tion. Of course, situations are never as clear cut and simple as this. In particular, one can identify two potential risk areas at this stage of the Product Development program; what can be called false positives and false negatives. The next example is a false positive. **(SLIDE 21)** From the earlier stages of the development program we determined a retrospective IVIVR based on a level C correlation at a number of time points. As we move to the next stage of scale-up, working with the same qualitative and quantitative formulation, this IVIVR predicted that based on the *in-vitro* profile of the new batch size we would not meet our target Cmax value which was critical for this product **(SLIDE 22)**. Because there had not been substantive change or modification from a processing perspective associated with the new batch size we decided to perform a confirmatory bioequivalence study which in fact showed that the new batch size product was acceptable and did meet the target Cmax value. When we relooked at the *in-vitro* method that had predicted the new batch size would not meet target we found it to be overly sensitive both to method conditions

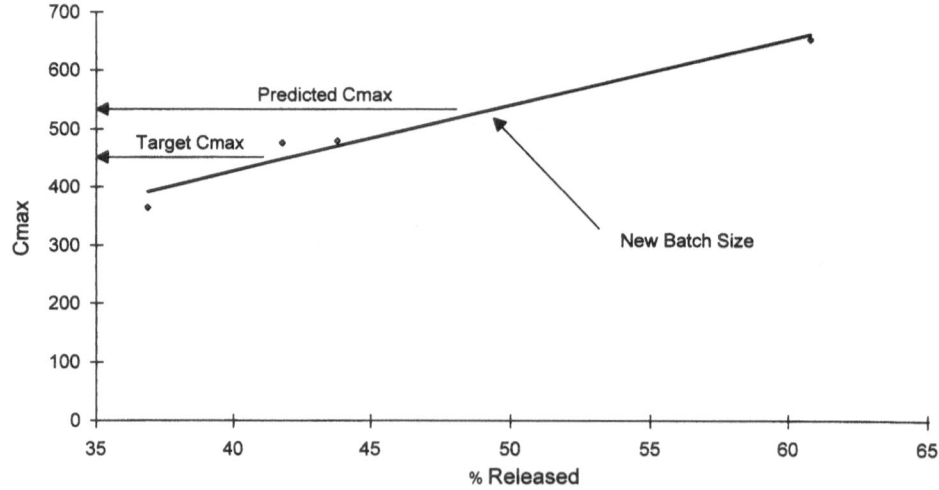

Slide 22. Retrospective IVIVR in vitro method Q2 level C.

and very minor formulation and process variations. **(SLIDE 23)** This lead us to revise the *in-vitro* method to be more robust and based on an ability to differentiate formulations that had been shown to be different previously but also now shown to predict the appropriate Cmax value of the new batch product. This was the method that was subsequently used for the prospective *in-vivo/in-vitro* study and that was used subsequently to guide pivotal batch manufacture, final product specifications etc.

The next example describes a false negative. As part of the development program of a tablet product, a number of different batches were tested *in-vivo* at different times. Ostensibly these tablets profiles were not different *in-vitro* based on the very reasonable dissolution method that had been used up to that point. However, on review it appeared that Cmax varied significantly across these tablet batches **(SLIDE 24).** When we examined the situation more closely it was clear that this particular product had a complex release system. The tablet itself was a rapidly disintegrating tablet. In turn there was a rapidly releasing component which released from the tablet matrix and was subject to both disintegration time and release characteristics from the tablet matrix. In turn there was a dominant portion of the dosage form which was available as a sustained release component from compressed coated beads which was the other portion of the tablet. On review, we believe that the critical *in-vitro* component impacting on the *in-vitro* performance was in fact the release characteristic from the coated beads. We therefore went back and used the in-process bead dissolution as a basis to establishing an IVIVR **(SLIDE 25).** We successfully established a level C correlation that predicted *in-vivo* Cmax. In this particular case therefore a huge amount of time and effort could have been expended trying to establish a perhaps complex dissolution system for the tablets in search of an *in-vivo/in-vitro* correlation when a greater understanding and more logical approach to the definition of what was critical within the formulation from an *in-vivo* perspective lead to an appropriate IVIVR that predicted *in-vivo* performance. Once a reliable IVIVR of course was established it allowed the next stages of the Product Development process such as scale-up pivotal manufacture and validation to go forward with a high level of confidence that changes or variations could be evaluated in terms of their likely *in-vivo* impact and appropriate

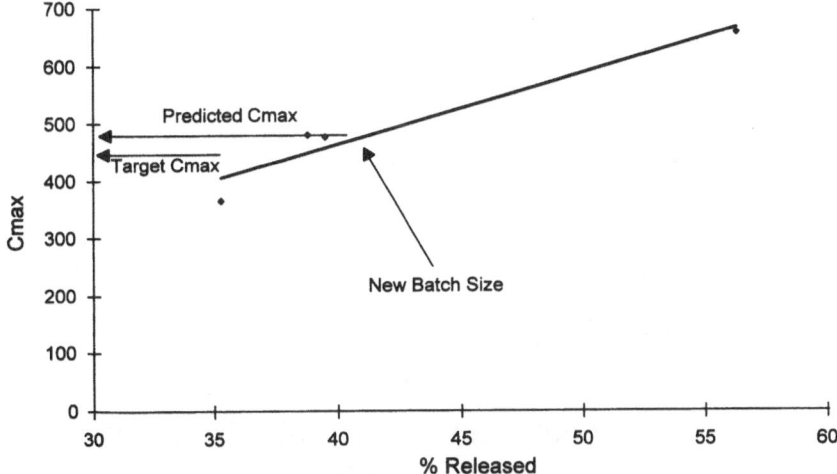

Slide 23. Q2 level C new IVIVR *in vitro* method.

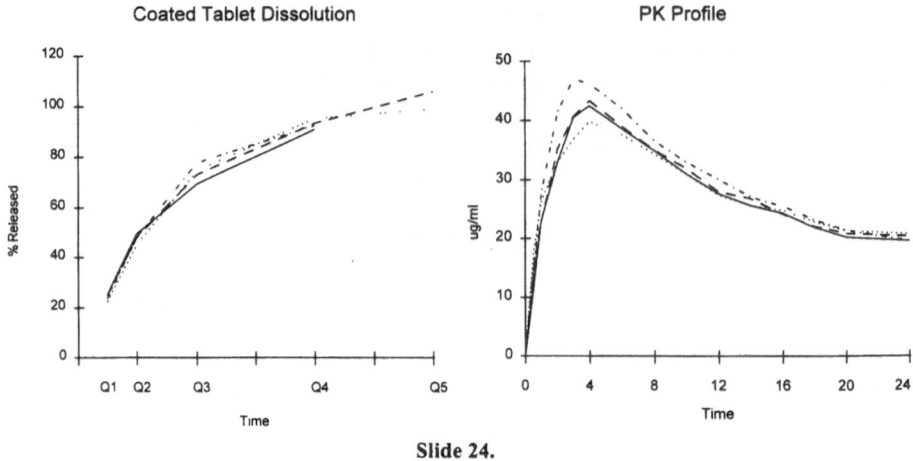

Slide 24.

judgements as to the acceptability or not of those steps made pre and post approval. **(SLIDE 26).**

To represent two later stages of the Product Development process, two final examples are presented. The first relates to establishing appropriate release specifications. In this particular case we are dealing with a level C *in-vivo/in-vitro* relationship and predicting Cmax as a parameter based on Q1 and Q2 **(SLIDE 27).** Three different lots were used to assess predictability as a basis to establishing the validity of the IVIVR. As can be seen across a predicted Cmax range ranging from 37mcg/ml through to 46mg/ml up to 65mcg/ml the prediction errors were relatively low and viewed as acceptable. Based on these correlations proposed specifications were arrived at for Q1 and Q2 with predicted Cmax values viewed as acceptable and likely to be consistent with a bioequivalence range **(SLIDE 28).** Because this development program had been relatively complex a number of full bioequivalence studies had been performed across a range of different batches and we had batches close to the low end and the high end of the proposed specification. By com-

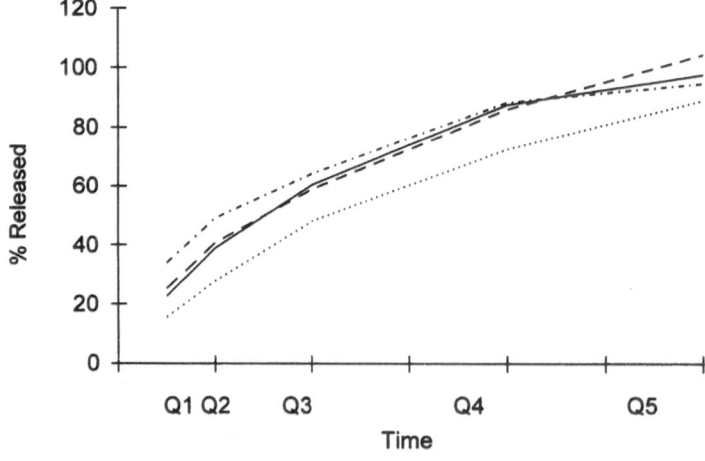

Slide 25. SR bead dissolution.

IVIVR ALLOWS

- JUDGEMENT OF IMPACT OF ANY
  FORMULATION/PROCESS ADJUSTMENTS
- PREDICTION OF PIVOTAL PK STUDY OUTCOME
- ESTABLISHMENT OF FINAL SPECIFICATIONS
  (RELEASE RATE)
- SUPPORT OF THE PROCESS VALIDATION PROGRAM

Slide 26. Scale-up/pivotal PK/validation.

bining the data we were able to assess, even though the batches were taken from different studies, the bioequivalence of the extremes. These ranges were consistent with bioequivalence and showed acceptable values within the 80–125% confidence interval range.

The last example is a situation which of course any R&D scientist hopes not to have to deal with but invariably does, and that is a situation where an extended release dosage form had been developed, characterised and indeed had entered clinical studies (**SLIDE 29**). As part of the development of that dosage form good *in-vivo/in-vitro* relationships were established. However, midway through the major development program we identified a need to change to an alternative formulation due to a long term shelf life issue specifically related to the beaded capsule formulation that we were working with. We used the established *in-vivo/in-vitro* relationship of that formulation to predict an acceptable *in-vivo* tablet profile where in particular peak and peak to trough levels were critical in terms of maintaining a comparable safety profile. With the original coated bead capsule formulation across 4 different lots, good level A type *in-vivo/in-vitro* relationships were estab-

Predictability of C_{max} using Level C IVIVR.
% DISS TC = % dissolution Q1 or Q2
PRD = predicted, OBS = observed

Lot 1 Time	% Diss.	Pred. Cmax	Obs. Cmax	% Error
Q 1	35.2	37.26	41.99	-11.26
Q 2	54.9	37.96		-9.61
Lot 2 Time	% Diss.	Pred. Cmax	Obs. Cmax	% Error
Q 1	25.2	45.92	44.53	3.12
Q 2	40.7	45.92		3.57
Lot 3 Time	% Diss.	Pred. Cmax	Obs. Cmax	% Error
Q 1	19.4	65.60	63.4	3.47
Q 2	32.4	65.25		2.91

Slide 27. Example establishing specifications.

TIME	SPECIFICATION (% DISSOLVED)	PREDICTED Cmax
Q1	15	42.08
	35	49.61
Q2	30	42.74
	50	49.06

Slide 28. Example establishing specifications. Predicited Cmax at extremes of the proposed specifications for Q1 and Q2 using the level C IVIVR.

lished **(SLIDE 30).** After reformulation, a tablet was developed with appropriate *in-vitro* characteristics and compared as follows with the capsule product *in-vitro*. From the *in-vitro* profile we anticipated and modelled an earlier absorption phase but also modelled Cmax and peak to trough fluctuation. The actual *in-vivo* performance was characterised **(SLIDE 31)** and the observed *in-vivo* absorption profiles match very closely what the *in-vitro* profile would have predicted. This is an example where an established IVIVR gave a high level of confidence that switching to a new dosage form would be acceptable and would not compromise the clinical/safety database already generated and allowed one to progress the development program rapidly with the tablet formulation.

Overall it is clear that IVIVR is a key element in the rational development of new and extended release dosage forms. It is also important to appreciate what the limitations of IVIVR methods are and indeed to acknowledge that in of itself, an IVIVR is of limited value and needs to be put in the context of a more integrated development program involving characterisation of the drug substance, understanding the mechanism of release of an extended release dosage form and integrated with the pharmacokinetic and indeed pharmacokinetic/pharmacodynamic characteristics. There are a number of open issues and future areas for development in the area of IVIVR particularly as it pertains to the Product Development process **(SLIDE 32).** Other chapters in this book have offered new ideas and concepts in the area of modelling, statistics and approaches to validation. The issue of false positive and false negatives has been described in this chapter. By its nature, linking

IVIVR USED AS A SURROGATE TO PREDICT MAJOR CHANGE

- IVIVR ESTABLISHED IN STAGE 2 FOR CAPSULE/COATED BEAD FORMULATION

- IDENTIFIED NEED TO CHANGE TO ALTERNATIVE PRESENTATION TO EXTEND SHELF LIFE

- USED IVIVR TO PREDICT "ACCEPTABLE" IN-VIVO TABLET PROFILE

- VERIFIED VALUE OF IVIVR AS A SURROGATE TO QUALIFY CHANGE

Slide 29. Qualifying change example.

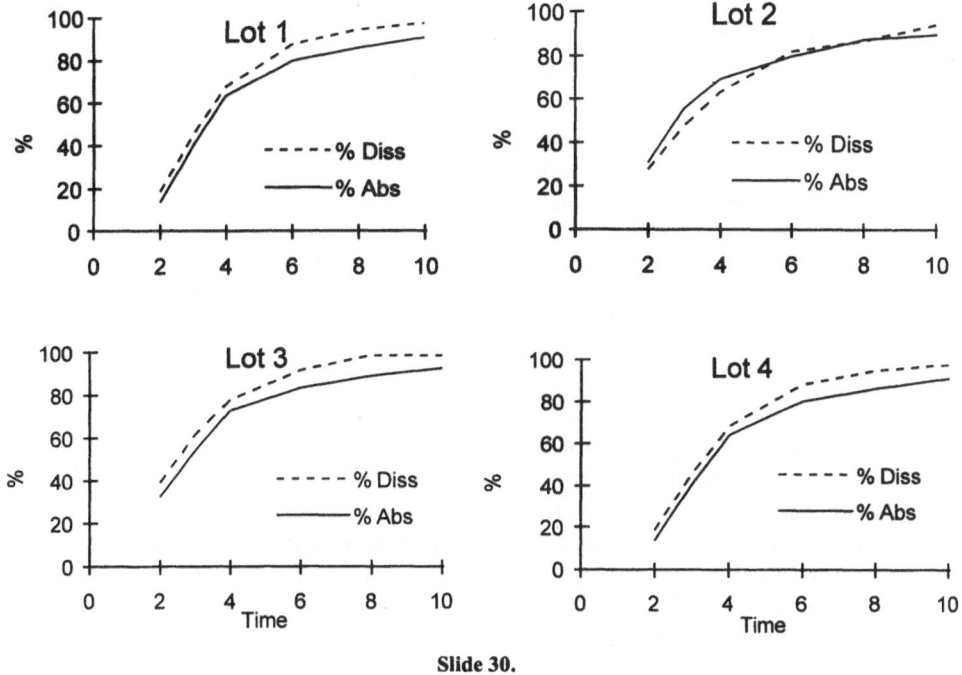

Slide 30.

as it does to what is usually a limited *in-vivo* database, the potential for false positives and false negatives will continue to be a concern until a wider experience is established and shared.

Clearly, as development scientists and also from a regulatory perspective one of our main concerns is to ensure that the products we develop, the specifications we set and the changes we make are constrained within acceptable bioequivalence limits. Linking *in-vivo/in-vitro* relationships based as they often are on limited study size and subject to the

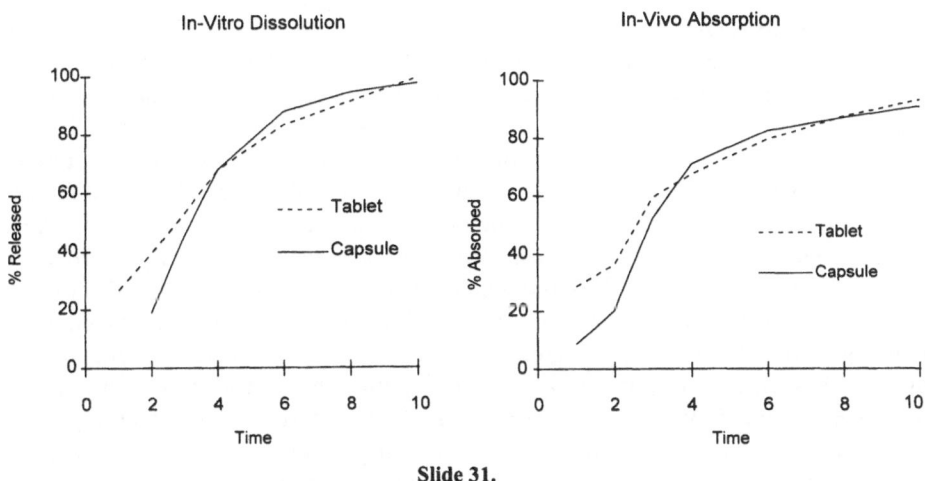

Slide 31.

- WHICH RELATIONSHIP
- WHAT MODELLING, STATISTICAL METHODS
- FALSE POSITIVES/FALSE NEGATIVES
- LINKING TO BE
- CROSS FORMULATIONS/TECHNOLOGIES
- STABILITY CHANGES
- "HEROIC" DISSOLUTION METHODS
- IN-VIVO STUDY DESIGN
- ENHANCED IN-VITRO/IN-VIVO CHARACTERISATION

Slide 32. Open issues/future developments.

inevitable inter and intra variability is an area of challenge. Obviously the ability to generalise IVIVR's across technologies and across formulations is desirable. Currently there is
a strong belief that IVIVR's should be viewed as formulation specific and perhaps even in
some circumstances process specific. Our ability to move from this conservative position
to a more general applicability within some combination of category of compounds and
technology approaches will obviously require a larger and stronger database built up over
the coming years. IVIVR as a basis to predicting the *in-vivo* consequences of stability
changes particularly in relation to release rate has not been significantly addressed. Shifts
in release rate on stability of extended release products is not an unknown phenomenon
but the mechanism of such shifts and changes maybe quite different and unrelated to the
formulation and process variation that was the basis for choosing the products upon on
which an IVIVR was originally established. To what extent an IVIVR based on such
methods can be used to predict stability changes remains to be determined. How far
should one go in pursuit of a level A correlation involving multiple media and multiple
condition changes and at the end of the day with no logical linkage to either the mechanism of release of the formulation or physiology is to my mind questionable. A level C
IVIVR that is logical, reasonable and based on more standard and rational methods may
prove at the end of the day to be more robust and relevant than what we refer to as heroic
efforts in terms of establishing correlated dissolution methods. The area of *in-vivo* design
and the choice of *in-vivo* design as a basis to establishing an IVIVR of course needs to be
product and situation specific. However, we do need to start developing some general concepts and guidelines in this area in order to standardise the data being generated and the
basis on which subsequent IVIVR's are developed and accepted. For a drug with non linear kinetics should IVIVR's be established based on steady state studies. Also if a narrow
therapeutic index and peak to through fluctuations is critical again should a steady state
study be considered. Should the traditional single dose IVIVR study always be fasted even
if the particular drug is recommended to be taken with or associated with food. These and
many more questions remain to be considered and resolved. We have already referred to a
need to understand and ideally link the mechanism of release of the dosage form into the
development of an IVIVR. Clearly methods which are described in another chapter of this
book using confocal microscopy to gain further insights into the actual mechanics and dy-
namics of drug release is potentially very powerful. These type of *in-vitro* techniques
linked to enhanced *in-vivo* characterisation such as offered by the techniques of the
gamma scintigraphy and other imaging systems can provide very powerful and enhanced
understanding and ultimately confidence in the robustness and reliability of both the dos-

CONCLUSIONS

IVIVR IN PRODUCT DEVELOPMENT

- RATIONAL TOOL IN DOSAGE FORM DEVELOPMENT
- IVIVR IS A DYNAMIC ELEMENT OF PRODUCT DEVELOPMENT PROCESS
- TECHNICAL ASPECTS OF IVIVR NEED TO CONTINUE TO BE DEVELOPED/ENHANCED
- MORE GENERAL APPLICABILITY/EXTRAPOLATION NEEDS TO BE EXPLORED
- CONSENSUS OF INDUSTRY, REGULATORS AND ACADEMIA NEEDS TO BE MAINTAINED AND BUILT ON

Slide 33.

age form and the corresponding *in-vitro* control methods used to control and predict *in-vivo* performance.

In conclusion therefore **(SLIDE 33)** we believe that IVIVR has evolved from being an interesting exercise to now a well founded rational tool in Product Development. The development of an IVIVR is a dynamic process that needs to be considered at the very earliest stages of the development program itself and then needs to be a constant thread that runs through the integrated formulation development, analytical testing and the *in-vivo* characterisation of the dosage form. Clearly there are pitfalls in the development of IVIVR as there are in all the other elements of a development program and for this reason the techniques, methods and approaches need to be constantly revised and critiqued. There are a number of areas that need to be explored in the future to further strengthen and enhance the applicability and general acceptance of IVIVR both to development scientists and to regulators. The consensus of academia, regulators and industry have built a firm foundation to the methodology that is now referred to as IVIVR and recognition of this is evidenced by the guidance that is now being established by the FDA and other regulatory agencies in the application of these methods in dosage form development and qualification of change.

24

THE ROLE OF *IN VITRO–IN VIVO* CORRELATIONS (IVIVC) TO REGULATORY AGENCIES

Henry J. Malinowski

Director Division of Pharmaceutical Evaluation I
Office of Clinical Pharmacology and Biopharmaceutics
Food and Drug Administration
1451 Rockville, Maryland 20852

1. INTRODUCTION

Dissolution testing remains a potentially powerful and nearly always useful method for obtaining data related to quality and, potentially, clinical performance of dosage forms, especially solid oral dosage forms. But, not surprisingly, dissolution is not always a surrogate for bioequivalence, which necessitates human testing for determination of bioequivalence in many instances. The key to confidence in dissolution testing is the strength of the relationship between dissolution and bioequivalence, in other words, the ability of dissolution testing to predict *in vivo* performance. Therefore, the availability of an IVIVC, as well as the type of dissolution testing conducted, are important considerations.

This discussion will put forward several examples of the use of IVIV relationships, including, but not limited to IVIVC, in various regulatory agencies around the world. Specific IVIVCs and the application of IVIVCs remains a relatively unusual circumstance. Such correlations seem most likely for some ER drug products which leaves a large body of dosage forms, both Immediate release (IR) and extended release (ER), for which IVIVC are not available, for which alterative approaches have been utilized. These approaches include concepts of side batches, the Biopharmaceutics Classification System and multi-media dissolution testing. Also available are the "FIP Guidelines for Dissolution Testing of Solid Oral Products" Final Draft (1) and the FDA draft Guidance "Extended Release Solid Oral Dosage Forms - Development, Evaluation and Application of In Vitro/In Vivo Correlations" (2).

While IVIVC for ER products remains the optimal goal, other approaches currently can provide means for increased confidence in dissolution testing as a surrogate for bioequivalence testing.

In Vitro–in Vivo Correlations, edited by Young *et al.*
Plenum Press, New York, 1997

2. EUROPE

One recent, primarily European, effort which provides insight into current thinking in the area of IVIVCs is the final draft "FIP Guidelines for Dissolution Testing of Solid Oral Products." This guideline is expected to be finalized in late 1996. An excellent summary of this Guideline was presented (3) by Martin Siewert. In the Guideline, a new term "*in vitro-in vivo* comparison" is used to identify a wider understanding than IVIVC or association. An *in vitro-in vivo* comparison is the process of comparing dissolution data to bioavailability data to determine what relationship exists between these two parameters. The stated purpose of these comparison studies is the "scientific verification of the *in vitro* test system and the respective specification limits for a given drug formulation." This comparison is applicable to both IR and ER dosage forms. It is useful to understand that, in the terminology of the Guideline, the comparison study may define a significant *in vitro-in vivo* association (IVIVC) but that useful information may still be obtained even when a correlation in the strict sense is not found.

For IR products, the comparison study suggested in the Guideline (Figure 1) For ER products, the Guideline endorses the categorization of correlation methods described in the USP, namely Levels A, B, and C. The type of correlation being attempted will determine how many batches should be included in the correlation study. It is suggested that a single batch may be sufficient for an acceptable IVIVC only for a Level A correlation for a drug product with dissolution completely independent of environmental conditions. It is suggested that a Level A correlation can be used as a surrogate for bioequivalence testing for changes in manufacturing site, minor formulation changes, scale-up considerations as well as for setting dissolution specifications.

For ER products, two alternative methods are suggested for verification of dissolution specifications, when a strict IVIVC cannot be developed. These are, situations where rank order correlation (Figure 2) is found and the concept of bioequivalence of side batches (Figure 3).

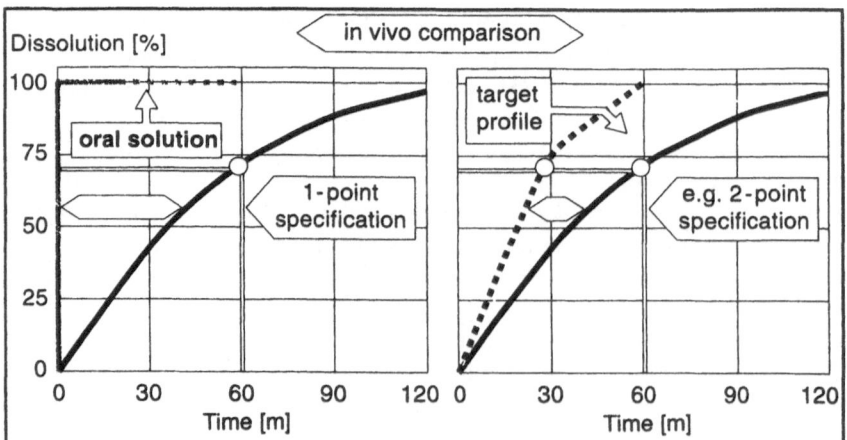

Figure 1. Specification type and verification study design concepts for immediate release products with Q specified for grater than 15 minutes (used with permission) consists of a two-way crossover study between an oral solution and a formulation which dissolves close to the dissolution specification limit.

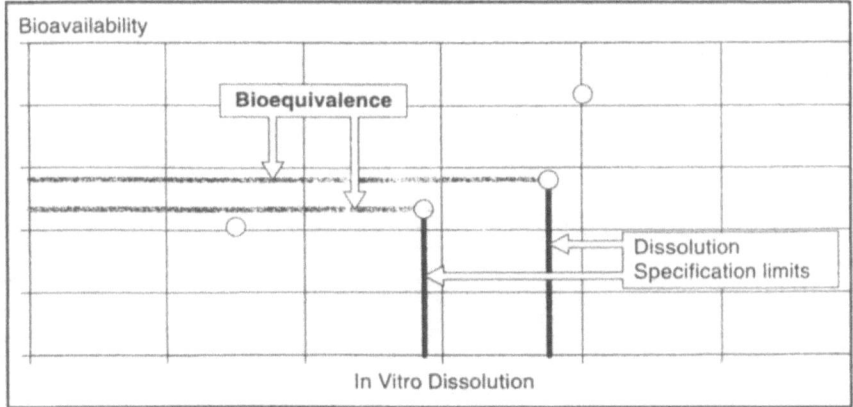

Figure 2. Application of a rank order correlation for verification of in-vitro dissolution specifications (used with permission).

In both of these situations, a separate bioequivalence study is suggested, to demonstrate bioequivalence of formulations with dissolution profiles near the upper and lower dissolution specifications.

3. JAPAN

For a Japanese view in this regard, I will refer a presentation (4) by Dr. Nobuo Aoyagi from the Ministry of Health and Welfare (MHW) in Japan. This presentation focuses on the role of dissolution tests for bioequivalence assessment and consistently emphasizes the use of several dissolution conditions particularly with regard to pH. This relates to concerns associated with individuals exhibiting achlorhydria.

(Figure 4) shows simulated expected performance for 2 products, one of which exhibits pH dependent dissolution, while the other's dissolution characteristics are pH inde-

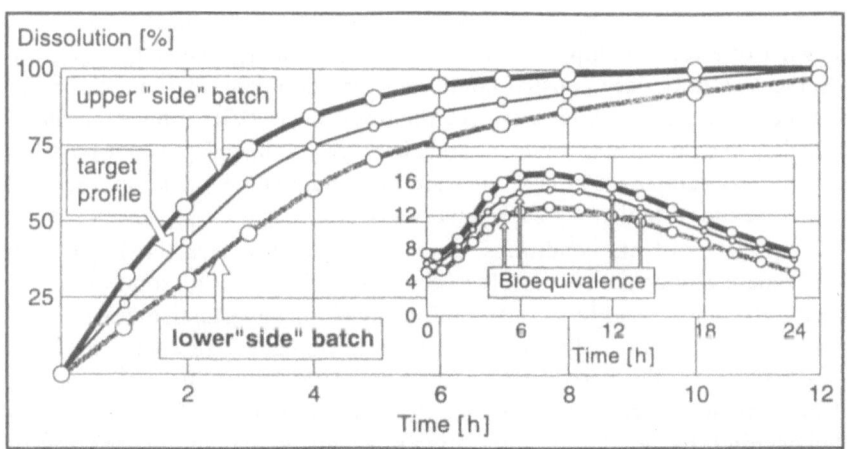

Figure 3. In-vivo verification of in-vitro test systems and specification based on the side-batch approach (example)(used with permission).

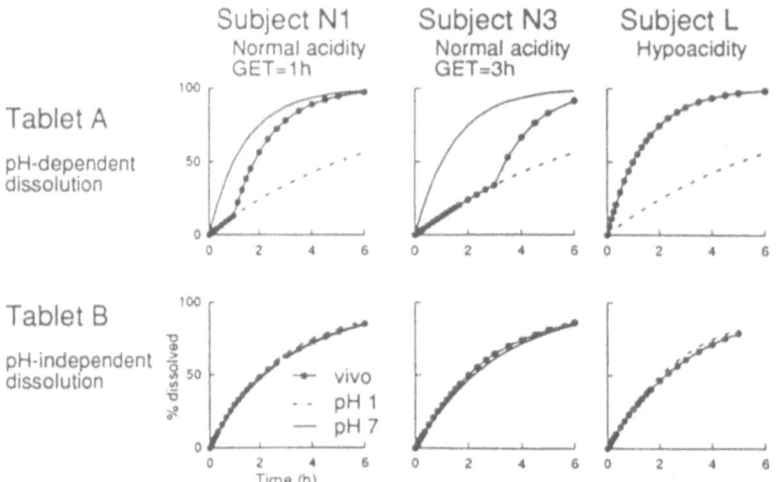

Figure 4. *In vitro/in vivo* dissolution profiles in three subjects with different gastric acidity and gastric emptying time for tablets A and B showing pH-dependent and independent dissolution (used with permission).

pendent. Shown are both dissolution and *in vivo* absorption data which indicate that the product with pH independent dissolution will more easily matches the expected bioavailability results in both patients with normal acidity or hypoacidity. For the pH dependent dissolution product, no one set of dissolution conditions will describe expected bioavailability for all patients. An important point to be made is that a product which shows pH independent dissolution *in vitro* is probably also pH independent, as far as dissolution, *in vivo*. This type of product is likely to be less susceptible to gastric acidity differences among patients.

Another aspect of the process establishing a relationship between dissolution and bioavailability is shown in Figure 5.

This shows bioavailability data for 2 subjects for 2 ER products. In one subject, the bioavailability profiles are very similar, in the other subject, the profiles are quite different. Somewhat of a relationship is shown for subject K using basket 150 rpm conditions, but, for Subject N, only very unusual conditions, paddle 0 rpm (1 minute stirring at each sampling time) resulted in a *in vitro in vivo* relationship (IVIVR). This example illustrates that *in vivo* data can be quite variable, as seen in these 2 subjects. It is not possible, or reasonable, to try to establish an IVIVR for each individual subject. Therefore, average data are normally used for establishing such relationships.

The importance of pH conditions, both *in vitro* and *in vivo*, is shown in Figure 6. These 2 diazepam products both dissolve very rapidly in pH1.2 dissolution medium. And in subjects with normal gastric acidity, both products have similar bioavailability. However, in subjects with hypoacidity, bioavailability differences are apparent and this is reflected in dissolution testing using pH 4.6 medium. This further illustrates the importance attached to studying several pH ranges for dissolution testing, related to concerns about gastric hypoacidity, by regulatory authorities in Japan. A detailed description of recommended dissolution testing may be found in the recent draft "Guideline for Bioequivalence Testing" (5) for generic drugs from the MHW.

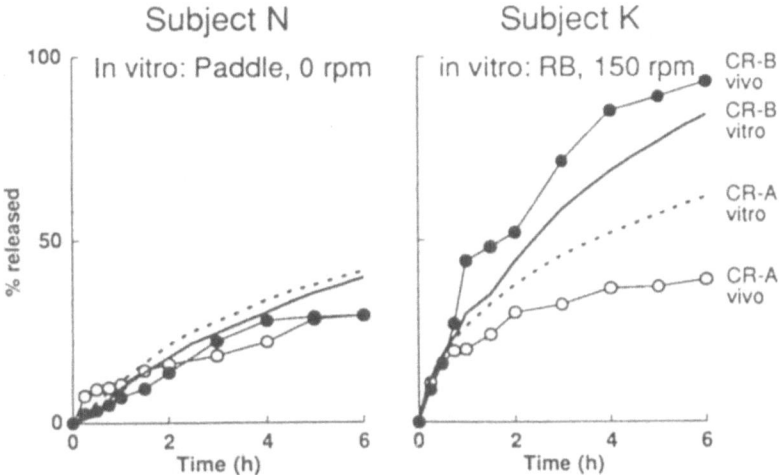

Figure 5. In vivo release of acetaminophen from CR-A and B in typical two subjects and *in vitro* release by JP paddle method a 0 rpm (1 min stirring at each sampling time) and rotating basket (RB) method at 150 rpm (used with permission).

Figure 7 illustrates the distribution of GI variables throughout the population. This is shown in relation to the degree of discrimination of tests for bioequivalence assurance, ranging from discriminatory (perhaps over discriminatory) conditions to nondiscriminatory conditions. It suggests that tests to assure bioequivalence generally are quite discriminatory, focused not on the average individual, but including nearly everyone in the population. While such typical conservative test conditions do provide a safety margin in detecting differences in measured parameters, they can also lead to situations where dif-

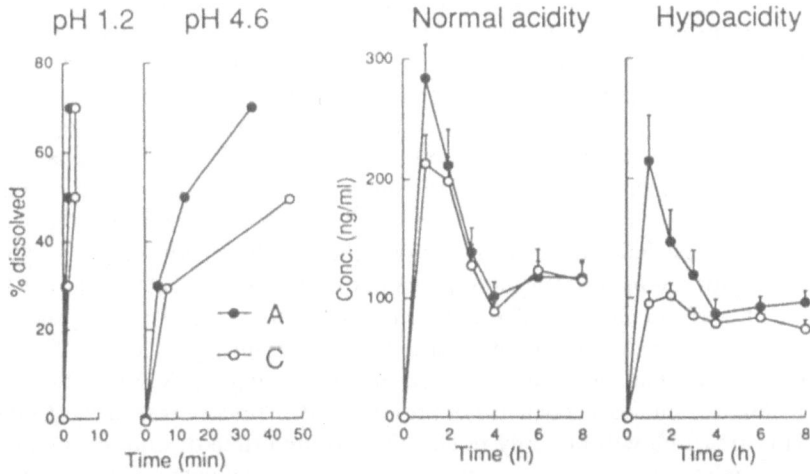

Figure 6. In vitro dissolution of diazepam from tablets A and C by a beaker method and serum concentration in two groups of subjects with normal and hypo-acidity of gastric fluid.

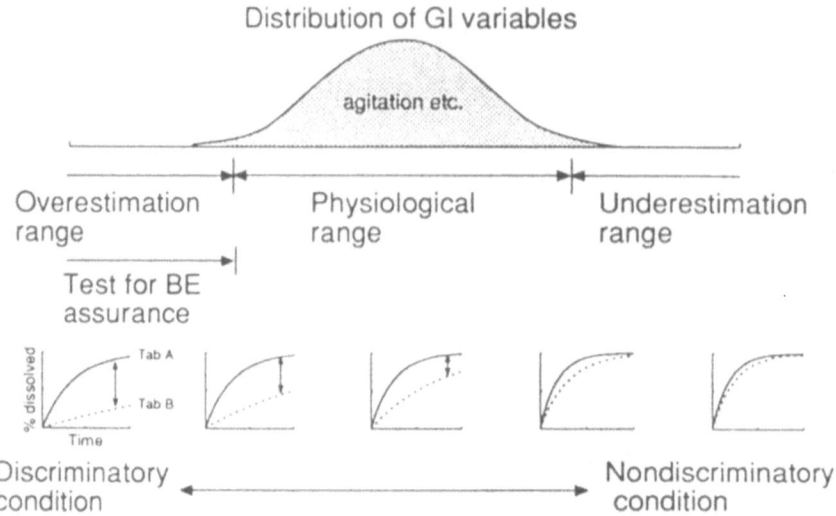

Figure 7. Physiological range of critical GI variables and *in vitro* testing condition of bioequivalence (BE) assurance

ferences are detected in, for example, dissolution testing, which do not relate to differences in bioavailability.

4. NORTH AMERICA

In the United States, a draft (2) Guidance for Development, Evaluation and Application of in Vitro/in Vivo Correlations was released on July 10, 1996. This guidance provides recommendations to pharmaceutical scientists related to various aspects of IVIVC for oral extended-release (ER) drug products particularly as utilized in the NDA/ANDA review process. It presents a comprehensive perspective on methods of developing IVIVC, appropriate means of evaluating the predictability of IVIVC, and relevant applications for IVIVC in the areas of changes (e.g., formulation, equipment, process, and manufacturing site) and setting dissolution specifications.

In vitro dissolution testing is important for: (1) providing necessary process control and quality assurance, (2) determining stability of the relevant release characteristics of the product, and (3) facilitating certain regulatory determinations and judgments concerning for example, minor formulation changes or change in site of manufacture. In addition, in certain cases, especially for ER formulations, the dissolution test can serve not only as a quality control for the manufacturing process but also as an indicator of how well the formulation will perform *in vivo*. Thus, the main objective of developing and evaluating IVIVC is to empower the dissolution test to serve as a surrogate marker for human bioequivalence studies. One additional purpose of establishing an IVIVC is to minimize the number of human studies needed for approving and maintaining a drug product on the market. This approach will not only reduce the cost and time of drug development by reducing the number of studies required to demonstrate adequate bioavailability, but will also facilitate the initial approval as well as scale-up and post-approval changes. However, for certain applications the adequacy of the *in vitro* method to act as a surrogate for *in vivo* testing must be demonstrated through an IVIVC for which predictability has been demonstrated.

4.1. Biowaivers

Regarding biowaivers, five categories of waivers are described in the FDA Guidance. These range from situations which are insignificant as far as expected effect on product performance, such as a change in manufacturing equipment where the new equipment has the same design and operating principles, to very significant changes for which biotesting would be required even if an IVIVC with good predictability has been developed. An example of a very significant change situation is the approval of another sponsor's ER product, for which an IVIVC has not been specifically developed, even with the same release controlling mechanism, where the reference product does have an IVIVC established.

Between these 2 categories are three categories for 1) non-narrow therapeutic index drugs, 2) narrow therapeutic index drugs and 3) ER drug products which have dissolution characteristics which are independent of dissolution test condition. Specific recommendations related to IVIVC for changes in site of manufacture, release-controlling and non-release-controlling excipients, manufacturing equipment and process, as well as approval of certain new strengths, are categorized and described in the draft Guidance, for each of these categories.

The criteria for granting biowaivers in circumstances where a IVIVC has been established are that the difference in predicted means of C_{max} and AUC is no more than 20% from that of the reference product and, where applicable, the new formulation meets the application/compendial dissolution specifications.

4.2. Setting Dissolution Specifications

One additional important use of IVIVC, as described in the draft Guidance, relates to using the IVIVC in the process of setting dissolution specifications. Also described is the situation of setting dissolution specifications where there is no IVIVC.

4.2.1. No IVIVC. In general, USP acceptance criteria for dissolution, regarding setting specifications, are utilized unless alternate acceptance criteria are justified. One less than optimal approach is to have the specifications established such that all lots pass at Stage 1 of testing. In other words, each individual dosage form among the dissolution data being utilized to determine appropriate specifications, passes the proposed specifications. Therefore, it is recommended that specifications should be established based on average dissolution data (Stage 2). Specification ranges of 20% or less are recommended. If justified, deviations from this criteria can be accepted, up to ranges of approximately 25%.

Specification ranges greater than 20% are generally acceptable only when supported by evidence that lots with mean dissolution profiles that are allowed by the upper and lower limit of the specifications are bioequivalent.

4.2.2. When an IVIVC Has Been Established. An important application of an IVIVC is in relation to setting appropriate dissolution specifications for the product. Ideally, the specifications should be established such that all lots that have dissolution profiles within the upper and lower limits of the specifications, are bioequivalent. Minimally, these lots should be bioequivalent to the lots used in the clinical trials or an appropriate reference standard. Specifications should be set on mean data using at least 12 individual dosage units per data set. Calculate the plasma concentration time profile for the upper and lower proposed dissolution specification profiles and using convolution techniques (or other appropriate modeling techniques) determine whether the lots with the fastest and slower release rates that are allowed by the dissolution specifications result in a maximal difference of 20% in the predicted C_{max} and AUC.

5. CONCLUSION

IVIVCs are not extensively used at this time by regulatory agencies around the world. However, there is general agreement that IVIVCs are very useful and desirable. And, IVIVRs, that is, a dissolution test which is meaningful in the sense that some relationship between the dissolution test results and expected bioavailability changes has been established, are universally accepted.

Recent draft Guidelines from Japan, the Federation Internationale Pharmaceutique (FIP), and the U.S. FDA can provide much useful information regarding current thinking in the major drug regulatory jurisdictions around the world regarding the role of *in vitro/in vivo* correlations as well as other *in vitro/in vivo* relationships.

REFERENCES

1. Guidelines for Dissolution Testing of Solid Oral Products Final Draft 1995, Pharmacopoeial Forum, 21 (5), 1371–1382, 1995.
2. Guidance for Industry - Extended-Release Solid Oral Dosage Forms: Development, Evaluation and Application of *in Vitro/in Vivo* Correlations, Final Draft July 1, 1996. Food and Drug Administration.
3. New FIP Guideline for Dissolution Testing of Solid Oral Products, Dissolution Technologies 3 (3), 3–6, 1996.
4. Use of Dissolution Tests for Bioequivalence Assessment in Japan, Nobuo Aoyagi, Abstract, FIP Bio International 1996, 132–135.
5. Guideline for Bioequivalence Studies of Generic Drugs, Draft July 1, 1996, Ministry of Health and Welfare, National Institute of Health Sciences, Japan.

25

DRAFT GUIDANCE FOR INDUSTRY EXTENDED-RELEASE SOLID ORAL DOSAGE FORMS

Development, Evaluation and Application of *in Vitro–in Vivo* Correlations

Henry Malinowski, Patrick Marroum, Venkata Ramana Uppoor,
William Gillespie, Hae-Young Ahn, Peter Lockwood, James Henderson,
Raman Baweja, Mohammad Hossain, Nicholas Fleischer, Lloyd Tillman,
Ajaz Hussain, Vinod Shah, Angelica Dorantes, Ray Zhu, He Sun, Kofi Kumi,
Stella Machado, Vijaya Tammara, Ting Eng Ong-Chen, Houda Mahayni,
Lawrence Lesko, and Roger Williams

I. PURPOSE OF DRAFT GUIDANCE

This draft guidance provides recommendations to pharmaceutical scientists related to various aspects of *in vitro/in vivo* correlations (IVIVC) for oral extended-release (ER) drug products particularly as utilized in the NDA/ANDA review process. It presents a comprehensive perspective on methods of developing IVIVC, appropriate means of evaluating the predictability of IVIVC, and relevant applications for IVIVC in the areas of changes (e.g., formulation, equipment, process, and manufacturing site) and setting dissolution specifications. To access the final guidance on the WWW, connect to theFDA home page at http://www.fda.gov/CDER/ and go to the "Regulatory Guidance" section.

II. DEFINITION OF TERMS

Application (As Applied to In Vitro/In Vivo Correlation). Use of an *in vitro/in vivo* correlation in setting dissolution specifications and/or as support for waiver of *in vivo* bioequivalence studies.

Batch. A specific quantity of a drug or other material produced according to a single manufacturing order during the same cycle of manufacture and intended to have uniform character and quality, within specified limits *(21 CFR 210.3(b)(2))*.

In Vitro–in Vivo Correlations, edited by Young *et al.*
Plenum Press, New York, 1997

Composition (Batch Formula). A complete list of the ingredients and their amounts to be used for the manufacture of a representative batch of the drug product. All ingredients should be included in the batch formula whether or not they remain in the finished product *(Guideline for submitting documentation for the manufacture of and controls for drug products, FDA, February 1987).*

Bioavailability. The rate and extent to which the active drug ingredient or therapeutic moiety is absorbed from a drug product and becomes available at the site of drug !action *(21 CFR 320.1(a)).*

Biobatch. The lot of drug product formulated for purposes of pharmacokinetic evaluation in a bioavailability or bioequivalence study. This lot is typically 10% as large as a production lot or at least 100,000 units.

Bioequivalent Drug Products. Pharmaceutical equivalents or pharmaceutical alternatives whose rate and extent of absorption do not show a significant difference when administered at the same molar dose of the therapeutic moiety under similar experimental conditions, either single dose or multiple dose. Some pharmaceutical equivalents or pharmaceutical alternatives may be equivalent in the extent of their absorption but not in their rate of absorption and yet may be considered bioequivalent because such differences in the rate of absorption are intentional and are reflected in the labeling, are not essential to the attainment of effective body drug concentrations on chronic use, or are considered medically insignificant for the particular drug product studied *(21 CFR 320.1(e)).*

Bootstrapping. A computer-based resampling method for estimating the precision of statistical estimates, e.g., standard errors and confidence intervals *(An Introduction to the Bootstrap by Bradley Efron and Robert J. Tibshirani).*

Convolution. Prediction of plasma drug concentrations using a mathematical model based on the convolution integral. For example, the following convolution integral equation may be used to predict the plasma concentration ($c(t)$) resulting from the absorption rate time course (r_{abs})

$$c(t) = \int_0^t c_\delta(t-u) \, r_{abs}(u) \, du$$

The function c represents the concentration time course that would result from the instantaneous absorption of a unit amount of drug and is typically estimated from i.v. bolus data.

Correlation. Having a connection to one another, or a mutual relationship. Typically for *in vitro/in vivo* correlation, a relationship is sought between *in vitro* dissolution rate and in vivo input (absorption) rate.

Cross-Validation. A statistical method for estimating prediction error *(An Introduction to the Bootstrap by Bradley Efron and Robert J. Tibshirani).*

Deconvolution. Estimation of the time course of drug input (usually *in vivo* absorption or dissolution) using a mathematical model based on the convolution integral. For example, the absorption rate time course (r_{abs}) that resulted in the plasma concentrations ($c(t)$) may be estimated by solving the following convolution integral equation for r_{abs}

$$c(t) = \int_0^t c_\delta(t-u) \, r_{abs}(u) \, du$$

The function c_d represents the concentration time course that would result from the instantaneous absorption of a unit amount of drug and is typically estimated from i.v. bolus data.

Development. Establishing an in vitro/in vivo correlation.

Drug Product. A finished dosage form, for example, tablet, capsule, or solution, that contains a drug substance, generally, but not necessarily, in association with one or more other ingredients *(21 CFR 314.3(b))*.

Hill Equation (D_{max} Function). In the context of this document, this is commonly used to model the relationship between time t and % dissolved D with a mathematical relationship as shown below:

$$D = \frac{D_{max} * t^Y}{D_{50}^Y + t^Y}$$

Extended-Release. Allows a reduction in dosing frequency as compared to that presented as a conventional dosage form (e.g., as a solution or an immediate release dosage form).

Formulation. A listing of the ingredients and composition of the dosage form.

In Vitro/In Vivo Correlation. A predictive mathematical model describing the relationship between an *in vitro* property of an extended release dosage form (usually the rate or extent of drug dissolution or release) and a relevant *in vivo* response (e.g., plasma drug concentration or amount of drug absorbed).

Level A Correlation. A predictive mathematical model for the relationship between the entire *in vitro* dissolution/release time course and the entire *in vivo* response time course (e.g., the time course of plasma drug concentration or amount of drug absorbed).

Level B Correlation. A predictive mathematical model for the relationship between summary parameters that characterize the *in vitro* and *in vivo* time courses, e.g., models that relate the mean *in vitro* dissolution time to the mean *in vivo* dissolution time, the mean *in vitro* dissolution time to the mean residence time *in vivo*, or the *in vitro* dissolution rate constant to the absorption rate constant.

Level C Correlation. A predictive mathematical model of the relationship between the amount dissolved in vitro at a particular time (or the time required for in vitro dissolution of a fixed percent of the dose, e.g., T %) and a summary parameter that characterizes the *in vivo* time course, e.g., C_{max} or AUC.

In Vivo Dissolution. This is the process of dissolution of drug in the gastro-intestinal tract.

In Vitro Release. Drug release from a dosage form as measured in an *in vitro* dissolution apparatus.

In Vivo Release. Release of drug from a dosage form as measured by pharmacokinetic studies in humans (patients or healthy volunteers).

Lot. A batch, or a specific identified portion of a batch, having uniform character and quality within specified limits or, in the case of a drug product produced by continuous process, a specific identified amount produced in a unit of time or quantity in a manner that assures its having uniform character and quality within specified limits *(21 CFR 210.3(b)(10)).*

Manufacturing Site Change. The relocation of the manufacturing process for a drug substance or dosage form from a noncontiguous building to other.

Meaningful Dissolution Method. An *in vitro* dissolution test which has been demonstrated to be directly correlatable with *in vivo* absorption.

Mean Absolute Prediction Error (PE). This is a prediction error metric calculated using the **abs** following equation:

$$PE_{abs} = \frac{1}{\sum_{k=1}^{N} n_k} \sum_{k=1}^{N} \sum_{i=1}^{n_k} |O(t_{ki}) - P(t_{ki})|$$

n = number of sampling times for k^{th} formulation k th
N = number of formulations
$O(t)$ = observed *in vivo* response to the kth formulation at the ith sampling time
$P(t)$ = predicted *in vivo* response to the kth formulation at the ith sampling time

Mean Absorption Time. This is the mean time required for drug to reach systemic circulation from the time of drug administration. This is commonly referred to as the mean time involved in the *in vivo* release and absorption processes as they occur in the input compartment and is estimated as $MAT = MRT_{oral} - MRT_{i.v.}$

Mean Dissolution Time. This reflects the mean time for drug to dissolve.

Mean in Vitro Dissolution Time. This is the mean time for the drug to dissolve under in vitro dissolution conditions. This is calculated using the following equation:

$$MDT_{vitro} = \frac{\int_o (M_\infty - M(t))\, dt}{M}$$

Mean in vivo Dissolution Time. For a solid drug product, mean *in vivo* dissolution time is: $\text{MDT}_{\text{solid}} = \text{MRT}_{\text{solid}} - \text{MRT}_{\text{solution}}$. This reflects the mean time for drug to dissolve *in vivo*.

Mean Residence Time. This describes the average time for the drug to reside in the body. MRT may also be considered as the mean transit time. MRT = AUMC/AUC.

Narrow Therapeutic Index Drugs. These drugs have for example less than a two-fold difference in the minimum toxic concentrations and the minimum effective concentrations *(21CFR 320.33 (c))*.

Non-Release Controlling Excipient (Non-Critical Compositional Variable). An excipient in the final dosage form which has no effect on the release of the active drug substance from the dosage form.

Predictability. Predictability of an IVIV model relates to verifying how well the model describes *in vivo* results from a test set of in vitro data as well as from the data that was used to develop the correlation.

Release Controlling Excipient (Critical Compositional Variable). An ingredient in the final dosage form whose primary function is to extend the release of the active drug substance from the dosage form.

Release Mechanism. The process by which the drug substance is released from the dosage form.

Release Rate. Fraction of dose in dosage form or amount of drug released per unit of time as defined by *in vitro* or *in vivo* testing.

Root Mean Square Prediction Error (PE_{rms}). This is a prediction error metric calculated using the following equation:

$$PE_{rms} = \sqrt{\frac{1}{\sum\limits_{k=1}^{N} n_k} \sum_{k=1}^{N} \sum_{i=1}^{n_k} \left[O(t_{ki}) - P(t_{ki})\right]^2}$$

n = number of sampling times for k^{th} formulation
N = number of formulations
$O(t_{ki})$ = observed *in vivo* response to the kth formulation at the i sampling time
$P(t_{ki})$ = predicted *in vivo* response to the kth formulation at the i sampling time

Statistical Moments. These are parameters that describe the characteristics of the time courses of plasma concentration (area, mean residence time and variance of mean residence time) and of urinary excretion rate *(Journal of Pharmacokinetics & Biopharmaceutics, vol. 6(6), 547, 1978)*.

Validation. "Establishing documented evidence which provides a high degree of assurance that a specific process will consistently produce a product meeting its predetermined specifications and quality attributes *(Guideline on General Principles of Process Validation, May 1987, Office of Compliance)*." "A validated process is one that has been proved to do what it purports or is represented to do" *(Program Guidance Manual for FDA Inspectors)*. In the context of *in vitro/in vivo* correlation

Validation/evaluation is a broad term encompassing experimental and statistical techniques used during development and evaluation of a correlation which aid in deciding whether the correlation is both predictive and of good quality.

Weibull Function. This is one of the models used to fit the dissolution profile, as shown in the equation below:

$$M = M_0 [1 - e^{-\frac{(t-T_0)^b}{a}}]$$

where, T = lag time; a = measure of rate of release; b = parametric description of curve form *(J. Appl. Mech., 18, 293, 1951)*.

III. INTRODUCTION

The concept of IVIVC, particularly for ER drug products, has been extensively discussed by pharmaceutical scientists. The ability to predict, accurately and precisely, expected bioavailability characteristics for an ER product from dissolution profile characteristics is a long-sought-after goal.

The "Report of the Workshop on CR Dosage Forms: Issues and Controversies" (1987) indicated that the state of science and technology at that time did not permit consistently meaningful IVIVC for ER dosage forms, and encouraged IVIVC as a future objective. Dissolution testing was suggested to be useful only for process control, stability, minor formulation changes, and manufacturing site changes.

A USP PF Stimuli Article in July, 1988 established the classification of IVIVC into Levels A, B and C, that are still in use today.

The Report "*In Vitro/in vivo* Testing and Correlation for Oral Controlled/Modified Release Dosage Forms" (1990) concluded that, while the science and technology may not always permit meaningful IVIVC, the development of IVIVC was stated to be an important objective on a product-by-product basis. Technologies for development, evaluation and application of IVIVC were described. Validation of dissolution specifications by way of a full bioequivalence study involving two batches of product with dissolution profiles at the upper and lower dissolution specifications was suggested.

USP Chapter <1088> similarly describes techniques appropriate for Level A, B and C correlations as well as methods for establishing dissolution specifications.

A further study of IVIVC may be found in "Workshop II Report: Scale-up of Oral Extended Release Dosage Forms" (1993). This report stated the objectives of an IVIVC as the use of dissolution as a surrogate for bioequivalency testing, and as an aid in setting dissolution specifications. It was concluded that dissolution may be used as a sensitive, reliable and reproducible surrogate for bioequivalence testing. The report gave support to the concepts of USP Chapter <1088> and further found that IVIVC may be useful for changes other than minor changes in formulation, equipment, process, manufacturing site, and batch size.

What is apparent from this review of IVIVC reports over the last 10 years is an increasing confidence in IVIVC as a sensitive means of using in vitro dissolution results as an indication of the *in vivo* bioavailability characteristics for the drug product being studied. In this regard, increased IVIVC activity in NDA submissions has been apparent. Still, the complete process of developing an IVIVC with high quality and predictability, and identifying specific applications for such correlations has not been well-defined.

As part of the process in developing this Guidance, surveys of NDA submissions for ER drug products were undertaken to ascertain the extent of IVIVC to be found therein. The first survey, of NDA submissions from 1982–1992, found 9 attempted IVIVCs among 60 such submissions. A more recent survey of NDA submissions from October 1994 to October 1995 found that IVIVCs were investigated in 9 out of 12 submissions. While this increased IVIVC activity is encouraging, there still is apparent the need to try to clarify and identify the entire process of developing correlations that are highly precise and predictable leading to the application of IVIVC in situations involving various changes in the ER drug products as well as for setting dissolution specifications.

It is for this reason that the Extended Release Dissolution Working Group was formed and this Guidance has been prepared. The Working Group has systematically studied available reports, compiled and considered published literature, and surveyed NDA submissions involving ER products.

The following Guidance is a result of these deliberations and it is recommended that sponsors developing an ER product always investigate the possibility of an IVIVC. Alternative scientifically valid procedures may be undertaken as deemed appropriate.

It is strongly believed that the availability of meaningful IVIVC of high quality and predictability will result in a significant positive impact on product quality of marketed ER products as well as reduced regulatory burden as the requirements for bioequivalency testing are lessened.

This Guidance puts forward current FDA recommendations regarding IVIVC for ER products. It describes the levels of correlations that can be established with varying degrees of robustness, important considerations for *in vivo* and *in vitro* experimentation, evaluation of the correlation by focusing on the critical feature of predictability, and finally, the practical applications that can be achieved using the IVIVC. It is intended to be the next step forward in the scientific community's efforts to delineate an approach to IVIVC that is meaningful, reasonable, and complete.

IV. CATEGORIES OF *IN VITRO/IN VIVO* CORRELATION

A. Level A

This is the most common type of correlation observed in NDAs submitted to the FDA. From a regulatory point of view it is considered to be the most useful. A correlation of this type which is usually estimated by a two stage procedure (e.g., deconvolution followed by comparison of the fraction absorbed to the fraction dissolved), is generally linear, and represents a point-to-point relationship between in vitro dissolution and the *in vivo* input rate (*in vivo* dissolution) of the drug from the dosage form. In such a correlation, the *in vitro* dissolution and *in vivo* input rate curves are either directly superimposable or may be made to be superimposable by the use of a scaling factor.

Alternative approaches to Level A IVIVC based on a convolution procedure model the relationship between in vitro dissolution and plasma concentration in a single step. Plasma concentrations predicted from the model and those observed are compared directly. For these methods the use of a reference treatment is desirable, but the lack of one does not preclude the ability to develop an IVIVC. Despite these methods being considered suitable, they have not been extensively utilized in NDA submissions to date.

Whatever the method used to establish a Level A IVIVC, the model should predict the entire *in vivo* time course from the in vitro data. In this context the model refers to the relationship between in vitro dissolution of an ER dosage form and an *in vivo* response (e.g., plasma drug concentration or amount of drug absorbed).

B. Level B

This approach utilizes the principles of statistical moment analysis. The mean in vitro dissolution time is compared to either the mean residence time or the mean *in vivo* dissolution time. Level B correlation, like Level A, utilizes all of the in vitro and *in vivo* data but is not considered to be a point-to-point correlation because it does not uniquely reflect the actual *in vivo* plasma level curve, since there are a number of different *in vivo* curves that will produce similar mean residence time values. This type of correlation is rarely seen in NDAs and in the first IVIVC survey for all approved NDAs between 1982 and 1992, accounted for one of the nine correlations and none in the 1995 survey.

C. Level C

This represents a single point correlation which relates one dissolution time point (e.g., $t_{50\%}$, $t_{90\%}$) to one pharmacokinetic parameter such as AUC, C_{max} or T_{max}. It does not reflect the complete shape of the plasma concentration time curve, which is the critical factor that defines the performance of ER products.

D. Multiple Level C

This involves correlation of one or several pharmacokinetic parameters of interest to the amount of drug dissolved at various time points of the dissolution profile. This type of IVIVC accounted for 2 of the 9 correlations submitted between 1982 and 1992 and none in the 1995 survey.

V. GENERAL CONSIDERATIONS

A. Human data is required for regulatory consideration of an IVIVC.
B. Bioavailability studies for IVIVC development may be performed with a small number of subjects. Crossover studies are preferred. Parallel studies or cross-study analyses are potentially acceptable if the *in vivo* results are normalized with a common reference treatment. An intravenous solution or aqueous oral solution of an approved immediate release product is the preferred reference treatment in most cases.
C. Any in vitro dissolution method can be utilized for obtaining the dissolution characteristics of the ER dosage form. Once the system is defined, it should be the same for all formulations tested.

1. Apparatus. In the majority of cases, it is expected that USP apparatus I (basket) or II (paddle) will be used at compendially recognized rotation speeds (e.g., 100 rpm for the basket and 50–75 rpm for the paddle). In other cases, the dissolution properties of some ER formulations may be determined with USP apparatus III (reciprocating cylinder) or IV (flow through cell). It is recommended that the Office of Pharmaceutical Sciences (OPS) be consulted prior to use of any other type of apparatus.

2. Medium. An aqueous medium, either water or a buffered solution preferably not exceeding pH 6.8, is recommended as the initial medium for development of an IVIVC. Sufficient data should be submitted to justify pH greater than 6.8. For poorly soluble drugs, addition of surfactant (e.g., 1% sodium lauryl sulfate) may be required. In general, nonaqueous and hydroalcoholic systems are discouraged unless all attempts with aqueous media are unsuccessful. In this case, it is recommended that the OPS be consulted.

3. Testing. The dissolution profiles of at least 12 individual dosage units from each batch must be determined. There must be a suitable distribution of sampling points to define adequately the profiles, and the coefficient of variation (CV) for mean dissolution of a single batch should be less than 10%.

D. Additional considerations

1. Among the three levels of IVIVC, Level A is the most informative, and hence is preferred and recommended.

2. Level C correlation can be useful in the early stages of formulation development when pilot formulations are being selected.

3. Level B correlation is least useful for the regulatory applications of IVIVC.

4. Multiple Level C correlations can be as useful as Level A correlations. However, in general, if it is possible to obtain a multiple Level C correlation, then a Level A correlation can also be developed for that case.

5. Rank order correlation is qualitative and is not very useful from a regulatory view point.

VI. DEVELOPMENT AND EVALUATION OF AN IN VITRO/ *IN VIVO* CORRELATION

A. Level A Correlations

1. Developing a Correlation.

   - Generally, the IVIVC relationship should be demonstrated consistently with two or more formulations with different release rates so as to result in differences in absorption profiles. Although an IVIVC can be defined with a minimum of two formulations with different release rates, it is recommended that formulations with three or more release rates be used.

   - Exceptions to this general requirement for 2 or more formulations with different release rates may be considered for formulations for which dissolution rate is independent of the dissolution test conditions (e.g., medium, agitation, pH) .

   - Ideally formulations should be compared in a single study with a crossover design. Alternatively, data obtained from separate studies, where all formulations being used to develop the IVIVC may have been studied in independent studies may be utilized if normalized appropriately. In cases where the correlation is attempted with three or more formulations and one of the formulations (highest or lowest release rate formulation) does not show the same relationship between in vitro dissolution and *in vivo* performance as the other formulations, the correlation can still be utilized within the range of release rates encompassed by the remaining formulations.

- The in vitro dissolution methodology should be discriminatory enough to distinguish between formulations. Dissolution testing can be carried out, during the formulation screening stage, using several methods. However, once the discriminatory system is developed, dissolution conditions should be the same for all formulations tested in the biostudy for development of the correlation and must be fixed before further steps towards correlation evaluation are undertaken.

a. Cases for which in vitro dissolution is either faster or slower than *in vivo* input rate: If percent dissolved in vitro is either faster or slower than fraction absorbed, one can consider changes in dissolution test conditions or time scaling.

   i. Adjustment of dissolution test conditions
      During the early stages of correlation development, dissolution conditions can be altered if the in vitro dissolution rate is faster or slower than *in vivo* dissolution. However, as stated above, once acceptable dissolution testing method and conditions have been selected, these must be fixed before proceeding any further and any manipulation of dissolution conditions after this stage is not appropriate.

   ii. Time scaling
      Another alternative for cases where the percent dissolved is either faster or slower than the fraction absorbed is time scaling. It is acceptable to carry out time scaling as long as the time scaling factor is the same across all formulations. It is inappropriate to use different time scales, and if a different time scale for each formulation becomes necessary, then this indicates that there is no IVIVC.

2. Evaluating the Predictability of a Level A Correlation.
   It is recommended that an IVIVC be evaluated to demonstrate that predictability of *in vivo* performance of a drug product from its in vitro dissolution characteristics is maintained over a range of in vitro dissolution release rates and manufacturing changes. Since the objective of an IVIVC is a predictive mathematical model describing the relationship between an in vitro property and a relevant *in vivo* response, the proposed evaluation approaches focus on the estimation of predictive performance or, conversely, prediction error. Depending on the intended use of an IVIVC, measures of prediction error internally or externally may be entailed. Evaluation of predictability with internal reference involves the use of the initial data used to define the IVIVC model. Evaluation of predictability with external reference generally involves the use of additional test data sets. Application of one or more of these procedures to the IVIV modeling process may be collectively defined as validation/evaluation.

      An important basic concept is that the less data available for initial IVIVC development and evaluation of predictability, the more additional data need to be evaluated to define completely the IVIVC's predictability. Some combination of 3 or more formulations with different release rates is considered optimal.

      Another significant factor is the range of release rates studied. The release rates, as measured by % dissolved, of formulations studied, should differ adequately (e.g., 10%) from each other. This should result in *in vivo* profiles that show a difference (e.g., 10%) in the pharmacokinetic parameters of interest (C_{max} or AUC) between each formulation.

Methodology for the evaluation of IVIVC predictability is an active area of investigation and a variety of methods are possible and potentially acceptable. Therefore, definitive recommendations regarding methods and criteria cannot be made at this time. Ideally, it is wished to determine that a correlation is accurately and consistently predictive of *in vivo* performance. Once this goal is achieved, in vitro dissolution could be used confidently as a surrogate for *in vivo* bioavailability of ER drug products.

a. Experimental Data Considerations

   i. Dosage Form Properties (Dependence of In Vitro Release on Experimental Conditions)

      pH Independent Dissolution: If in vitro dissolution is shown to be independent of dissolution conditions such as pH and agitation, and the in vitro dissolution profile is shown to be equal to the *in vivo* absorption or *in vivo* dissolution profile, then the results for a single formulation (one release rate) may be sufficient. No further evaluation of the IVIVC is necessary for most applications. For narrow therapeutic index drugs, external predictability needs to be evaluated as described below in VI A2aiii.

      pH Dependent Dissolution: In all other submissions which present an IVIV model, results from a single formulation (one release rate) are not sufficient and some separate measure of the predictability of the IVIVC is desirable.

   ii. Internal versus External Predictability

      Two distinct aspects of predictability can be considered, however both aspects are not recommended for all submissions.

      Estimation of Prediction Error Internally The first aspect relates to evaluating how well the model describes the data used to define the IVIVC relationship and is appropriate for all submissions.

      If formulations with 3 or more different release rates are used to develop the IVIVC model, no further evaluation beyond this initial estimation of prediction error may be necessary for non-narrow therapeutic index drugs (Category 2 applications). However, depending on the results of this prediction error calculation, determination of prediction error externally may be appropriate.

      If only 2 formulations with different release rates are utilized, the application of the IVIVC is further limited to Category 2a applications and determination of prediction error externally is recommended for complete evaluation and subsequent full application of the IVIVC.

      Estimation of Prediction Error Externally The second aspect relates to how well the relationship predicts data when one or more additional test data sets are utilized which differ from those used to define the correlation. This is appropriate for some submissions, particularly when only 2 formulations with different release rates are used to develop the IVIV model, when calculation of prediction error internally is inconclusive or for narrow therapeutic index drugs.

      The additional test data sets used for external prediction error estimation may have several differing characteristics compared to the data sets used in IVIVC development. While formulations with different release rates provide the optimal test of the IVIVC's predictability, it is not expected that a formulation would be specially prepared solely for this purpose. In the absence of such a formulation, data from other types of formulations may be considered. In each case, bioavailability data must be available for the data set under consideration.

The following represent, in decreasing order of preference, formulations that can be used to estimate prediction error externally.

1) Formulation with a different release rate than those used in IVIVC development.

2) Formulation with the same or similar release rate but involving some change in manufacture of this batch (e.g., composition, process, equipment, site)

3) Formulation with the same or similar release rate obtained from another batch/lot, with no changes in manufacturing.

iii. Pharmacologic Properties of the Drug (Therapeutic Index)

Narrow Therapeutic Index Drugs If an IVIV model is to be used in estimating the *in vivo* performance of formulations of narrow therapeutic index drugs, the model's predictability should be further tested with a data set which differs from those used to define the correlation, in other words, the external predictability of the correlation should be evaluated. The release rate of the test formulation can either be within or outside the range used to define the IVIVC relationship.

Non-Narrow Therapeutic Index Drugs If an IVIV model is to be used in estimating the *in vivo* performance of formulations of non-narrow therapeutic index drugs, it is desirable although not always necessary to test further the model's predictability with a data set which differs from that used to define the correlation.

Note - If unsure about the classification of a drug as a narrow therapeutic index drug, CDER should be consulted.

b. Methods for Evaluation of Predictability.

It is not possible at this time to describe completely exact methods appropriate for the determination of predictability of an IVIVC. However, it is hoped that it will be useful to identify concepts that appear to have validity related to evaluation of predictability of correlations.

The objective of IVIVC evaluation is to estimate the magnitude of the error in predicting the *in vivo* results from in vitro dissolution data (for a new formulation with a possibly different in vitro release profile). This objective should guide the choice and interpretation of evaluation methods.

i. Metrics of Prediction Error A variety of prediction error metrics are possible and have yet to be critically assessed for IVIVC applications. Listed below are two under consideration:

a) Root Mean Square Prediction Error (PE_{rms})

b) Mean Absolute Prediction Error (PE_{abs})

ii. Prediction Error Estimation Methods Following is a summary of several methods recommended for further consideration.

a) Cross-Validation (Internal Prediction): Cross-validation is a resampling technique that is widely used for estimating prediction error. It is particularly useful for internal estimation of prediction error. Briefly, the application of cross-validation to the evaluation of IVIVC is as follows: - Leave the data from one formulation out, develop the IVIV model with the remaining data from the other (N-1) formulations, and calculate the prediction error of the fitted model when predicting the *in vivo* response profile (e.g., absorption or plasma concentration profile) of the nth formulation based on its in vitro dissolution profile. - Do the above for n = 1, 2,.....N and combine the N estimates of prediction error.

b) Naive Prediction Model: The predictability of the IVIVC may be demonstrated by showing the improvement in prediction error relative to the model which assumes no relationship between in vitro and *in vivo* data. The greater the difference in measures of prediction error, the better the predictability. One possible naive prediction model is the mean of the *in vivo* responses at each time for the formulations used to develop the IVIVC, i.e.,

$$P_{naive}(t) = \frac{1}{N} \sum_{k=1}^{N} O_k(t)$$

N = number of formulations
$P(t)$ = naive model prediction of *in vivo* response at time t *naive*
$O(t)$ = Observed *in vivo* response to the k formulation at time t k th

The IVIVC model and the naive prediction model are then compared with respect to one of the prediction error metrics.

c) Precision and Comparison of Estimated Prediction Error: Formal inferences regarding prediction errors, such as comparison of the IVIVC and naive model predictions, should consider the precision of the prediction error estimates. Such precision might be expressed in terms of standard errors or confidence interval estimates of the prediction error. Bootstrapping approaches may be used to estimate such quantities and to evaluate hypothesis tests relevant to testing the existence and quality of the IVIVC.

B. Level C Correlations

A single point Level C correlation can only be useful in setting dissolution specifications at that specific time point. From a regulatory perspective, it may not be possible to obtain *in vivo* biowaivers that would usually require an IVIVC, based on this single point correlation.

It is possible to obtain biowaivers based on a multiple Level C correlation. However, the correlation should be established over the whole dissolution profile with some of the pharmacokinetic parameters of interest. This could be achieved by correlating the amount dissolved at various time points with either C_{max} , AUC or any other suitable parameter. A relationship should be demonstrated at each time point with the same parameter such that the effect on the *in vivo* performance of any change in dissolution could be assessed. However, if such a multiple Level C correlation is achievable, then the development of a Level A correlation should also be feasible.

Such a relationship between the amount of drug dissolved and the same pharmacokinetic parameter is needed at a minimum of three time points in order to consider a multiple Level C correlation useful in obtaining *in vivo* bioavailability waivers. Preferably, these time points should cover the early, middle and late stages of the dissolution profile.

The recommendations for assessing the predictability of Level C correlations will also depend on the type of application for which the correlation is to be used. These requirements are similar to those for a Level A correlation (refer to VI A2).

C. Level B Correlations

From a regulatory perspective, Level B correlations are not useful for obtaining any type of *in vivo* bioavailability waivers.

VII. APPLICATIONS OF IVIVC

In vitro dissolution testing is important for: (**1**) providing necessary process control and quality assurance, (**2**) determining stability of the relevant release characteristics of the product, and (**3**) facilitating certain regulatory determinations and judgments concerning for example, minor formulation changes or change in site of manufacture. In addition, in certain cases, especially for ER formulations, the dissolution test can serve not only as a quality control for the manufacturing process but also as an indicator of how well the formulation will perform *in vivo*. Thus, the main objective of developing and evaluating IVIVC is to empower the dissolution test to serve as a surrogate marker for human bioequivalence studies. One additional primary purpose of establishing an IVIVC is to minimize the number of human studies needed for approving and maintaining a drug product on the market. This approach will not only reduce the cost and time of drug development by reducing the number of studies required to demonstrate adequate bioavailability, but will also facilitate the initial approval as well as scale-up and post-approval changes. However, for certain applications that are outlined below, the adequacy of the in vitro method to act as a surrogate for *in vivo* testing must be shown through an IVIVC for which predicability has been demonstrated.

This section describes how an IVIVC can be used in obtaining a waiver for demonstrating *in vivo* bioavailability and how an IVIVC can be used in setting dissolution specifications that would assure consistent *in vivo* performance.

A. Biowaivers for Changes in Drug Product Manufacturing

  1. Categories of Waivers to Be Granted without an IVIVC.

    These types of waivers can be granted for both narrow and non-narrow therapeutic index drugs.

    a. Situations in which such waivers can be granted:

    Pre and post-approval:

      i. Situations described in SUPAC MR that do not require either a biostudy or an IVIVC (refer to SUPAC MR Guidance document)

      ii. For situations described in SUPAC MR that require a biostudy, waiver of *in vivo* testing for such changes made on lower strengths can be granted provided that:

        a) the strengths are compositionally proportional or qualitatively the same

        b) bioequivalence (required in SUPAC MR) has been demonstrated on the highest strength and dose proportionality has been demonstrated for this

drug product, or bioequivalence (required in SUPAC MR) has been demonstrated on the highest and lowest strengths and the drug exhibits linear
kinetics.

iii. For beaded formulations in capsules, approval of additional strengths provided bioavailability data is available on the highest strength. The only difference is in the number of beads per dosage strength.

For cases (VII A 1 a) i - iii above, waivers can be granted without IVIVC
provided that dissolution data is submitted in five media (e.g., water, 0.1 N
HCl, USP buffer media at pH 4.5, 6.8 and 7.5), and similar dissolution behavior between old and new formulations is demonstrated in all five media.

b. Criteria for granting biowaivers:

i. Comparison of dissolution profiles

The sponsor should apply appropriate statistical testing with justifica-
tions (e.g., the f2 equation) for comparing dissolution profiles.

Dissolution profiles may be compared using the following equation that
defines a similarity factor (f_2):

$$f_2 = 50 \text{ LOG } \{[1+1/n \sum_{t=1}^{n} (R_t - T_t)^2]^{-0.5} \times 100\}$$

where R_t and T_t are the percent dissolved at each time point. An f_2 value between 50 and 100 suggests the two dissolution profiles are similar. Also, the
average difference at any dissolution sampling time point should not be
greater than 15% between the changed drug product and the biobatch or
marketed batch (unchanged drug product) dissolution profiles. An appropriate reference for this comparison should represent an average dissolution
profile derived from at least three consecutive recent batches of the current
formulation. Finally, the dissolution data obtained under the applica-
tion/compendial dissolution test conditions (e.g., media, agitation), on both
the changed drug product and the biobatch or marketed batch (unchanged
drug product) dissolution profiles must be within the application/compendial
specifications.

2. Categories of Waivers for Non-Narrow Therapeutic Index Drugs.

a. IVIVC developed with 2 formulations/release rates

Pre and post-approval:

i. Level III manufacturing site changes of SUPAC MR

ii. Level III non-release controlling excipient changes of SUPAC MR This does
not include complete removal of or replacement of excipients.

b. IVIVC developed with 3 formulations/release rates (or developed with 2 formu-
lations/release rates and external predictability has been evaluated)

Pre and post-approval:

i. Level III process changes of SUPAC MR

ii. Complete removal of or replacement of non-release controlling excipients
without affecting the extended release mechanism

iii. Level III changes in the release controlling excipients of SUPAC MR

iv. Approval of new strengths, lower than the highest strength, within the investigated dosing range for all formulations other than beaded formulations provided that these new strengths are compositionally proportional or
qualitatively the same, are made with the same type of equipment, are manu-

factured using the same process, and are at the same site as other strengths that have been tested in a biostudy. The efficacy and safety of these new strengths should have been demonstrated in clinical trials. Such changes for beaded formulations are already covered under category 1.

c. Waivers for lower strengths.

If an IVIVC is developed with the highest strength, waivers for changes made on the highest strength and any lower strengths can be granted provided that these strengths are compositionally proportional or qualitatively the same, and the in vitro dissolution profiles of all the strengths are similar.

d. Changes in release controlling excipients in the formulation should be within the range of release controlling excipients of the established correlation.

e. Criteria for granting 2a, 2b, and 2c biowaivers

The difference in predicted means of C_{max} and AUC is no more than 20% from that of the reference product and where applicable the new formulation meets the application/compendial dissolution specifications.

3. Categories of Waivers for Narrow Therapeutic Index Drugs.

If external predictability of IVIVC is evaluated, all of the following waivers can be granted if a minimum of 2 formulations/release rates have been studied for the development of the the IVIVC.

a. Situations in which such waivers can be granted:

Pre and post-approval

i. Level III non-release controlling excipient changes of SUPAC MR

ii. Complete removal of or replacement of non-release controlling excipients without affecting the extended release mechanism

iii. Level II release controlling excipients changes of SUPAC MR

iv. Level III release controlling excipients changes of SUPAC MR

v. Level III process changes as of SUPAC MR

vi. Level III site change as of SUPAC MR

vii. Approval of new strengths, lower than the highest strength, within the investigated dosing range for all formulations other than beaded formulations provided that these new strengths are compositionally proportional or qualitatively the same, are made with the same type of equipment, are manufactured using the same process, and are at the same site as other strengths that have been tested in a biostudy. The efficacy and safety of these new strengths should have been demonstrated in clinical trials. Such changes for beaded formulations are already covered under category 1.

b. Waivers for lower strengths.

If an IVIVC is developed with the highest strength, waivers for changes made on the highest strength and any lower strengths can be granted provided that these strengths are compositionally proportional or qualitatively the same, and the in vitro dissolution profiles of all the strengths are similar.

c. Changes in release controlling excipients in the formulation should be within the range of release controlling excipients of the established correlation.

d. Criteria for granting 3a and 3b biowaivers.

The difference in predicted means of C_{max} and AUC is no more than 20% from that of the reference product and where applicable the new formulation meets the application/compendial dissolution specifications.

4. Categories of Waivers When in Vitro Dissolution Is Independent of Dissolution Tes T Conditions.

    a. Situations in which such waivers can be granted for both narrow and non-narrow therapeutic index drugs:

        Category 2 and Category 3 applications (not Category 5) can be granted in this case with an adequately developed IVIVC with one formulation/release rate. Category 1 waivers are also allowable.

        For non-narrow therapeutic index drugs, no further evaluation of the IVIVC is necessary provided dissolution data is submitted in five media (e.g., water, 0.1 N HCl, USP buffer media at pH 4.5, 6.8 and 7.5), and similar dissolution behavior is demonstrated in all five media. For narrow therapeutic index drugs, external predictability of the IVIVC should be evaluated.

    b. Criteria for granting biowaivers:

       i. In vitro dissolution should be shown to be independent of dissolution test coditions even after changes are made in drug product manufacturing.

      ii. Comparison of dissolution profiles

        The sponsor should apply appropriate statistical testing with justifications (e.g., the f_2 equation) for comparing dissolution profiles.

        Dissolution profiles may be compared using the following equation that defines a similarity factor (f_2):

$$f_2 = 50 \ \text{LOG} \ \{[1+1/n \textstyle\sum_{t=1}^{n} (R_t - T_t)^2]^{-0.5} \times 100\}$$

        where R_t and T_t are the percent dissolved at each time point. An f_2 value between 50 and 100 suggests the two dissolution profiles are similar. Also, the average difference at any dissolution sampling time point should not be greater than 15% between the changed drug product and the biobatch or marketed batch (unchanged drug product) dissolution profiles. An appropriate reference for this comparison should represent an average dissolution profile derived from at least three consecutive recent batches of the current formulation. Finally, the dissolution data obtained under the application/compendial dissolution test conditions (e.g., media, agitation), on both the changed drug product and the biobatch or marketed batch (unchanged drug product) dissolution profiles must be within the application/compendial specifications.

     iii. The difference in predicted means of C_{max} and AUC should be no more than 20% from that of the reference product and where applicable the new formulation should meet the application/compendial dissolution specifications.

5. Categories of Waivers for Which an Externally Evaluated IVIVC Cannot Be Used for Approval.

    a. Approval of a new product/formulation of an already approved drug but the new product/formulation has a different release mechanism

    b. Approval of a dosage strength beyond the limits of what was tested in clinical trials

    c. Approval of another sponsor's product even with the same release controlling mechanism

B. Dissolution Specifications

    It is an accepted scientific principle to establish the in vitro dissolution specifications based on the performance of the bio/clinical lots. Moreover, the dissolution specifications are sometimes set such that the scale-up lots as well as the stability lots pass these specifications. This suggests that more variable formulations may need wider dissolution specifications. This is due to the fact that the in vitro dissolution test is sometimes considered to be a quality control test without any *in vivo* significance even though in certain cases the rate limiting step in the absorption of the drug is the dissolution of the active drug substance from the formulation as is the case with many extended release formulations.

The aim of establishing an IVIVC is to take a quality control test and to use it as a surrogate for *in vivo* bioavailability. The goal to achieve is that the in vitro dissolution test becomes a predictor of *in vivo* performance of the formulation, and thus, dissolution specifications can be established to minimize the possibility of releasing lots that would be different in their *in vivo* performance.

1. Case with No IVIVC.

    Presently, the USP acceptance criteria for dissolution specifications are adopted unless alternate acceptance criteria are specified in the NDA. One less than optimal approach is to have the specifications established such that all lots pass at Stage 1 of testing. This may result in specifications which are very permissive and would not prevent the release of lots that might not have optimal *in vivo* performance.

    a. Maximum suggested range at any dissolution time point specification is ± 10% of label claim deviation from the mean dissolution profile obtained from the clinical/bio lots. However, in certain cases, deviations from the ± 10 % criteria can be accepted provided that the range at any time point does not exceed 25 %.

    b. Inter-lot variability should not be a primary consideration in setting the specifications.

    c. Specifications greater than 20 % are generally acceptable only when the sponsor submits evidence that lots with mean dissolution profiles that are allowed by the upper and lower limit of the specifications are bioequivalent (for e.g., lots that have the fastest and the slowest dissolution rates).

    d. A minimum of three time points are required to set the specifications. Preferably, these time points should cover the early, middle and late stages of the dissolution profile. The last time point should be the time point where at least 80% of the label claim has dissolved. If the maximum amount dissolved is less than 80 %, then the last time point should be the time where the plateau in the dissolution profile has been reached.

    e. USP acceptance criteria can be used. Specifications should be established based on average dissolution data for each lot under study (equivalent to passing at Stage 2 level of testing).

2. Case Where an IVIVC Has Been Established.

    a. Whether the IVIVC is externally validated or not, it should still be used to establish the specifications for in vitro drug release. The use of an established IVIVC for setting dissolution specifications even though not evaluated for predictability, represents an improvement over dissolution specifications that are based solely on the in vitro performance. A well established IVIVC may allow for the setting of wider dissolution specifications. This would be dependent on the predictions of the IVIVC (viz., 20% differences in the predicted AUC and C_{max}).

  b. Ideally, the specifications should be established such that all lots that have dissolution profiles within the upper and lower limits of the specifications, are bioequivalent. Minimally, these lots should be bioequivalent to the lots used in the clinical trials or an appropriate reference standard.

  c. When there is a point to point correlation (Level A) the following aspects should be considered

    i. In the case where an IVIVC exists, specifications should be set on mean data using at least 12 individual dosage units per data set.

    ii. A minimum of three time points are required to establish the specifications. Preferably, these time points should cover the early, middle and late stages of the dissolution profile. The last time point should be the time point where at least 80 % of the label claim has dissolved (for oral dosage forms only) or the earliest time where the maximum possible amount that could possibly dissolve for that product (if 80% of label claim is not reasonable).

    iii. Calculate the plasma concentration time profile using convolution techniques (or other appropriate modelling techniques) and determine whether the lots with the fastest and slowest release rates that are allowed by the dissolution specifications result in a maximal difference of 20% in the predicted C_{max} and AUC.

3. Release Rate Specifications.

  a. If the release characteristics of the formulation can be described by a truly zero-order process for some period of time (e.g., 5%/hr from 4 to 12 hours) and the dissolution profile appears to fit a linear function for that period of time, then a release rate specification could be established to describe the dissolution characteristics of that formulation.

  b. This release rate specification may be in addition to the specifications established on the cumulative amount dissolved at the chosen time points.

  c. Alternatively, this release rate specification may be the only specification in addition to the specification for time where at least 80% of the label claim has dissolved.

4. Another Approach for Future Consideration.

    If the dissolution profile can be described by a certain mathematical model (e.g., Weibull function or Hill equation), then the dissolution specification limits may be established on the model parameters of interest. A 90% CI is constructed in a way that the extreme in dissolution profiles permitted by the lower and upper limit of the CI will not result in more than a 20% difference in the AUC and C_{max} predicted by convolution of these extreme dissolution profiles.

5. Multiple Level C Correlation.

  a. If a multiple point Level C correlation has been established, then set the specifications at each time point such that there is a maximum of 20% difference in the predicted AUC or C_{max}. If there is a correlation with two or more parameters, then the specifications should be set with the parameter that would result in tighter dissolution limits.

  b. Additionally, the last time point should be the time point where at least 80% of label claim has dissolved (for oral dosage forms only).

6. One Time Point Correlation (Level C).

    This one time point may be used to establish the specification where there is a maximum of 20% difference in the predicted AUC and C_{max}; further, the other time points should not be more than 20% wide with the clinical/bio lots profiles considered to be the target profile to be achieved. The last time point should be the time point where at least 80% of label claim has dissolved.

VIII. REFERENCES

1. "An Introduction to the Bootstrap", B.Efron and R.J.Tibshirani, Chapman & Hill, New York, 1993, pp. 237–257.
2. "Draft Guidance for Industry: Modified Release Solid Oral Dosage Forms; Scale-Up and Post-Approval Changes: Chemistry, Manufacturing and Controls, In Vitro Dissolution Testing, and *in vivo* Bioequivalence Documentation", U.S. Department of Health and Human Services, Food and Drug Administration, July 1996.
3. "Mathematical comparison of curves with an emphasis on dissolution profiles", J.W.Moore and H.H.Flanner, presented at the AAPS National Meeting, November 1994. Personal Communication from AAI Inc., Wilmington, NC 28405.
4. "Report of the Workshop on CR Dosage Forms: Issues and Controversies", J.P. Skelly et al., Pharmaceutical Research, 4(1), 1987, pp. 75 - 78.
5. "In-vitro/In-Vivo Correlation for Extended-Release Oral Dosage Forms", USP PF Stimuli Article, July 1988, pp. 4160–4161.
6. "Report of Workshop on In vitro and *in vivo* Testing and Correlation for Oral Controlled/Modified-Release Dosage Forms", J.P. Skelly et al., Journal of Pharmaceutical Sciences, 79(9), September 1990, pp. 849–854.
7. "In Vitro *in vivo* Evaluation of Dosage Forms", USP XXIII chapter <1088>, United States Pharmacopoeal Convention, Inc., pp. 1927–1929.
8. Workshop II report "Scale-up of Oral Extended Release Dosage Forms", J.P.Skelly et al., Pharmaceutical Research, 10(12), 1993, pp. 1800–1805.

26

ISSUES IMPORTANT TO *IN VITRO–IN VIVO* CORRELATION

IVIVR Workshop Open Discussion

Although this book addresses numerous topics in the area of *in vitro-in vivo* correlation (IVIVC), this chapter summarizes some of the scientific and regulatory issues important to IVIVC and the Draft FDA Guidelines on IVIVC. This chapter was based on the open discussion sessions at the IVIVR Workshop in Baltimore, Maryland on September 4–6, 1996. The chapter represents a compilation of comments and the consensus of the attendees at the meeting. This chapter is organized into five major topics with each issue listed under the topic and the consensus response from the attendees. For those few times that a consensus could not be reached, the different points of view are provided.

GASTROINTESTINAL PHYSIOLOGY

1. Our knowledge about GI physiology is greater than ever before but more research is needed.

   The consensus was that this statement is true and a better understanding of GI physiology would provide a better understanding as to the limits of developing and applying IVIVCs.

2. The use of animal models for IVIVC investigation needs to be considered.

   The consensus was that animal models should not be used in the regulatory decision process for oral or extended release products. However, they may be a useful tool for product development.

3. *In vitro* and in situ animal and human systems are important to IVIVC.

   The consensus was that these systems, in principal, have a role in IVIVC research but the role is not well defined except for permeability.

4. New *in vitro* and in situ animal and human models need to be developed and correlated to man.

   The consensus was that additional research on permeability and regional permeability will be valuable for the development of ER products.

In Vitro–in Vivo Correlations, edited by Young *et al*.
Plenum Press, New York, 1997

5. Investigation of factors affecting BA.

This is an important area to investigate but the consensus was that it is outside the realm of this conference on IVIVC.

DISSOLUTION

1. *In vitro* dissolution can be used for QC and as a surrogate for *in vivo* (IVIVC).

It was agreed that *in vitro* dissolution can be used for both QC and as a surrogate for *in vivo*. More importantly, however, was the agreement that the systems are not *required* to be the same. If they are not the same, a major concern was that the QC dissolution specifications would be based on batch history and not the anticipated *in vivo* consequences to the formulations.

A concern by a number of individuals was the appropriateness of using a different dissolution test for QC release vs. IVIVC. The problem is that the QC specifications from a dissolution system different from the IVIVC system does not ensure comparable *in vivo* response for different formulations.

In a similar vein, a number of attendees stated, "how can the dissolution specifications be developed from a different system than the IVIVC dissolution system which is a surrogate for the *in vivo*, if we are concerned about *in vivo* response to the product?" It was agreed that there may be financial and/or logistical reasons why a company would not be able to use the dissolution system from the IVIVC, in this case the ideal situation would be to relate the two dissolution systems in some manner (e.g., mapping the formulations from one dissolution system to another).

2. We must define and further investigate the methodology needed to determine differences in dissolution.

It was agreed that further research in this area is appropriate.

3. We must define what is an acceptable vs. unacceptable difference.

For biowavers which require IVIVC, the acceptance criteria for the comparison of curves should be based on *in vivo* consequences not just the *in vitro* curves.

4. The guidance states "If *in vitro* dissolution is shown to be independent of dissolution conditions such as pH and agitation....". We must define *independent*.

The consensus was that the variables and range of variables to be investigated need to consider the characteristics of the product. The variables one might consider are: pH, agitation, ionic strength, lipids/surfactants, apparatus.

5. Optimization of dissolution systems to obtain IVIVC is important.

The attendees agreed that the optimization process can go to the extreme. A better approach would be to describe the IVIVC by more complex mathematical/statistical models without making heroic efforts to change dissolution. This assumes that the existing dissolution system is discriminating.

6. We should be developing mechanistic (mimicking physical/biological systems) as well as empirical dissolution systems, but this is not a regulatory issue.

   The consensus was that theses dissolution systems may need to be developed and the FDA must be open to accepting these systems for IVIVC and specifications.

7. The dissolution system used for IVIVC should be a discriminatory system based on the critical manufacturing variables that will affect *in vivo* response.

   The attendees agreed with this statement and emphasized the fact that the critical manufacturing variables should be "critical" based on *in vivo* consequences not *in vitro* consequences.

8. Dissolution specifications should be developed from IVIVC models.

   The attendees agreed with this statement but again emphasized that, if the QC dissolution system is different, QC specifications would not be based on IVIVC. Some type of relationship (possibly mapping) may be required to relate QC dissolution to IVIVC dissolution. The attendees believed that the specifications should be based on IVIVC models, if possible, but that the use of an IVIVC model should not be a requirement.

9. Should the formulations be bioequivalent at the extremes of the dissolution specifications or should the extremes of the dissolution specifications be bioequivalent to the pivotal batch?

   The consensus was that the upper/lower dissolution specification limits should be *equivalent* as a product but studies should not be required to verify that the top and bottom are equivalent. There was not, however, an agreement about the definition or criteria of equivalent. It was agreed that equivalent is a broader term than bioequivalent.

   Differing opinions existed as to the criteria for equivalent. Some of the comments are provided below: 1) The development of the upper and lower dissolution specifications should be based on a desire to have the formulations bioequivalent following our present criteria (if we do not desire bioequivalence at our present criteria, the science is moving backwards), 2) A 20% mean range for Cmax and AUC is a reasonable position at this time for ER products, 3) A 20% mean range for Cmax and AUC is arbitrary and is a step backward in our science, 4) A 10% difference in the mean Cmax and AUC between the pivotal batch and upper or lower specification is reasonable, 5) a 10% difference in the mean Cmax and AUC is appropriate for the upper and lower specifications, 6) The innovator must be bioequivalent across all batches otherwise new generic products may target the top or bottom of the innovator and therefore produce a wide range of generic products which may not be bioequivalent to all batches of the innovator, 7) Bioequivalence standards that are drug specific may be the most appropriate and the criteria should depend on the drug product itself rather than 1 criteria for all products.

   There are really two situations: 1) specifications developed from IVIC and 2) specifications developed from side batches when an IVIVC does not exist. In the second situation, a clinical study may be needed to confirm equivalence.

STUDY DESIGN

1. Clinical studies should be conducted as early as possible after choosing the dosage form.

   This was strongly recommended by attendees.

2. Food studies may play an important role in IVIVC and should be performed when necessary.

   If an IVIVC developed in the fasted state is capable of predicting the *in vivo* characteristics in the fed state, no additional IVIVC work under fed conditions is required. If not, more investigation may be required and one might even consider performing an IVIVC study under fed conditions.

3. Crossover studies are preferable to parallel designed studies. If parallel or crossover studies are used in IVIVC, a common reference is needed.

   The attendees recommended that the statement "a common reference is needed" in the Draft IVIVC Guidelines be changed to "a common reference may be useful". There was no recommendation on the nature of the common reference but further investigations may be needed to determine the optimal choice (solution, IR, IV) for specific circumstances.

4. How many formulations are needed to develop an IVIVC model? What factors does it depend on?

   Multiple formulations should be used with the exception of one formulation as stated in the Draft IVIVC Guidelines. An exact number could not be defined.

5. How many subjects are needed for an IVIVC study?

   The number of subjects required for IVIVC study needs to be determined after the modelling method and a statistical criteria are established.

6. One should consider the potential site of absorption and release from the ER formulation when choosing the unit impulse (solution vs. IR solid dosage form vs. IV). The unit impulse should be chosen to match the sites of absorption as much as possible.

   The attendees agreed with this statement.

MODELLING

1. The present deconvolution and convolution methods have provided a starting point for IVIVC but further research in these two area is required (e.g., Verotta, Gillespie).

The consensus was that more research is needed in these areas and that a user friendly deconvolution and convolution computer program is needed.

2. Both empirical and mechanistic models (e.g., accounting for *in vivo* release, mechanism of release, dosage form-GI interaction) need to be investigated and evaluated further.

The consensus was that more research is required for both types of modelling approaches.

3. If the *in vivo* or *in vitro* profiles is faster, the dissolution system should be changed or alternative models (e.g., time scaling) developed.

The consensus was both approaches were appropriate.

4. We should be developing Level A correlations with the entire curves. What about early vs. late points?

The consensus was that the *in vitro* dissolution should predict the entire *in vivo* curve up to the post-absorption phase, which needs to be defined. If, for example, there is site limited absorption then it may not be necessary to use the entire *in vitro* curve.

5. All formulations need to fit the same model.

Since all formulations within the range of the formulation and manufacturing specifications should fit the IVIVC, a single IVIVC should be developed for multiple formulations.

6. We should account for correlation of data within a subject.

New approaches which account for the correlation of data is required to appropriately model the IVIVC but a user friendly program is needed in order to make the approach practical.

7. We should account for variability in the *in vitro* dissolution.

New approaches are needed to account for the variability in the *in vitro* dissolution but the mathematical and statistical methods need to be further developed and implemented in a user friendly manner in order to make the approaches practical.

8. Do we develop a Level A for 1 formulation or multiple formulations (similar to cross formulation approach in Level C)?

Since all formulations within the range of the formulation and manufacturing specifications should fit the IVIVC, multiple formulations should be used when developing an IVIVC.

The exception is that stated in the guidelines when the dissolution remains the same for dissolution systems that have differences in, for example, pH, agitation, system. In this case, the study is considered a confirmatory study, not a modelling study since the model is based on the *in vitro* dissolution results.

9. Clarify internal vs. external predictions and when is one better than another?

One major difference between internal vs. external predictions is that the external procedure predicts the *in vivo* response for formulation(s) that are in no way used in the development of the IVIVC model. Internal prediction, on the other hand, uses formulation(s) that are used to develop the model. An example of the internal procedure is cross-validation where one formulation is left out of the IVIVC model and predicted from the IVIVC model developed from the other formulations. This is repeated for each formulation.

The external prediction procedure will generally prove to be the best but the conditions where it performs better than the internal prediction procedure has not been defined. Further research is needed to define the various conditions where the two predictions will result in false positive and false negative findings.

EVALUATION / VALIDATION

1. An investigation of the appropriate prediction error metrics needs to be undertaken.

   The attendees agreed with this.

2. More research into the area of model validation or the criteria for IVIVC acceptability needs to be undertaken.

   The attendees strongly agreed with this and stated this should be a major area of future research.

3. A definition of what is acceptable and what is not acceptable needs to be developed.

   The attendees strongly agreed with this and stated that it is extremely difficult for the industry to implement validation procedures without a criteria for acceptable and not acceptable. The acceptance criteria for a valid IVIVC may be different for drugs with a narrow therapeutic index.

OTHER ISSUES

1. The biopharmaceutical classification for ER products needs to be continued.

   The biopharmaceutical classification for IR formulations has been beneficial and an expansion into ER formulations would be helpful. Attendees were concerned about the ability to develop such a classification.

2. Additional research on the IVIVCs for highly variable drugs is required.

   There was a consensus that further research in this area is required.

3. Methods need to be developed to deal with the following: Truncated absorption (change in F) and saturable presystemic metabolism.

   There was a consensus by the attendees that the modelling procedures for these areas is required.

4. IVIVC models need to be developed early to assist in product development.

There was a strong consensus that early development of IVIVC models would be beneficial to product development.

5. Define the role of IVIVC in generic drugs

IVIVC may be important for generic drugs (e.g., in the development of dissolution specifications), but further research in the general area of IVIVC is required. The same principles for developing IVIVC models for NDA and ANDA applications should apply.

6. What is the cost / benefit ratio for IVIVC development?

IVIVC can be extremely useful in product development, product optimization, and post-approval changes. The cost/benefit ratio, however, could not be defined but the attendees felt that it was the sponsors decision to determine the cost/benefit ratio for developing IVIVC models.

7. Can this be extended to products administered other than po.

It was agreed that the principals of IVIVC can be extended to products administered by other routes such as IM.

8. Explain further what is a formulation.

For IVIVC, a formulation is not defined based on the type or quantity of polymer and/or excipients within the formulation, but a more important issue is the choice of the range of differences between the formulations. Formulations may have different release rates but they should posses the same mechanism of release.

9. Semantics within the guidelines

It is important for the FDA to realize that although many guidelines use the terms "desires" as something less than "required", the pharmaceutical industry interprets the term "desires" as equal to required.

INDEX